Marine Glycobiology: Principles and Applications

Marine Glycobiology: Principles and Applications

Editor

Mikolaj Tosto

scitus academics

Marine Glycobiology: Principles and Applications
Edited by **Mikolaj Tosto**

ISBN: 978-1-68117-689-5
Library of Congress Control Number: 2016958861

SCITUS Academics LLC,
616, Corporate Way, Suite 2, 4766,
Valley Cottage, NY 10989

www.scitusacademics.com

Notice

Printed in United States of America on Acid Free Paper

Preface

In the field of medicine and science, glycobiology is a new and exciting place. For all natural systems, glycobiology, which is the study of the structure and function of carbohydrates and carbohydrates containing molecules, is crucial. Progress in glycobiology has shed light on the selection of complex natural processes associated with cell and molecular communication, immunology and disease, and developmental biology. Marine glycobiology: applications and principles revolutionize the idea of marine glycobiology, focusing on the latest principles and applications and relationships of marine glycobiology. Glycobiology actually combines all branches of biomedical research and reveals the various and vast types of carbohydrate structures in nature. Structural progress has enabled scientists to study the function of complex carbohydrates much more thoroughly and to identify the work they do in a wide range of natural processes. Research in glycobiology in marine systems has focused largely on reproduction, especially with regard to chemical communication between gamets. There is an acknowledged, if rather modest, tradition of glycobiology research in marine systems that focus mainly on reproduction, biofouling and chemicals.

This title is intended to emphasize the fundamental value of glycobiology in many natural procedures in marine creatures and is based on highly demarcated systems. The current position of research in marine glycobiology is mainly eloquent with very limited progress in structural clarification and the consequent definition of specific functional roles beyond some classical examples, e.g. induction of the acrosome reaction in echinoderms. However, with the latest advances in analytical instruments, marine glycoconjugates can now be characterized, many of which have potential biomedical and biotechnological applications.

Table of Contents

1

CHAPTER

Glycobiology of Reproductive Processes in Marine Animals: The State of the Art

Alessandra Gallo and Maria Costantini *

Laboratory of Animal Physiology and Evolution, Stazione Zoologica Anton Dohrn, Villa Comunale, Naples 80121, Italy

ABSTRACT

Glycobiology is the study of complex carbohydrates in biological systems and represents a developing field of science that has made huge advances in the last half century. In fact, it combines all branches of biomedical research, revealing the vast and diverse forms of carbohydrate structures that exist in nature. Advances in structure determination have enabled scientists to study the function of complex carbohydrates in more depth and to determine the role that they play in a wide range of biological processes. Glycobiology research in marine systems has primarily focused on reproduction, in particular for what concern the chemical communication between the gametes. The current status of marine glycobiology is primarily descriptive, devoted to characterizing marine glycoconjugates with potential biomedical and biotechnological applications. In this review, we describe the current status of the glycobiology in the reproductive processes from gametogenesis to fertilization and embryo development of marine animals.

Keywords: embryo development; ferilization; glycobiology; oogeneesis; spermatogenesis

1. INTRODUCTION

Glycoconjugate chemistry and glycobiology were rapidly evolving as scientific disciplines in the 1980s. Their impact, however, on understanding fundamental biological processes was not immediately forthcoming. In recent years, there has been resurgence in the field with major discoveries leading to powerful new insights into the complex role of glycoconjugates in biological processes. Progress in glycobiology has shed light on a range of complex biological processes associated with, for example, disease and immunology, molecular and cellular communication and developmental biology [1]. The diversity of complex carbohydrates has fascinated and frustrated glycobiologists for years. The most studied for their biological properties are mammalian sulfated polysaccharides or glycoconjugates constituted by glycosaminoglycans (GAGs) composed of negatively charged acidic chains, most of them covalently linked to proteins [2]. Glycoconjugates are also widely diffused in marine environment. In fact, they are responsible for a remarkably diverse array of biological activities in marine organisms ranging from immunity, sperm-egg binding, thrust generation in nonflagellar cyanobacteria, electroreception in sharks, and a multitude of cell-cell and cell-molecular recognition events [1]. For these reasons, there is a considerable interest in characterizing the glycan structures that mediate many of these processes, as many will have potential medical and biotechnological applications. Moreover, manipulating oligosaccharide composition in the embryo promises new insights into their developmental functions [3]. For example, *O*-linked *N*-acetylglucosamine (*O*-GlcNAc) is a highly dynamic post-translational modification of cytoplasmic and nuclear proteins. Although the function of this abundant modification is yet to be definitively elucidated, considering also that all *O*-GlcNAc proteins are phosphoproteins. Further, the serine and threonine residues substituted with *O*-GlcNAc are often sites of, or close to sites of, protein phosphorylation. This implies that there may be a dynamic interplay between these two post-translational modifications to regulate protein function. It has been demonstrated that several cellular function and developmental regulation might be affected by changes in *O*-GlcNAc levels [4]. In this respect, one of the most fascinating properties of complex carbohydrates moieties is their enormous information potential, considering glycocoding as an information management system in reproduction and embryonic development. In fact, reproduction is an evolutionary imperative and a fundamental feature of all known life. Understanding of the mechanism(s), by which the spatial/temporal regulation of early development is managed, represents a very topical issue. The "transitions" which occur during cell to cell cluster, cell cluster to early organ architecture, and early organ architecture to functional organ development are fundamental and mirror the evolutionary process of biological information management. All of these "transitions" involve increasing complexity, the development of hierarchies of information management systems (integrated bidirectionally), and spatial/temporal regulation, which relies on historical events to map future structure and function.

Carbohydrates appear to be important functional groups in reproductive biology with variation in glycosylation products contributing toward reproductive functionality in male and female animals [5,6]. Carbohydrates are known to play essential role in various biological processes including development.

In this review, we will present the relevance of glycobiology to gametes physiology, fertilization and embryo development.

2. THE GLYCOBIOLOGY IN OOGENESIS

During oogenesis, immature oocytes are arrested at the prophase I of the meiotic process and are characterized by a large nucleus that is referred as the "germinal vesicle". Along the maturation process oocytes grow and acquire the competence for fertilization. During the growth, large amounts of glycoconjugates, in particular glycoproteins, are synthesized and represent the molecular constituents of cortical vesicles, vitelline envelope and yolk granules of the fully grown oocyte of marine organisms [7,8].

Cortical alveoli are membrane-limited round structures located in the oocyte cortex containing neutral glycoproteins, carboxylated glycoconjugates, neutral glycoproteins plus sialic acid-rich glycoprotein, and neutral glycoproteins plus sulfated glycoconjugates [9].

The major constituent of cortical alveoli of fish eggs is the hyosophorin (HSP), a highly glycosylated protein in which carbohydrates account for 80% to 90%. Polysialoglycoprotein (PSGP) is the first HSP isolated from the unfertilized eggs of rainbow trout (*Salmo gairderi*) [10]. Homologs were subsequently found in salmon [11,12] and flounder [13] and they are collectively denoted as hyosophorins, which respond to the following criteria: high carbohydrate content and tandem repeats of an identical peptide sequence.

PSGP from rainbow trout eggs has a low protein content (about 15% by weight) and a high sialic acid content (about 60% by weight) most of which occurs in polysialic acid (polySia) chains linked to *O*-glycosidic carbohydrate units [14]. PSGP consists of tandem repeats of tridecaglycopeptide carrying three *O*-linked glycans with polySia chain which is an α2,8-linked polySia with chain length of up to 25 sialic acid residues [15]. These polySia chains are capped at the non-reducing ends by deaminated neuraminic acid, 3-deoxy-D-glycero-D-galacto-nonulosonic acid (KDN) [16] and their biosynthesis is developmentally regulated and occurs at later stage of oogenesis [17]. A species-specific structural diversity has been revealed in polySia chain of salmonid egg PSGP.

In rainbow trout and in *Oncorhynchus* fish species, PSGP sialic acid residues are exclusively *N*-glycolylneuraminic acid (Neu5Gc) while *Salmo* and *Saluelinus* fish species contain both *N*-acetylneuraminic acid (Neu5Ac) and Neu5Gc residues [14,15,18]. Hyosophorin isolated from flounder has a neutral fucosylated pentaantennary glycan chain [13] that in *herring*and *Fundulus heteroclitus* is sialylated.

In fish, cortical alveoli contain a considerable cellular heterogeneity of glycoconjugate sugar residues. The glycoconjugate pattern is specie specific, being cortical alveoli rich in α-*N*-acetyl-D-glucosamine (α-GlcNAc), sialic acids, α-*N*-Acetyl-D-galactosamine (α-GalNAc), α- or β-GalNAc, β-D-Galactosyl-(1-3)-*N*-acetyl-D-galactosamine (β-D-Gal(1-3)-GalNAc) in swordfish [8]; GalNAc and α-galactose (αGal) in flatfish [19]; GlcNAc, GalNAc, Gal and sialic acid in protogynous teleost*Epinephelus marginatus* [20] and GlcNAc, GalNAc and sialic acid in bluefin tuna [7].

The cortical alveoli of fishes, therefore, are considered homologous to the cortical granules present in invertebrates [21,22] and in other vertebrate species [23,24,25]. Several studies on the contents of cortical granules in sea urchins oocytes demonstrated that they contain diverse repertoire of molecules that includes enzymes, such as an ovoperoxidase, a glycoprotein containing both mannose (Man) and GlcNAc moieties [26], a protease, and a glycosidase; structural proteins such as SFE9 (soft fertilization envelopes clone 9), proteoliaisin, and SFE1 (soft fertilization envelope clone one); glycosaminoglycans; and perivitelline molecules such as glucanase and 330 kDa fibrillar glycoprotein hyaline [27].

In Crustaceans, mature oocytes are characterized by the presence of rod-like bodies, called cortical rods (CRs), arranged radially around the periphery of the oocyte plasma membranes [28,29,30,31,32] and located in the extracellular crypts formed by the invagination of the oolemma into the egg cortex. During spawning, CRs are released upon contact of the eggs with seawater and form a jelly investment around the eggs in many penaeiod shrimp species but is lacking in other crustaceans [30,31,33]. Biochemical characterization of the isolated CRs has revealed that they contain approximately 25%–30% carbohydrates and 70%–75%

proteins by weight [29]. Histochemical evidence has also consistently demonstrated that CRs are glycoprotein-based materials [34]. To date, two CRs glycoproteins have been characterized: the shrimp ovarian peritrophin (SOP) (Figure 1) [35,36] and the cortical rod protein (CRP) [37,38]. SOP and CRP are high-Man glycoproteins that also contain a minor proportion of GlcNAc, sialic acid and fucose (Fuc) residues, some of which are additionally sulfated, playing a key role in modulating shrimp sperm acrosomal reaction response [39].

One of the most important aspects of oogenesis is the formation and storage of the yolk proteins, a process known as vitellogenesis. In oviparous animals, yolk proteins are the most important source of nutrients for developing embryos, constituting 60%–90% of the total egg proteins. Oocyte yolk proteins derive from the enzymatic cleavage of a common precursor called vitellogenin (Vg), which is a glycolipophosphoprotein showing similar characteristics in vertebrates and invertebrates [40].

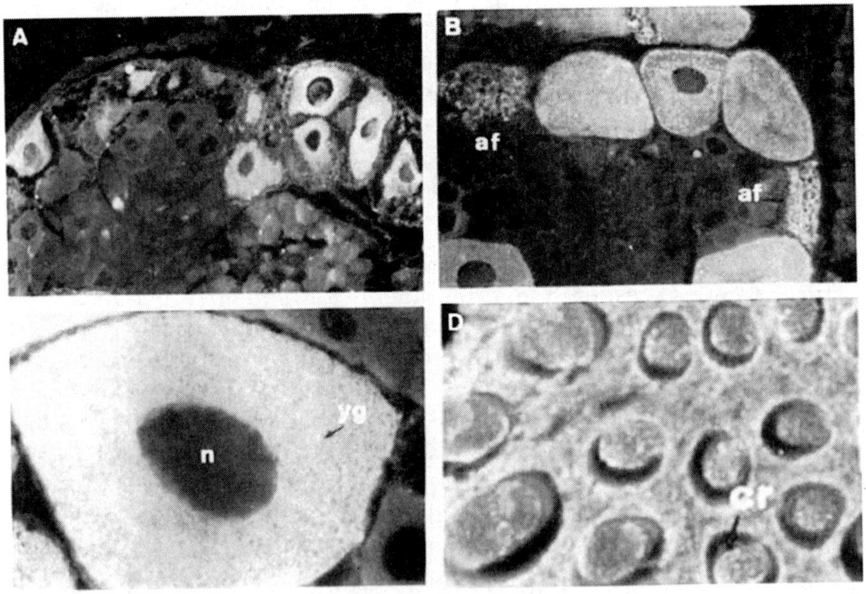

Figure 1. Immunolocalization of shrimp ovarian peritrophin (SOP) in the shrimp ovarian sections at different stages of oocyte development. Immunofluorescence labeled previtellogenic (**A**) and vitellogenic oocytes (**B**) localized at the periphery of ovarian lobe. Higher magnification revealed that SOP was specifically present in the oocyte cytoplasm from vitellogenic (**C**) and late vitellogenic (**D**) ovarian stages; no label was detected in the nucleus or in the surrounding follicular cells. In vitellogenic oocytes, SOP immunoreactivity was also detected in CRs (**D**) (Modified from [35]).

In all model systems studied, Vg is uptake by the growing oocyte, in which is proteolytically cleaved into a characteristic suite of yolk proteins: a high-density lipoglycoprotein called lipovitellins, highly phosphorylated phosvitins, and a β-component, stored within membrane-bound structures called yolk platelets, major source of nutrients for eggs and developing embryo [8]. The primary degradation products of Vg have also been shown to play a role in regulating oocyte hydration [41] and buoyancy of eggs [42]. In addition, the ion-binding properties of Vg serve as a major supply of minerals to the oocytes [43].

The structure of the glycan moieties of Vg has been discovered only in decapod crustaceans. Recent study of the composition of oligosaccharides attached to the Vg of the crayfish *Cherax quadricarinatus* showed that Vg, which is produced in the hepatopancreas and secreted to the hemolymph [44], is posttranslationally modified by *N*-linked oligosaccharides. These *N*-glycan moieties are composed of monoglucosylated and highly mannosylated glycans, ranging from Glc1Man9GlcNAc2 (one glucoses, nine mannoses, and two GlcNAc) to $Man_5GlcNAc_2$ (five mannoses and two GlcNAc). Vg is a large hydrophobic protein containing glucose-capped oligosaccharides. It has been suggested that glycosylation of Vg plays an important role in the folding and subunits assembly to achieve the mature protein in the hemolymph and ovary [45]. Glycosylation is known to increase solubility of proteins [46] so *N*-glycans might have a significant role in keeping this large, hydrophobic protein in the hemolymph to improve its transport in the ovary [45]. The uptake of Vg into the oocytes is known to be mediated by receptors [47,48]. In this respect, the glycan moiety might play a role in recognition and receptor-mediated endocytosis. Once taken into the oocytes, it might also have a role in packaging and compacting its products in yolk bodies [45].

In several sea urchin species, Vg is synthetized in the intestine [49] as a 195-kDa glycoprotein and secreted into celomic fluid of the adult. From the coelom, this glycoprotein is absorbed by the nutritive phagocytes (accessory cells) of the ovary, where it is stored. Then it is transported to the growing oocytes to be accumulated as a 180-kDa glycoprotein, termed major yolk protein (MYP) [50,51,52]. This decrease in molecular weight from Vg to MYPs seems to be a common phenomenon in sea urchin. It is possible that Vg is slightly modified in molecular structure after its incorporation into the gonads. Shyu *et al.*[49] presumed that this modification takes place in the oocytes, but it has also been suggested that it occurs in the nutritive phagocytes of both sexes immediately after incorporation [51].

Glycosphingolipid (GSL) is composed of a sugar chain and a ceramide consisted of a fatty acid and sphingoid base [53]. GSLs are ubiquitous on the outer surface of the plasma membrane in animal cells and play an essential role in intercellular interaction and recognition [54,55,56,57]. However, GSL functions have not yet been clarified because of the structural complexity of the sugar chain and ceramide moiety [58,59]. According to the structure of the carbohydrate moiety, GSLs have been divided in two main groups: neutral glycosphingolipids (NGSLs) and acidic glycosphingolipids.

Gangliosides are more complex glycosphingolipids in which oligosaccharide chains, containing sialic acid, are attached to a ceramide. In the eggs of some sea urchins, the chemical structures of the major gangliosides have been identified as NeuGcα2-6Glcβ1-1Cer (M5) and HSO3-NeuGcα2-6Glcβ1-1Cer (T1). M5 ganglioside constitutes more than 90% of the egg gangliosides and 0.8% or more of the egg dry weight [60,61,62]. In unfertilized egg ganglioside M5 has been shown to be localized in the plasma membrane and in yolk granule, where it is associated with yolk lipoproteins and is involved in the uptake of yolk lipoproteins into the growing oocytes during oogenesis and transported from yolk granules to other cellular components during embryogenesis. M5 ganglioside associating with yolk lipoproteins in yolk granules may be considered a significant stored material to be utilized for carly embryogenesis [63,64,65].

Sea urchin oocytes contain a variety of NGLs such as glucosylceramide, melibiosylceramide, ceramide trihexoside, and GL-5 [61,66,67]. The chemical structures of these glycolipids was determined showing that ceramide moieties of these are almost identical.

Glucosylceramide (Glcβl-1Cer) and melibiosylceramide (Galα 1-6Glcβ1-1Cer) have the same long-chain base compositions that are very characteristic, all of them are phytosphingosines. The fatty acid compositions of these glycolsphingolipids also resemble each other [68].

The chemical structure of the ceramide trihexoside was determined (Galβl-6Galβ1-6Glcβ1-lCer) and its carbohydrate structure is a novel trisaccharide. Glucosylceramide resembles ceramide trihexoside in ceramide structure suggesting that it is synthesized from glucosylceramide [66].

GL-5 chemical structure is Fucα1-3GalNAcβ1-4(Fucαl-3)GlcNAcβl-4Glcβ1-lCer. GL-5 has a unique saccharide sequence: the reducing terminal disaccharide core is GlcNAcβl-4Glc. The defucosylated core structure (GalNAcβ1-4GlcNAcβl-4Glcβ1-) is a novel trisaccharide chain. Fuc directly binds to GalNAc and the sugar structure is one of the shortest saccharide chains so far found among the difucosylated glycolipids [67,69]. Kubo *et al.* [68] speculate on the possible biological functions of these glycolsphingolipids suggesting that they could participate to sperm egg interaction process.

Another feature of oocyte development is the formation of an extracellular eggshell, or vitelline envelope [70], that in some species is encompassed by a jelly-like layer. The most common molecules on the surface of oocyte envelope are glycoproteins [71,72,73,74] that are provided by the liver [75] or by the ovary [76,77,78] or by both of them [79]. The vitelline envelope possesses many functions, such as prevention of polyspermy, protection of the growing oocyte and the developing embryo, uptake of nutrients and other molecules during oogenesis, and guidance of the spermatozoa to the oocyte.

The vitelline envelope of fish eggs is composed of an outer layer rich in glycolipids and of an inner layer, called zona radiate, containing neutral glycoproteins [7]. The zona radiata is composed of 3–4 glycoproteins derived from glycoproteins precursors, known as choriogenins [8]. During oogenesis, choriogenin is synthesized in either the ovary or the liver and incorporated into the egg envelope [80].

Although the biological function of these glycoconjugates is unknown, it has been suggested that they are involved in the hormones binding, in the transport of metabolites and ions across plasmalemma, and in the sperm-egg interaction process [81].

From the vitelline envelope of the unfertilized eggs of rainbow trout (*Salmo gairdneri*), it was isolated, for the first time, a new acid glycoprotein that is designated as "KDN -glycoprotein" because it contains only KDN but no sialic acid as the acid carbohydrate moieties. Other carbohydrate components of KDN-glycoprotein are Gal and GalNAc [18]. KDN-glycoprotein consists of 500 kDa polypeptide chain, to which a number of *O*-linked glycans are attached. These oligosaccharide chains have a core trisaccharide Galβ1-3GalNAcα1-3GalNAc in which the terminal Gal residue is blocked by a single residue of KDN and the proximal GalNAc residue is linked to α2,8-linked oligo(KDN) chains with different degrees of polymerization [82]. KDN-glycoprotein was shown to contain also *N*-linked complex-type glycan chains as minor components. KDN-glycoprotein is the unique glycoprotein of the rainbow trout egg envelope but it is also the major component of vitelline envelope of chum and kokanee salmon. KDN-glycoprotein is located in second layer of vitelline envelope but exposed to the outer surface of vitelline envelope around the micropyle through which spermatozoa can penetrate the oocyte at fertilization. Although, little is known about the functional roles of these glycan chains, probably they act as cell surface receptor for spermatozoa in fertilization [83]. An analogous family of mucin-type glycoprotein that contains Neu5Gc instead of KDN is the major component of cherry salmon vitelline envelope [84].

Glycoconjugates forming zona radiate of different teleost fish eggs have been well investigated. In the flatfish *Solea senegalensis*, the presence of glycans with terminal GalNAc and/or αGal and with terminal/internal αMan is reported. As oogenesis proceeds, the glycan pattern of zona radiata decreases drastically and in the last phase of maturation only few βGalNAc residues are detected. Zona radiate of dusky grouper *Epinephelus marginatus* is characterized by a higher content of GlcNAc and sialic acid [20] as in swordfish zona radiate in which α-L-Fuc residues are also detected [8]. During oocyte development, the glycan pattern changes and this may reflect the different activity of zona radiate during different phases of oogenesis [19].

Studies carried out on different species of Crustaceans, such as rock shrimp *Rhynchocinetes types* and peneaid shrimp*Sycionia ingentis*, have characterized oligosaccharides on the glycoproteins present on oocyte envelope [85,86]. They show a high concentration of glucose (Glc) and α1-3mannose (α1-3Man) on the oocyte at all levels of maturation. Gal, Fuc, GlcNAc and GalNAc are others monosaccharides detect in lowest concentration on the oocyte envelope. Fuc, Gal and GalNAc exclusively have a structural function

In glycoconjugates while α-GlcNAc has great importance, together with Glc and Man, during gamete interaction process, in spite of its small concentration.

Differently from Crustaceans, α-GlcNAc is highly expressed on the oocyte envelope of ascidian *Phallusia mammillata.* This monosaccharide, as well as GalNAc, is predominant with 86% by weight of total sugar content and Fuc, Man and Glc accounted for the remaining 14% [87]. GalNAc, α-GlcNAc and Man, with Gal are also the major components of the vitelline coat of another ascidian *Halocinthia roretzi*, in which monosaccharide analysis of glycans reveals also a high content of arabinose, xylose and rhamnose, but any information about their plausible biological significance is not available. Glycans are *O*-linked and Gal and xylose residues are detected at reducing termini. Several analysis suggest that 1,4-linked xylose, 1,3-linked Gal and GalNAc residues constitutes the core structures of anionic glycans [88]. Rhamnose is also one of the monosaccharides detected in the vitelline coat of another ascidian *Ciona intestinalis* [89]. The vitelline coat of the mature oocyte of *Ciona intestinalis* contains three glycoproteins and the most abundant carbohydrate is GalNAc, followed by Fuc and Gal [90].

In different sea urchin species, the oocyte jelly coat was shown to contain two major acid glycoproteins, sialic acid-rich glycoprotein and fucose sulfate glycoconjugate (FSG) [91,92]. The sialic acid-rich glycoproteins have a low protein content and contain many *O*-linked sialyglycan chains attached to threonine residues on the core protein. The polySia chains in polysialic acid-glycoprotein contain Neu5Gc residues and are composed of Fuc, Gal, and GalNAc residues. Most interestingly, the inner residue linkages in the polySia chains contain unique (→5-O_{glycolyl}-Neu5Gcα2→)$_n$, linkages [93]. This structure represents the first naturally occurring polySia as an integral part of a glycoprotein, which potentiates the FSG-induced acrosome reaction in a dose dependent manner [94]. FSGs are the major macromolecules present on the egg jelly coat responsible for inducing the sperm acrosome reaction [95,96,97]. They are linear polysaccharides with a repeated unit composed of sulfated 1–4 hexose moieties. It was found that, among sea urchin species, they differ in sugar composition (L-Fuc or L-Gal), glycosidic linkage (α1→3 or α1→4) and in the sulfation pattern (2- or 4-*O*-sulfate) [98,99]. These differences are responsible of species-specific induction of the acrosomal reaction [94,98].

Starfish egg jelly coat contains two glycoproteins: the first one is a high Man glycoprotein with a molecular mass of 80 kDa and an unique saccharide structures, but its function is not well clarified [100,101]; the second one is a highly sulfated Fuc-rich glycoprotein, a unique glycoprotein playing a key role in the trigger of sperm acrosome reaction, named acrosome reaction-inducing substance (ARIS) [101,102].

The biological activity of ARIS resides in one of the sugar fragments called fragment 1 [94,103]. The molecular size of this glycan is about 10 kDa and does not contain amino acid residues. Its structure was revealed as ten or so repeats of the following pentasacharide unit; [→4)-β-D-Xyl*p*-(1→3)-α-D-Gal*p*-(1→3)-α-LFuc*p*-4(SO_3^-)-(1→3)-α-L-Fuc*p*4(SO_3^-)-(1→4)-α-L-Fuc*p*-(1→] [104]. This sugar chain links to the peptide part by *O*-glycosidic linkage trough another sugar chain with different structure from fragment 1 [94].

The inner sugar portion of ARIS has been isolated as fragment 2. It is mostly composed of sulfated glycans and retains about 10% (w/w) of the protein part. Fragment 2 has a molecular mass of 400 kDa and its glycans are *O*-linked. It is composed of the heptasaccharide units of [→3)-Gal*p*-(1→3)-Fuc*p*-(1→3)-Gal*p*-(1→4)-GalNAc*p*-(1→4)-GlcNAc*p*-6(SO_3^-)-(1→6)-Gal*p*4(SO_3^-)-(1→4)-GalNAc*p*-(1→] [105]. Polymerizations of outer glycans are necessary to form the layer. ARIS has another acid sugar chain composed of Man, GalNAc and GlcNAc, although it is not important for the biological activity [106].

The oocytes of a tropical abalone *Haliotis asinina* have two protective barriers: the egg jelly coat and the vitelline envelope. The egg jelly contains two major glycoproteins of 107 kDa and 178 kDa, whereas the vitelline envelope contains a broad spectrum of protein bands ranging from 15 to 200 kDa. Glc is the major sugar residue of both egg jelly and vitelline envelope glycoproteins, whereas minor proportions of arabinose, fructose, Gal, and Fuc are present in both the egg jelly and vitelline envelope [107].

3. THE GLYCOBIOLOGY IN SPERMATOGENESIS

Spermatogenesis is a developmental process during which spermatogonial stem cells produce highly differentiated spermatozoa. It starts with the mitotic proliferation of spermatogonia and formation of primary spermatocytes, passing then through meiosis into secondary spermatocytes, which differentiate into spermatids and finally through spermiogenesis into mature spermatozoa [108,109,110,111].

During spermatogenesis, glycoprotein composition significantly changes suggesting the important role of these components in this process [112,113,114]. In sperm development, some complex and not well known germ cell-somatic cell interactions are involved and several reports suggest that glycoconjugates are involved in cell adhesion during this process [115,116].

Spermatogenesis in fishes is characterized by a cystic mode that consist of a set of Sertoli cells surrounding a group of synchronously differentiating germ line cells through the respective developmental stages until spermatogenesis is completed and mature spermatozoa are formed. The development of spermatogenic cells requires a special microenvironment that is created by the Sertoli cells [117]. Through cytoplasmic projections, these cells surround clones of primordial germ cells forming cysts in the seminiferous tubules [118].

The testes of the teleosts have two compartments: a tubular compartment constituted of Sertoli's cells and germinative cells forming spermatogenic cysts; and an interstitial compartment that is formed of connective tissue and Leydig's cells [119].

The identification and localization of glycoprotein oligosaccharide sequences in the testis of teleost fish *Solea senegalesis*have been well investigated [120]. Glycoproteins are localized in both compartments of the testis, the interstitial and germinal compartments, but also in the basal lamina separating them. In the interstitial stroma, a very complex glycoprotein composition, including asialo-, as well as sialoglycans, in *N*- and *O*- linked oligosaccharides and *N*-linked glycans, was discovered. It has been also demonstrated the glycosylation pattern of the interstitial stroma is depending on the testicular region analyzed. In fact Neu5Acα2,3Galβ1,4GlcNAc and GalNAcα1,3(L-Fuc α 1,2)Galβ1,3/4GlcNAcβ1 are more abundant in the medullar region than in the cortex. This different pattern is due to a differential glycoprotein compound trafficking patterns from the vascular system to each testis region. This occurs because germ cell proliferation and differentiation take place in the cortical region of the testis while final sperm maturation in medullar region, in order to support distinct glycoprotein requirements in different phases of developmental process.

The interstitial compartment of fish species is characterized by the presence of melano-macrophage centers, that contain glycans terminating with Galβ1,3GalNAc but their role is not known, and Leydig cells, that unlike from other teleost species, does not exhibit glycoprotein oligosaccharide sequences [121].

The basal lamina is characterized by glycans with terminal/internal Man, internal βGlcNAc, terminal Neu5Acα2,6Gal/GalNAc, Neu5acGalβ1,3GalNAc, Galβ1,3GalNAc, Galβ1,4GlcNAc, GalNAc, αGal and αL-Fuc. This complex glycosylation pattern is related to the presence of meshwork composed of several glycoconjugate components, responsible for biological function of basal laminas.

In the germinal compartment, Sertoli cells express sialoglycans terminating with Neu5Acα2,3Galβ1,4GlcNAc that may be involved in the establishment of interaction between adjacent spermatocysts playing a key role in the organization of spermatogenic cysts.

Different from *Senegalese sole*, the Sertoli cells of other fish species, such as spotted ray and Nile tilapia, display a more complex glycosylation pattern, but sialoglycans seem to be lacking in those species [121,122].

Glycoprotein oligosaccharide sequences are absent in spermatogonia while primary spermatocytes express, in cytoplasm and nucleus, glycans terminating with Galβ1,3GalNAc and αGalNAc, respectively [120]. Since germ cell nuclear glycoproteins are associated with the chromatin, it has been suggested that they play a role in the regulation of transcription factors and in the control of cell cycle [123]. Spermatid cytoplasm exhibits *N*-linked glycans, containing high-Man residues, as well as oligosaccharides terminating with α/βGalNAc, αGal and αL-Fuc. These glycoproteins were also found in fine granular structures located in the cytoplasm of spermatids.

The presence of glycoconjugates has been also demonstrated in germ cells and somatic cells, Leydig and Sertoli cells, in cartilaginous fish testis. In particular, germ and somatic cells undergo extensive modifications regarding the composition of glycoside residues at the level of surface, cytoplasm, and nucleus. On the surface of germ and Sertoli cells, it has been identified cadherins, glycoprotein superfamily containing mannosyl chains. Cadherin play a key role in establishing interactions between germ and somatic cells, implied in cyst and spermatoblast formation. In addition, germ cells and Sertoli cells undergoing apoptosis change the cellular surface composition, specifically overexpressing on the cell surface GalNAc and Gal that make them susceptible to phagocytosis. During cyst maturation, also Leyding cells change their surface glycoconjugate composition due to a modification of their activity during spermatogenesis. Germ cells express new sugar residues during their differentiation, the presence of glycoconjugates is limited to the Golgi zone in spermatocytes, to acrosome formation region in spermatids, and to the acrosome in spermatozoa [121]. This finding suggests that the acrosome glycosylation pattern depends on stoichiometric and spatial distribution of binding sites and/or the activity of glycolytic enzymes and glycosyltransferases [121,124]. Glycoconjugates are also expressed in the germ cell nucleus where they are involved in the control of cell cycle [121].

Studies on sea urchin testis have revealed the presence of the yolk protein precursor, Vg, in immature male gonad as well as in the female one, suggesting that male Vg could be a precursor of MYP incorporated in the testicular nutritive phagocytes as a nutrient source for spermatogenesis [51]. As spermatogenesis proceeds, MYP decreases in quantity; in fact, it is utilized as material for synthesizing new substances that participate to the formation of spermatozoa and are metabolized as an energy source during this period [125].

The immature sea urchin testis also contains a large quantity of polysaccharides, most of which is probably glycogen in the form of granules [126,127]. The polysaccharide content decreases as gametogenesis proceeds in both sexes, since possibly they are used as an energy source, as suggested for *Strongylocentrotus intermedius* [127].

In sea urchin testis, it has also been demonstrated the expression of a heavily sialylated glycoprotein, named flagellasialin. Flagellasialin has a molecular mass ranging from 40 to 80 kDa and contains a unique polySia sulfated α2,9-linked polyNeu5Ac [128]. Flagellasialin is exclusively located to the flagellum where probably α2,9-linked polySia regulates voltage sensitive sodium and calcium channels influencing the intracellular calcium and sperm motility [129,130]. In the flagellum but also in the sperm head, it has been demonstrated the expression of a 190 kDa glycoprotein linked to α2,8-linked polyNeu5Ac structures. The biological function of this glycoprotein is not yet clarified. The conformational differences between α2,9- and α2,8- linked polySia structures and their co-localization in the same sperm might reflect functional differences of these two polySia-containing glycoprotein at fertilization. The α2,8-linked polyNeu5Ac structure occurs also in glycolipids, localized in the sperm head in order to facilitate rearrangement of the membrane proteins on the sperm surface upon sperm activation [131,132].

In Decapoda it has been shown that immature germ cells and spermatozoa have different glycoconjugate composition. In*Aristaeomorpha foliacea*, immature germ cells only express *N*-linked oligosaccharides which contain terminal and internal α-D-Man, internal β-D-GlcNAc, NeuAcα2,6Gal/GalNAc and terminal NeuAcα2,3Galβ1,4GlcNAc. Spermatozoa express in the cytoplasm both *N*- and *O*-linked oligosaccharides. The *N*-linked oligosaccharides consist of terminal GlcNAc while *O*-linked oligosaccharides terminated with β-Gal(1-3)-GalNAc and/or α-GalNAc in the cytoplasm, whereas they ends with sialic acid linked to β-Gal(1-3)-GalNAc in the nucleus [133]. The extracellular matrix in which the spermatozoa are embedded consists of neutral and acid glycoconjugate. Acid polysaccharides have been reported in *Albunea symnista* [134] and *Panulirus homarus* [135]. Neutral glycoprotein and acid polysaccharide have been found in *M.rosebergii* [136]. Differences in glycoconjugate composition of extracellular matrix between immature and mature hemispermatophores have been observed: the former expresses both *N*- and *O*-linked glycoconjugates while mature hemispermatophores express only *O*-linked oligosaccharides and contain more sialyl-glycoconjugates than immature ones [133]. The role of extracellular matrix in decapods is not well known. It has been suggested that the acid mucopolysaccharide of spermatophores may act as a cementing agent or an antimicrobial agent [137] or in the maintenance of spermatozoa during the storage within female spermathecae [133,138].

Previuos study on *A. foliacea* also demonstrates that sperm glycoprotein pattern seasonally changes and spermatozoa undergo maturative changes in glycoconjugate compotion during their transit from testis to hemispermatophore [139].

4. THE GLYCOBIOLOGY IN FERTILIZATION

Fertilization is a highly specialized process of cell-cell interaction that marks the creation of a new individual. Although the details of fertilization vary between species, it generally consists of five major events. The first event is the recognition and the species-specific interaction between sperm and the egg coat. This interaction allows spermatozoa to undergo the acrosome reaction, an exocytosis of the acrosomal vesicle located on the tip of the spermatozoa head, which releases a lytic agent. These events enable spermatozoa to cross the extracellular matrix and reach the oocyte plasma membrane where binding between the two cells occurs and causes the fusion of the genetic material of spermatozoa and egg leading to the activation of egg metabolism, mitosis and the beginning of development [140,141,142].

Glycoconjugates play a number of pivotal roles in multifaceted processes during fertilization [1,143]. The initial interaction between oocytes and spermatozoa involves glycoproteins of oocyte envelope and complementary sperm surface receptors [144]. In particular, the terminal oligosaccharides of egg envelope glycoprotein are responsible for spermatozoa binding through recognition of species-specific polypeptide chains on the spermatozoa [87,145,146].

The first evidence supporting the function of complex glycoconjugates in fertilization was found in sea urchins [91]. Two carbohydrates are involved in sea urchin fertilization. The first carbohydrate is egg jelly coat FSG that binds a glycoprotein receptor, named suREJ located exclusively in the plasma membrane just on the sperm acrosomal vesicle at the anterior apex of the spermatozoa head [147,148]. suREJ contains one epidermal growth factor module and two *C*-type lectin carbohydrate-recognition modules [147]. After this binding, the spermatozoa undergo consecutive morphological and biochemical changes called the acrosomal reaction. In response to signals transduced by sperm receptor, acrosomal vesicles fuse with the plasma membrane and actin polymerizes to form the acrosomal process. The exocytosis of acrosomal vesicles expels the protein bindin that coats the spermatozoa acrosomal process [149]. The second carbohydrate involved in sea urchin fertilization is a glycoprotein receptor for bindin in the vitelline layer [150]. This receptor, namely EBR1, is a 350 kDa glycoprotein containing both *N*- and *O*-linked oligosaccharide chains with different biological activity, being inactive and active respectively. The active oligosaccharide is linked to serine or threonine via an *O*-glycosidic bond to GalNAc, which contain a Gal bound β(l→3) to the linkage sugar. The remainder of the structure consists of GlcNAc and Fuc residues, with the sulfate moieties on the Fuc residues. The *O*-linked oligosaccharide chains show different level of sulfation correlated to their activity, being greatly enhanced with increasing sulfation. Two-step models involving carbohydrate and protein chains have been proposed for sperm-egg interaction in sea urchin. The first step is postulated to be a low-affinity ionic interaction of the sulfated *O*-linked oligosaccharide chains of the receptor with spermatozoa that is not species specific. This is followed by a high affinity, species-specific interaction of one or more domains of the polypeptide chain and bindin on the acrosomal process [151]. The species specificity of this interaction is determined by differences in the position of attachment of the oligosaccharide chains on the protein backbone [152].

In starfish, the acrosome reaction is induced by the combination of ARIS and other two components of the egg jelly coat: sulfated steroid saponin namely Co-ARIS and an oligopeptide known as asterosap [153,154]. When spermatozoa reach the egg jelly coat, the first binding occurs between ARIS and its receptor, specifically located to the anterior region of spermatozoa heads [155].

The tertiary structure of the ARIS is important for its activity. The signal for a correct folding is derived from fragment 1. The fragment 1 has a double helix and a compact carbohydrate core with sulfates protruding in pairs away from the center of the helix. It has been demonstrated that the double helix structure of ARIS saccharide chain is important for the induction of acrosome reaction [94]. The binding gates calcium channels inducing an intracellular calcium increase. This signal causes exocytosis by which spermatozoa expose the devices essential for penetration through the egg coats and for fusion with the egg plasma membrane.

In sea water, ARIS requires Co-ARIS to induce acrosome reaction [156] and microdomain changes [103]. Co-ARIS is composed of sulfated steroid and a pentasaccharide chain. Its activity requires sulfate moiety and mainly depends on the structure of steroidal side chain, while not necessarily requiring a specific structure of the saccharide chain [157]. Spermatozoa do not have a specific receptor for Co-ARIS. It has been suggested that steroid ring and side chain of Co-ARIS may infiltrate or be inserted into spermatozoa plasma membrane and the sulfated group and sugar chain contribute to keep it in the right position and orientation to interact correctly with other components of spermatozoa plasma membrane [101].

Asterosap is a glutamine-rich tetratriacontapeptide with a 25-residue ring formed by a disulfide linkage that is essential for its biological activity [158]. It transiently increases the intracellular pH and calcium via the activation of asterosap receptor that is a guanylyl cyclase located in the sperm flagellar plasma membrane [159,160,161]. Calcium and pH increase are essential to trigger acrosome reaction.

In archaeogastropods, the molecules analogous to sea urchin bindin and EBR1 are lysin and its receptor, most extensively studied in abalone. The abalone spermatozoa swims easily through the egg jelly coat and reaches the vitelline envelope. The contact between the spermatozoa and the vitelline envelope induces the polymerization of actin to generate acrosomal process and the exocytotic of acrosomal vesicle that release protein lysin onto the surface of the vitelline envelope. Lysin is a 16 kDa nonenzymatic, cationic protein that forms a dimer, which is able to bind a giant 1000 kDa glycoprotein named Vitelline Envelope Receptor for Lysin (VERL) [151]. VERL is a long, unbranched, fibrous glycoprotein that is composed of at least 50% saccharide, containing Glc and Man residues, and comprises 30% of the vitelline envelope mass [162]. Dimer binding results in monomerization and the tight, species-specific binding of lysin to VERL that involves the carbohydrate moieties of VERL linked to serine or threonine via an *O*-glycosidic bond. Upon binding lysin monomers, the fibrous VERL molecules lose cohesion and splay apart, through which the spermatozoa passes to reach the egg cell membrane and the tip of its acrosomal process fuses with the egg plasma membrane [163,164]. Other vitelline envelope glycoproteins of 30–50 kDa, even if do not binding lysin, could be involved in mediating species selectivity based on their interaction with VERL [162,165].

In ascidians, sperm-egg binding is mediated by an enzyme-substrate complex established between a specific spermatozoa surface glycosidase and corresponding glycans on the surface of the vitelline coat [166,167]. In the alkaline seawater, this complex remains stable, since the glycosidase has an acid pH optimum [167,168,169]. In *Ascidia nigra* and *Phallusia mamillata*,*N*-acetyl-glucosaminidase in sperm membranes recognizes terminal α-GlcNAc residues on the oocyte envelope [168,170], while in *Ciona intestinalis* and *Halocynthia roretzi*, the sulfated, Fuc-containing glycans of vitelline coat glycoproteins are responsible for the spermatozoa binding to the vitelline coat. In particular, the sperm glycosidase, α-L-fucosidase, binds to terminal L-Fuc residues of the vitelline layer [94,101,160,171]. The degree of sulfation and the proper spacing of sulfate groups seem essential for biological activity [88]. When spermatozoa bind vitelline envelope, acrosome reaction occurs, the acrosomal outer membrane fuses with the plasmalemma enclosing the acrosome, resulting in exocytosis of the acrosomal substances. After the acrosome reaction, apical processes protrude mainly from the peripheral region of the apex of the spermatozoa head [172,173,174]. Although the chemical nature and the precise role of the acrosomal substances remain to be elucidated, it has been proposed that these acrosomal substances are responsible for membrane fusion between the apical processes and the egg plasma membrane [173,174]. The sperm-egg vitelline interaction activates a spermatozoa lysine system. In particular, three proteases are involved in spermatozoa penetration of the vitelline coat: two trypsin-like proteases, acrosin and spermosin, and one chymotrypsin-like protease. The chymotrypsin-like activity is involved in spermatozoa penetration of the vitelline coat, but spermosin and acrosin both function to increase the rate of fertilization [175,176,177].

In decapod crustaceans, the first sperm-egg interaction is established between the apical end of a sperm's appendage (spike) and the outermost egg envelope. In particular, in *Rhynchocinetes typus* it has been demonstrated the presence of a lectin-like molecule on the tip of spermatozoa spike that recognized specific carbohydrates on the oocyte envelope such as GlcNAc, Glc and Man [86,178]. At this point, the tip of the rigid spike exerts a lytic effect upon vitelline envelope causing a perforation through which the spermatozoa passage to reach the plasma membrane oocyte [179,180,181].

Polyspermy is the fusion of more than one spermatozoa with an egg, which is a lethal condition to the embryos of most organisms [182]. At fertilization, rapid and slow changes occur to block polyspermy. The rapid changes are related to the modification of egg membrane potential. The slow changes are due to the modifications of egg vitelline envelope, which transform it to a hard layer called the "fertilization membrane". The hardening of vitelline envelope is trigged by the contents of cortical granules that are released upon fertilization. Fertilization membrane forms a protective barrier that repels additional spermatozoa, but also bacteria and small eukaryotic invaders [27]. Several studies report that glycoconjugates are implied in to block polyspermy.

In sea urchins, fertilization envelope formation is initiated by trypsin-like proteases that cleave both a sperm-binding protein [183,184,185,186] removing supernumerary spermatozoa and preventing further spermatozoa binding, and the proteins that connect the vitelline envelope to the plasma membrane [27,187]. At the same time, sugars that are released from the cortical granules attract water into the perivitelline space allowing the vitelline envelope lifts off the egg plasma membrane that is also due to the formation of hyaline layer [27]. Hyalin is a filamentous molecule that tends to form aggregates forming an amorphous layer [188]. In particular, its carbohydrate residues act as receptors for hyaline-hyalin and hyaline-cellinteraction [27]. Then, ovoperoxidases harden the vitelline envelope that becomes resistant to both mechanical and enzymatic modifications [27]. Ovoperoxidase released in the perivitelline space interacts with another cortical granule protein, the proteoliaisin. When this complex is associated with the nascent fertilization envelope, ovoperoxidase catalyzes the covalent cross-linking of juxtaposed tyrosine residues in adjacent polypeptide chains, including SFE 1 and SFE 9, to form a stable, macromolecular complex [189]. Proteoliaisin targets the ovoperoxidase to the nascent vitelline layer/fertilization envelope [190] and protects this enzyme from proteolytic digestion [191].

Fish spermatozoa lack an acrosome and they attach directly to the egg plasma membrane through the micropyle, a single, small pore in the chorion [148]. Following sperm-egg fusion, cortical alveoli fuse with the plasma membrane and discharge their contents, PSGP, hyaline, proteases and transglutaminases into the presumptive perivitelline space [15,182,192,193,194]. Once released from the cortical alveoli, proteases cleave the 200 kDa PSGP (L-PSGP) into small 9 kDa PSGP (H-PSGP) [15,192,193]. Both amino acid and carbohydrate compositions of L-PSGP and H-PSGP are identical. H-PSGP is still too large to permeate the chorion. This establishes a colloid osmotic pressure, which causes an influx of mostly external water that swells the perivitelline space. The perivitelline space cushions the embryo and bathes it in a special medium containing the 9 kDa PSGP, lipids, carbohydrates, and ions as well as providing a sink for nitrogenous wastes [195,196,197]. The perivitelline space also provides room for the free movement and growth of the embryo, which, due to its hypertonicity, absorbs water osmotically and slowly swells during embryogenesis [198]. With the establishment of the perivitelline space, the rise in hydrostatic pressure is aided by the hardening of the chorion, which contributes to the closure of the micropyle and thereby reduces the probability of polyspermy and microbial infection [199,200]. The hardening process is catalysed by transglutaminases activated of PSGP that have proteinase activity [201]. Thus, activated transglutaminase forms covalent ε-(γ-glutamyl)-lysine crosslinks constituents of adjacent proteins or glycoproteins of chorion [202].

Ascidian eggs lack cortical granules and prevent polyspermy by first releasing a large quantity of glycosidase, followed by an electrical modification of the egg plasma membrane [87,203,204,205]. Sperm-egg binding triggers eggs to release large quantities of glycosidase that rapidly bind the vitelline-coat surface as spermatozoa surface glycosidase blocking the binding of supernumerary spermatozoa [74,101,204].

5. THE GLYCOBIOLOGY IN EMBRYO DEVELOPMENT

Glycoconjugates have relevance for embryo development, tissue and cell specialization and organogenesis. Several studies reported that glycoconjugates are involved in embryo development of marine invertebrates. Early studies on inhibition of glycoprotein synthesis and embryonic development of sea urchin revealed that tunicamycin (a mixture of homologous antibiotics that blocks *N*-glycosylation of proteins, inhibiting *N*-acetylglucosamine transferases) blocked the process of gastrulation [206,207]. Subsequently, it was established that the sensitivity of the gastrulation process to tunicamycin could be related to a marked increase in the level of *N*-linked glycoprotein synthesis, which occurs just before gastrulation [208]. This requirement for glycoprotein synthesis was consistent with the finding that gastrulation could also be blocked by inhibiting synthesis of dolichyl phosphate [209,210], the lipid carrier required for oligosaccharide chain assembly in *N*-linked glycoprotein synthesis. These early studies of the effect of tunicamycin on development also revealed that this drug blocked spiculogenesis when added at the late gastrula stage [206]. More recently, this apparent requirement for *N*-linked glycoprotein synthesis during spicules formation has received greater attention. Carson *et al.* [211] established that addition of a monoclonal antibody (mAb 1223) to a culture of primary mesenchyme cells caused a block in spiculogenesis. Immunofluorescence studies showed that the 1223 antigen was primary mesenchyme cell-specific. Immunoblot analysis revealed that one of the proteins containing the 1223 epitope was a 130-kD polypeptide whose level of expression correlated with the acquisition of the ability of primary mesenchyme cells to accumulate calcium during spicule formation. These observations, coupled with subsequent studies establishing that the 1223 antigen was a glycoprotein, and that the mAb 1223 was directed toward a carbohydrate chain on the glycoprotein [212,213], led to study both the distribution of this epitope in the embryo and the nature of the oligosaccharide moiety in more detail. These studies have established that the epitope on the glycoprotein recognized by mAb 1223 is a complex, *N*-linked oligosaccharide chain [213] that is found in the cortical granules of the egg, disappears after fertilization and reappears in association with primary mesenchymal cells before spiculogenesis [212,214]. Moreover, some authors [215] demonstrated that two other independently generated monoclonal antibodies, 1G8 [216] and B2C2 [215] also recognized an oligosaccharide group on the same 130-kD protein, suggesting that it is a highly immunogenic mesenchymal cell marker.

Given the complex nature of the carbohydrate chain of the 1223 antigen and its apparently essential role in a step in spiculogenesis, it seemed likely that inhibitors of the processing of oligosaccharide chains would block spicule formation. Kabakoff and Lennarz [217] reported the results of biochemical and morphological studies with such inhibitors using both intact embryos and primary mesenchyme cells in culture. In order to ascertain whether the processing of high mannose oligosaccharides to complex oligosaccharides is necessary for spiculogenesis, intact embryos and cultures of spicule-forming primary mesenchyme cells were treated with two glycoprotein processing inhibitors, deoxymannojirimycin (1-MMN) and deoxynojirimycin (1-DNJ). In both cases, normal embryonic development between gastrula and prisma stage was affected by impairing or completely blocking spicule formation (Figure 2).

These inhibitors did not affect gastrulation in whole embryos or filopodial network formation in cell cultures. It was also demonstrated that (a) 1-MMN entered the embryos and blocked glycoprotein processing in the 24-h period before spicule formation as assessed by a twofold increase in endoglycosidase H sensitivity among newly synthesized glycoproteins upon addition of 1-MMN; (b) 1-MMN did not affect general protein synthesis until after its effects on spicule formation were observed; (c) Immunoblot analysis with an antibody directed towards the polypeptide chain of the 130-kD protein (mAb A3) showed that 1-MMN did not affect the level of the polypeptide that is known to be synthesized just before spicule formation; (d) 1-MMN and 1-DNJ almost completely abolished (>95%) the appearance of mAb 1223 reactive complex oligosaccharide moiety associated with the 130-kD glycoprotein.

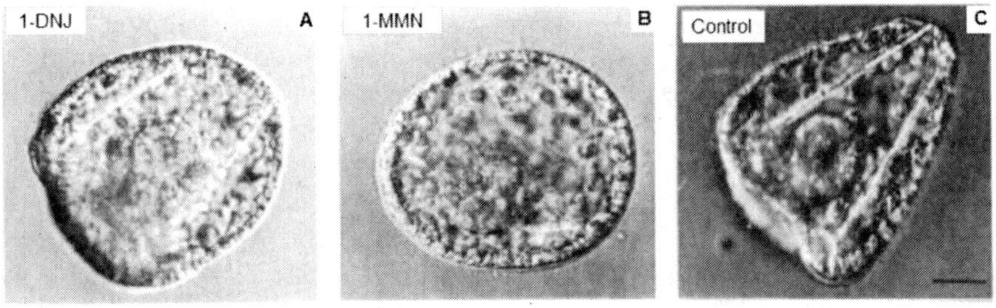

Figure 2. Effects of glycoprotein processing inhibitors. Eggs were fertilized and embryos cultured until 72 h post fertilization, when they were treated with (**A**) 4 mM 1-deoxynojirimycin (1-DNJ, which blocks both glucosidase I and II; (**B**) 2 mM 1-deoxymannojirimycin (1-MMN, which blocks mannosidase I); (**C**) untreated controls. The samples were fixed in 2% glutaraldeyde. Scale bar, 5 μm. (Modified from [217]).

These results indicate that the conversion of high mannose oligosaccharides to complex oligosaccharides is required for spiculogenesis in sea urchin embryos and they suggest that the 130-kD protein is one of these essential complex glycoproteins. In fact, a correlation between a block in processing of the oligosaccharide chain of glycoproteins and impairment of spicule formation was established. Moreover, the conversion of high mannose ologosaccharides to complex oligosaccharides was required for spiculogenesis in sea urchin embryos, suggesting that the 130-kD protein was one of these essential complex glycoproteins.

Gangliosides play important roles also during embryo development. In fact, revealing the distribution of M5 during embryogenesis is an important step in evaluating the significance of gangliosides in early development. Nezuo *et al.* [62] reported that ganglioside M5 was secreted during embryogenesis and localized in the extracellular matrix (ECM). Since M5 exists in unfertilized eggs in yolk granules, it was transported from the yolk granules to the ECM and/or the plasma membrane after fertilization. M5, however, is quantitatively constant during embryogenesis [218]. Using fluorescent-labeled ganglioside NBD-M5, similar thick layers of staining were observed in 2-, 4-, 8-cell stage and blastula embryos (see Figure 3). A number of studies have been shown the role of gangliosides as mediators in the interaction of various cells with ECM in vertebrates, showing also a high affinity to fibronectin [219,220]. Since the sea urchin embryo is surrounded by several ECM components, such as molecules similar to vertebrate fibronectin, collagen and laminin [221], M5 may be involved in the organization of these molecules, although it remains to be determined whether M5 actually interacts with these proteins.

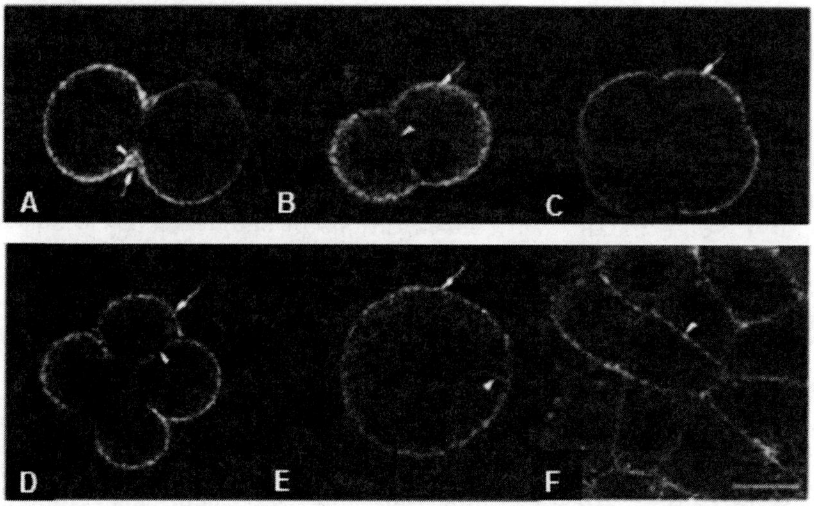

Figure 3. Detection of NBD-M5 in early development with a confocal microscope. To observe the localization of NBD-M5 in living embryos, fertilized eggs pre-labeled with NBD-M5 were grown in artificial sea water without dye. At the end of mitosis (first cell division), the thick ECM (hyaline layer) near the cleavage furrow was stained (**A**); similar tick layers of staining were observed in 2- (**B**), 4- (**C**), 8-cell stage (**D**) and blastula before hatching (**E** and **F**). (**F**) is an higher magnification of the cortical region of (**E**). The localization of NBD-M5 did not change. Scale bar, 10 μm. (Modified from [62]).

Dermatan sulfate is a macromolecule member of a class of natural, structurally complex, sulfated, linear polymers named glycosaminoglycans (GAGs). There are many types of GAGs generally grouped into four categories: (1) hyaluronic acid or hyaluronan; (2) keratan sulfate; (3) chondroitin sulfate (CS)/DS; and (4) heparan sulfate (HS)/heparin. They are biosynthesized as polysaccharides of repeating disaccharides with an *N*-acetylhexosamine, *N*-acetylgalactosamine (GalNAc), or *N*-acetylglucosamine as one of the sugars. The alternating sugar is glucuronic acid (GlcA) with the exception of keratan sulfate, which instead contains galactose. The hyaluronic acid is not further modified, whereas the other classes are modified by (1) the addition of *O*-sulfate groups on various hydroxyls (the three classes); (2) 5-epimerization of some GlcA residues to form iduronic acid (IdoA) residues (DS, HS, heparin), and (3) the removal of acetyl residues from some hexosamines replaced with *N*-sulfates (HS and heparin) [222].

Unfertilized eggs of the sea urchin *Strongylocentrotus purpuratus* are surrounded by a gelatinous layer rich in sulfated fucan. Shortly after fertilization this polysaccharide disappears, but 24 h later, the embryos synthesize high amounts of dermatan sulfate concomitantly with the mesenchyme blastula-early gastrula stage when the larval gut is forming [223]. This glycosaminoglycan has the same back-bone structure [4-α-L-IdoA-1→3-β-D-GalNAc-1]$_n$ as the mammalian counterpart but possesses a different sulfation pattern. It has a high content of 4-O and 6-*O*-disulfated galactosamine units. In addition, chains of this dermatan sulfate are considerable longer than those of vertebrate tissues. Adult sea urchin tissues contain high concentrations of sulfated polysaccharides, but dermatan sulfate is restricted to the adult body wall where it accounts for 20% of the total sulfated polysaccharides, indicating that it retains a biological role in the adult stage. Vilela-Silva *et al.* [223] concentrated their studies on the characterization of the sulfated polysaccharide synthesized by the larvae of the sea urchin*S. purpuratus*. Shortly after fertilization, the sulfated fucan of egg jelly disappears, and by 24 h of development the embryos begin to synthetize large amounts of dermatan sulfate. In addition, sulfation at the 4-*O*-position decreases markedly in the dematan sulfate from adult sea urchin when compared with the glycan from larvae. The importance of dermatan sulfate for the development of sea urchin embryos was emphasized in several experiments where inhibitors of sulfation, chain elongation, or proteoglycan formation were added to the medium and inhibited embryonic development [207,224,225,226,227].

Interesting data were reported on the very attractive process during embryo development, the transdifferentiation, also known as lineage reprogramming. In fact transdifferentiation means the conversion of cells from one differentiated cell type to another. Reber-Muller *et al.* [215] demonstrated that transdifferentiation at high rates is also initiated by treatment of the isolated tissue fragments with the monoclonal antibody (mAb)19 developed against the ECM of the medusa *Podycoryne carnea*. This antibody interferes strongly with cell-ECM interactions. Moreover, oxidation of carbohydrates with sodium-meta-periodate in enzyme linked immunosorbent assays (ELISAs) demonstrated that mAb19 is directed against a carbohydrate epitope. The results suggested that in striated muscle of the medusa cell adhesion and the maintenance of the differentiated state depend on carbohydrate mediated cell-ECM interactions. If these interactions were disturbed by mAb19 the striated muscle cells will undergo DNA-replication and transdifferentiate. In conclusion, medusa cells depend on the presence of species-specific carbohydrates for cell adhesion and spreading [216].

6. CONCLUSIONS

Marine environment offers a tremendous biodiversity and original polysaccharides, presenting a great chemical diversity that is largely species-specific. Marine polysaccharides present an enormous variety of structures, presenting a real potential for natural product drug discovery and for the delivery of new marine derived products for therapeutic applications. They are involved in a range of complex biological processes associated with diseases, immunology, molecular and cellular communications, and mainly with developmental biology. In fact, most recent findings showed that glycobiology of reproductive processes in marine organisms represent a rapidly advancing field and permits an innovative view on the carbohydrate-based interactions that lead to gametogenesis, to fertilization and to embryo development. In conclusion, this study is of special interest for general scientists in reproductive biology and medicine, for glycobiologists and for clinicians working in the fields of human and animal fertility.

ACKNOWLEDGMENTS

We thank Elisabetta Tosti for helpful discussions.

CONFLICT OF INTEREST

The authors declare no conflict of interest.

REFERENCES

1. Caldwell, G.S.; Pagett, H.E. Marine glycobiology: Current status and future perspectives. *Mar. Biotechnol.* **2010**, *12*, 241–252.

2. Shriver, Z.; Raguram, S.; Sasisekharan, R. Glycomics: A pathway to a class of new and improved therapeutics. *Nat. Rev. Drug Discov.* **2004**, *3*, 863–873.

3. Shur, B.D. Glycobiology: The beginning of a sweet tale. *Curr. Biol.* **1994**, *4*, 996–999.

4. O'Donnell, N. Intracellular glycosylation and development. *Biochim. Biophys. Acta* **2002**, *1573*, 336–345.

5. Morris, H.R.; Dell, A.; Easton, R.L.; Panico, M.; Koistinen, H.; Koistinen, R.; Oehninger, S.; Patankar, M.S.; Seppala, M.; Clark, G.F. Gender-specific glycosylation of human glycodelin affects its contraceptive activity. *J. Biol. Chem.* **1996**, *271*, 32159–32167.

6. Dell, A.; Morris, H.R.; Easton, R.L.; Patankar, M.; Clark, G.F. The glycobiology of gametes and fertilization. *Biochim. Biophys. Acta* **1999**, *1473*, 196–205.

7. Sarasquete, C.; Cardenas, S.; de Gonzalez, C.M.; Pascual, E. Oogenesis in the bluefin tuna, *Thunnus thynnus* L.: A histological and histochemical study. *Histol. Histopathol.* **2002**, *17*, 775–788.

8. Ortiz-Delgado, J.B.; Porcelloni, S.; Fossi, C.; Sarasquete, C. Histochemical characterisation of oocytes of the swordfish*Xiphias gladius*. *Sci. Mar.* **2008**, *72*, 549–564.

9. Patino, R.; Sullivan, C.V. Ovarian follicle growth, maturation, and ovulation in teleost fish. *Fish. Physiol. Biochem.* **2002**,*26*, 57–70.

10. Inoue, S.; Iwasaki, M. Isolation of a novel glycoprotein from the eggs of rainbowtrout: Occurrence of disialosyl groups on all carbohydrate chains. *Biochem. Biophys. Res. Commun.* **1978**, *83*, 1018–1023.

11. Kitajima, K.; Sorimachi, H.; Inoue, S.; Inoue, Y. Comparative structures of the apopolysialoglycoproteins from unfertilized

and fertilized eggs of salmonid fishes. *Biochemistry* **1988**, *27*, 7141–7145.

12. Song, Y.; Kitajima, K.; Inoue, Y. New tandem-repeating peptide structures in polysialoglycoproteins from the unfertilized eggs of kokanee salmon. *Arch. Biochem. Biophys.* **1990**, *283*, 167–172.

13. Seko, A.; Kitajima, K.; Iwasaki, M.; Inoue, S.; Inoue, Y. Structural studies of fertilization-associated carbohydrate-rich glycoproteins (hyosophorin) isolated from the fertilized and unfertilized eggs of flounder, *Paralichthys olivaceus*. Presence of a novel penta-antennary *N*-linked glycan chain in the tandem repeating glycopeptide unit of hyosophorin. *J. Biol. Chem.* **1989**, *264*, 15922–15929.

14. Iwasaki, M.; Inoue, S. Structures of the carbohydrate units of polysialoglycoproteins isolated from the eggs of four species of salmonid fishes. *Glycoconj. J.* **1985**, *2*, 209–228.

15. Kitajima, K.; Inoue, Y.; Inoue, S. Polysialoglycoproteins of Salmonidae fish eggs. Complete structure of 200-kDa polysialoglycoprotein from the unfertilized eggs of rainbow trout (*Salmo gairdneri*). *J. Biol. Chem.* **1986**, *261*, 5262–5269.

16. Nadano, D.; Iwasaki, M.; Endo, S.; Kitajima, K.; Inoue, S.; Inoue, Y. A naturally occurring deaminated neuraminic acid, 3-deoxy-D-glycero-D-galacto-nonulosonic acid (KDN). ts unique occurrence at the nonreducing ends of oligosialyl chains in polysialoglycoprotein of rainbow trout eggs. *J. Biol. Chem.* **1986**, *261*, 11550–11557.

17. Asahina, S.; Sato, C.; Matsuno, M.; Matsuda, T.; Colley, K.; Kitajima, K. Involvement of the alpha2,8-polysialyltransferases II/STX and IV/PST in the biosynthesis of polysialic acid chains on the *O*-linked glycoproteins in rainbow trout ovary. *J. Biochem.* **2006**, *140*, 687–701.

18. Inoue, S.; Kanamori, A.; Kitajima, K.; Inoue, Y. KDN-glycoprotein: A novel deaminated neuraminic acid-rich glycoprotein isolated from vitelline envelope of rainbow trout eggs. *Biochem. Biophys. Res. Commun.* **1988**, *153*, 172–176.

19. Accogli, G.; Zizza, S.; Garcia-Lopez, A.; Sarasquete, C.; Desantis, S. Lectin-binding pattern of the Senegalese sole Solea senegalensis oogenesis. *Microsc. Res. Tech.* **2012**, *75*, 1124–1135.

20. Mandich, A.; Massari, A.; Bottero, S.; Marino, G. Histological and histochemical study of female germ cell development in the

dusky grouper *Epinephelus marginatus* (Lowe, 1834). *Eur. J. Histochem.* **2002**, *46*, 87–100.

21. Longo, F.J. Fertilization: A comparative ultrastructural review. *Biol. Reprod.* **1973**, *9*, 149–215.

22. Schuel, H. Secretory Functions of Egg Cortical Granules. In *Biology of Fertilization*; Metz, C.B., Monroy, A., Eds.; Academic Press: San Diego, CA, USA, 1985; Volume 3, pp. 1–44.

23. Zamboni, L. Ultrastructure of mammalian oocytes and ova. *Biol. Reprod.* **1970**, *2*, 44–63.

24. Wolf, D.P. On the contents of the cortical granules from *Xenopus* laevis eggs. *Dev. Biol.* **1974**, *38*, 14–29.

25. Gilkey, J.C. Mechanisms of fertilization in fishes. *Am. Zool.* **1981**, *21*, 359–375.

26. Deits, T.; Farrance, M.; Kay, E.S.; Medill, L.; Turner, E.E.; Weidman, P.J.; Shapiro, B.M. Purification and properties of ovoperoxidase, the enzyme responsible for hardening the fertilization membrane of the sea urchin egg. *J. Biol. Chem.***1984**, *259*, 13525–13533.

27. Wessel, G.M.; Brooks, J.M.; Green, E.; Haley, S.; Voronina, E.; Wong, J.; Zaydfudim, V.; Conner, S. The biology of cortical granules. *Int. Rev. Cytol.* **2001**, *209*, 117–206.

28. Anderson, S.L.; Chang, E.S.; Clark, W.H.J. Timing of postvitellogenic ovarian changes in the ridgeback prawn *Sicyonia ingentis* (Penaeidae) determined by ovarian biopsy. *Aquaculture* **1984**, *42*, 257–271.

29. Lynn, J.W.; Clark, W.H.J. Physiological and biochemical investigations of the egg jelly release in *Penaeus aztecus*. *Biol. Bull.* **1987**, *173*, 451–460.

30. Pongtippatee-Taweepreda, P.; Chavadej, J.; Plodpai, P.; Pratoomchart, B.; Sobhon, P.; Weerachatyanukul, W.; Withyachumnarnkul, B. Egg activation in the black tiger shrimp *Penaeus monodon*. *Aquaculture* **2004**, *234*, 183–198.

31. Rankin, S.M.; Davis, R.W. Ultrastructure of oocytes of the shrimp, *Penaeus vannamei*: Cortical specialization formation.*Tissue Cell* **1990**, *22*, 879–893.

32. Kruevaisayawan, H.; Vanichviriyakit, R.; Withyachumnarnkul, B.; Chavadej, J.; Sobhon, P. Oogenesis and formation of cortical rods

in the black tiger shrimp, *Penaeus monodon*. *Aquaculture* **2010**, *301*, 91–98.

33. Clark, J.W.H.; Yudin, A.I.; Lynn, J.W.; Griffin, F.J.; Pillai, M.C. Jelly layer formation in Penaeoidean shrimp eggs. *Biol. Bull.* **1990**, *178*, 295–299.

34. Tan-Fermin, J.D.; Pudadera, R.A. Ovarian maturation stages of the wild giant tiger prawn, *Penaeus monodon* Fabricius.*Aquaculture* **1989**, *77*, 229–242.

35. Khayat, M.; Babin, P.J.; Funkenstein, B.; Sammar, M.; Nagasawa, H.; Tietz, A.; Lubzens, E. Molecular characterization and high expression during oocyte development of a shrimp ovarian cortical rod protein homologous to insect intestinal peritrophins. *Biol. Reprod.* **2001**, *64*, 1090–1099.

36. Avarre, J.C.; Khayat, M.; Michelis, R.; Nagasawa, H.; Tietz, A.; Lubzens, E. Inhibition of *de novo* synthesis of a jelly layer precursor protein by crustacean hyperglycemic hormone family peptides and posttranscriptional regulation by sinus gland extracts in *Penaeus semisulcatus* ovaries. *Gen. Comp. Endocrinol.* **2001**, *124*, 257–268.

37. Kim, Y.K.; Kawazoe, I.; Tsutsui, N.; Jasmani, S.; Wilder, M.N.; Aida, K. Isolation and cDNA cloning of ovarian cortical rod protein in kuruma prawn *Marsupenaeus japonicus* (Crustacea: Decapoda: Penaeidae). *Zool. Sci.* **2004**, *21*, 1109–1119.

38. Kim, Y.K.; Tsutsui, N.; Kawazoe, I.; Okumura, T.; Kaneko, T.; Aida, K. Localization and developmental expression of mRNA for cortical rod protein in kuruma prawn *Marsupenaeus japonicus*. *Zool. Sci.* **2005**, *22*, 675–680.

39. Kruevaisayawan, H.; Vanichviriyakit, R.; Weerachatyanukul, W.; Withyachumnarnkul, W.; Sobhon, P. Biochemical characterization and physiological role of cortical rods in black tiger shrimp, *Penaeus monodon*. *Aquaculture* **2007**, *270*, 289–298.

40. Matozzo, V.; Gagne, F.; Marin, M.G.; Ricciardi, F.; Blaise, C. Vitellogenin as a biomarker of exposure to estrogenic compounds in aquatic invertebrates: A review. *Environ. Int.* **2008**, *34*, 531–545.

41. Fabra, M.; Raldua, D.; Bozzo, M.G.; Deen, P.M.; Lubzens, E.; Cerda, J. Yolk proteolysis and aquaporin-1o play essential roles to regulate fish oocyte hydration during meiosis resumption. *Dev. Biol.* **2006**, *295*, 250–262.

42. Matsubara, T.; Ohkubo, N.; Andoh, T.; Sullivan, C.V.; Hara, A. Two forms of vitellogenin, yielding two distinct lipovitellins, play different roles during oocyte maturation and early development of barfin flounder, *Verasper moseri*, a marine teleost that spawns pelagic eggs. *Dev. Biol.* **1999**, *213*, 18–32.

43. Arukwe, A.; Goksoyr, A. Eggshell and egg yolk proteins in fish: Hepatic proteins for the next generation: Oogenetic, population, and evolutionary implications of endocrine disruption. *Comp. Hepatol.* **2003**, *2*, 4.

44. Abdu, U.; Davis, C.; Khalaila, I.; Sagi, A. The vitellogenin cDNA of Cherax quadricarinatus encodes a lipoprotein with calcium binding ability, and its expression is induced following the removal of the androgenic gland in a sexually plastic system. *Gen. Comp. Endocrinol.* **2002**, *127*, 263–272.

45. Khalaila, I.; Peter-Katalinic, J.; Tsang, C.; Radcliffe, C.M.; Aflalo, E.D.; Harvey, D.J.; Dwek, R.A.; Rudd, P.M.; Sagi, A. Structural characterization of the *N*-glycan moiety and site of glycosylation in vitellogenin from the decapod crustacean *Cherax quadricarinatus*. *Glycobiology* **2004**, *14*, 767–774.

46. Jaenicke, R. Protein stability and protein folding. *Ciba Found. Symp.* **1991**, *161*, 206–221.

47. Raikhel, A.S.; Dhadialla, T.S. Accumulation of yolk proteins in insect oocytes. *Annu. Rev. Entomol.* **1992**, *37*, 217–251.

48. Warrier, S.; Subramoniam, T. Receptor mediated yolk protein uptake in the crab *Scylla serrata*: Crustacean vitellogenin receptor recognizes related mammalian serum lipoproteins. *Mol. Reprod. Dev.* **2002**, *61*, 536–548.

49. Shyu, A.B.; Raff, R.A.; Blumenthal, T. Expression of the vitellogenin gene in female and male sea urchin. *Proc. Natl. Acad. Sci. USA* **1986**, *83*, 3865–3869.

50. Ozaki, H.; Moriya, O.; Harrington, F.E. A glycoprotein in the accessory cell of the echinoid ovary and its role in vitellogenesis. *Dev. Genes Evol.* **1986**, *195*, 74–79.

51. Unuma, T.; Suzuki, T.; Kurokawa, T.; Yamamoto, T.; Akiyama, T. A protein identical to the yolk protein is stored in the testis in male red sea urchin, *Pseudocentrotus depressus*. *Biol. Bull.* **1998**, *194*, 92–97.

52. Harrington, F.E.; Ozaki, H. The major yolk glycoprotein precursor in echinoids is secreted by coelomocytes into the coelomic plasma. *Cell Differ.* **1986**, *19*, 51–57.

53. Degroote, S.; Wolthoorn, J.; van Meer, G. The cell biology of glycosphingolipids. *Semin. Cell Dev. Biol.* **2004**, *15*, 375–387.

54. Fishman, P.H. Recent advances in identifying the functions of gangliosides. *Chem. Phys. Lipids* **1986**, *42*, 137–151.

55. Hakomori, S. Glycosphingolipids in cellular interaction, differentiation, and oncogenesis. *Annu. Rev. Biochem.* **1981**, *50*, 733–764.

56. Hakomori, S. Bifunctional role of glycosphingolipids. Modulators for transmembrane signaling and mediators for cellular interactions. *J. Biol. Chem.* **1990**, *265*, 18713–18716.

57. Hakomori, S. Structure and function of sphingoglycolipids in transmembrane signalling and cell-cell interactions.*Biochem. Soc. Trans.* **1993**, *21*, 583–595.

58. Hakomori, S.I. Cell adhesion/recognition and signal transduction through glycosphingolipid microdomain. *Glycoconj. J.* **2000**, *17*, 143–151.

59. Kojima, H.; Tohsato, Y.; Kabayama, K.; Itonori, S.; Ito, M. Biochemical studies on sphingolipids of *Artemia franciscana*: Complex neutral glycosphingolipids. *Glycoconj. J.* **2012**.

60. Yamamoto, T.; Teshima, T.; Saitoh, U.; Hoshi, M.; Shiba, T. Synthesis of ganglioside M5 from sea urchin egg.*Tetrahedron Lett.* **1994**, *35*, 2701–2704.

61. Kubo, H.; Irie, A.; Inagaki, F.; Hoshi, M. Gangliosides from the eggs of the sea urchin, *Anthocidaris crassispina*. *J. Biochem.* **1990**, *108*, 185–192.

62. Nezuo, M.; Shogomori, H.; Hoshi, M.; Yamamoto, T.; Teshima, T.; Shiba, T.; Chiba, K. Developmental changes in localization of the main ganglioside during sea urchin embryogenesis. *Glycobiology* **2000**, *10*, 1243–1247.

63. Shogomori, H.; Chiba, K.; Kubo, H.; Hoshi, M. Non-plasmalemmal localisation of the major ganglioside in sea urchin eggs. *Zygote* **1993**, *1*, 215–223.

64. Shogomori, H.; Chiba, K.; Hoshi, M. Association of the major ganglioside in sea urchin eggs with yolk lipoproteins.*Glycobiology* **1997**, *7*, 391–398.

65. Kubo, H.; Hoshi, M. Immunocytochemical study of the distribution of a ganglioside in sea urchin eggs. *J. Biochem.*1990, *108*, 193–199.

66. Kubo, H.; Jiang, G.J.; Irie, A.; Morita, M.; Matsubara, T.; Hoshi, M. A novel ceramide trihexoside from the eggs of the sea urchin, *Hemicentrotus pulcherrimus*. *J. Biochem.* **1992**, *111*, 726–731.

67. Kubo, H.; Jiang, G.J.; Irie, A.; Suzuki, M.; Inagaki, F.; Hoshi, M. A novel difucosylated neutral glycosphingolipid from the eggs of the sea urchin, *Hemicentrotus pulcherrimus*: I. Purification and structural determination of the glycolipid. *J. Biochem.* **1992**, *112*, 281–285.

68. Kubo, H.; Irie, A.; Inagaki, F.; Hoshi, M. Melibiosylceramide as the sole ceramide dihexoside from the eggs of the sea urchin, *Anthocidaris crassispina*. *J. Biochem.* **1988**, *104*, 755–760.

69. Inagaki, F.; Tate, S.; Kubo, H.; Hoshi, M. A novel difucosylated neutral glycosphingolipid from the eggs of the sea urchin, *Hemicentrotus pulcherrimus*: II. Structural determination by two-dimensional NMR. *J. Biochem.* **1992**, *112*, 286–289.

70. Dumont, J.N.; Brummett, A.R. Egg envelopes in vertebrates. *Dev. Biol.* **1985**, *1*, 235–288.

71. Glabe, C.G.; Grabel, L.B.; Vacquier, V.D.; Rosen, S.D. Carbohydrate specificity of sea urchin sperm bindin: A cell surface lectin mediating sperm-egg adhesion. *J. Cell Biol.* **1982**, *94*, 123–128.

72. Focarelli, R.; Rosa, D.; Rosati, F. Differentiation of the vitelline coat and the polarized site of sperm entrance in the egg of *Unio elongatulus* (Mollusca, Bivalvia). *J. Exp. Zool.* **1990**, *254*, 88–96.

73. Wikramanayake, A.H.; Clark, W.H. Two extracellular matrices from oocytes of the marine shrimp *Sicyonia ingentis* that independently mediate only primary or secondary sperm binding. *Dev. Growth Differ.* **1994**, *36*, 89–101.

74. Honegger, T.G.; Koyanagi, R. The ascidian egg envelope in fertilization: Structural and molecular features. *Int. J. Dev. Biol.* **2008**, *52*, 527–533.

75. Murata, K.; Sugiyama, H.; Yasumasu, S.; Iuchi, I.; Yasumasu, I.; Yamagami, K. Cloning of cDNA and estrogen-induced hepatic gene expression for choriogenin H, a precursor protein of the fish egg envelope (chorion). *Proc. Natl. Acad. Sci. USA* **1997**, *94*, 2050–2055.

76. Chang, Y.S.; Hsu, C.C.; Wang, S.C.; Tsao, C.C.; Huang, F.L. Molecular cloning, structural analysis, and expression of carp ZP2 gene. *Mol. Reprod. Dev.* **1997**, *46*, 258–267.

77. Mold, D.E.; Kim, I.F.; Tsai, C.M.; Lee, D.; Chang, C.Y.; Huang, R.C. Cluster of genes encoding the major egg envelope protein of zebrafish. *Mol. Reprod. Dev.* **2001**, *58*, 4–14.

78. Kanamori, A. Systematic identification of genes expressed during early oogenesis in medaka. *Mol. Reprod. Dev.* **2000**,*55*, 31–36.

79. Hyllner, S.J.; Westerlund, L.; Olsson, P.E.; Schopen, A. Cloning of rainbow trout egg envelope proteins: Members of a unique group of structural proteins. *Biol. Reprod.* **2001**, *64*, 805–811.

80. Lee, C.; Jeon, S.H.; Na, J.G.; Park, K. Sequence analysis of choriogenin H gene of medaka (*Oryzias latipes*) and mRNA expression. *Environ. Toxicol. Chem.* **2002**, *21*, 1709–1714.

81. Miller, D.J.; Ax, R.L. Carbohydrates and fertilization in animals. *Mol. Reprod. Dev.* **1990**, *26*, 184–198.

82. Kanamori, A.; Inoue, S.; Iwasaki, M.; Kitajima, K.; Kawai, G.; Yokoyama, S.; Inoue, Y. Deaminated neuraminic acid-rich glycoprotein of rainbow trout egg vitelline envelope. Occurrence of a novel alpha-2,8-linked oligo(deaminated neuraminic acid) structure in *O*-linked glycan chains. *J. Biol. Chem.* **1990**, *265*, 21811–21819.

83. Tezuka, T.; Taguchi, T.; Kanamori, A.; Muto, Y.; Kitajima, K.; Inoue, Y.; Inoue, S. Identification and structural determination of the KDN-containing *N*-linked glycan chains consisting of bi- and triantennary complex-type units of KDN-glycoprotein previously isolated from rainbow trout vitelline envelopes. *Biochemistry* **1994**, *33*, 6495–6502.

84. Inoue, S.; Inoue, Y. Fish Glycoprotein. In *Glycoprotein II*; Montreuil, J., Vliegenthart, J.F.G., Schachter, H., Eds.; Elsevier Science BV: Amsterdam, The Netherlands, 1997; Volume 29, pp. 143–161.

85. Pillai, M.C.; Clark, W.H., Jr. Development of cortical vesicles in *Sicyonia ingentis* ova: Their heterogeneity and role in elaboration of the hatching envelope. *Mol. Reprod. Dev.* **1990**, *26*, 78–89.

86. Dupré, E.; Gòmez, D.; Araya, A.; Gallardo, C. Role of egg surface glycoconjugate in the fertilization of the rock

shrimp*Rhynchocinetes typus* (Milne-Edwards, 1837). *Lat. Am. J. Aquat. Res.* **2012**, *40*, 22–29.

87. Litscher, E.; Honegger, T.G. Glycoprotein constituents of the vitelline coat of *Phallusia mammillata* (Ascidiacea) with fertilization inhibiting activity. *Dev. Biol.* **1991**, *148*, 536–551.

88. Baginski, T.; Hirohashi, N.; Hoshi, M. Sulfated *O*-linked glycans of the vitelline coat as ligands in gamete interaction in the ascidian, *Halocynthia roretzi*. *Dev. Growth Differ.* **1999**, *41*, 357–364.

89. De Santis, R.; Pinto, M.R. Gamete Interaction in Ascidians: Sperm Binding and Penetration through the Vitelline Coat. In *Mechanism of Fertilization: Plants to Humans*; Dale, B., Ed.; Springer-Verlag: Berlin, Germany, 1990; pp. 297–304.

90. Rosati, F.; Cotelli, F.; de Santis, R.; Monroy, A.; Pinto, M.R. Synthesis of fucosyl-containing glycoproteins of the vitelline coat in oocytes of *Ciona intestinalis* (Ascidia). *Proc. Natl. Acad. Sci. USA* **1982**, *79*, 1908–1911.

91. SeGall, G.K.; Lennarz, W.J. Chemical characterization of the component of the jelly coat from sea urchin eggs responsible for induction of the acrosome reaction. *Dev. Biol.* **1979**, *71*, 33–48.

92. Shimizu, T.; Kinoh, H.; Yamaguchi, M.; Suzuki, N. Purification and characterization of the egg jelly macromolecules, sialoglycoprotein and fucose sulfate glycoconjugate, of the sea urchin *Hemicentrotus Pulcherrimus*. *Dev. Growth Differ.* **1990**, *32*, 473–487.

93. Kitazume, S.; Kitajima, K.; Inoue, S.; Troy, F.A., II.; Cho, J.W.; Lennarz, W.J.; Inoue, Y. Identification of polysialic acid-containing glycoprotein in the jelly coat of sea urchin eggs. Occurrence of a novel type of polysialic acid structure. *J. Biol. Chem.* **1994**, *269*, 22712–22718.

94. Hirohashi, N.; Kamei, N.; Kubo, H.; Sawada, H.; Matsumoto, M.; Hoshi, M. Egg and sperm recognition systems during fertilization. *Dev. Growth Differ.* **2008**, *50*, S221–S238.

95. Mulloy, B.; Ribeiro, A.C.; Alves, A.P.; Vieira, R.P.; Mourao, P.A. Sulfated fucans from echinoderms have a regular tetrasaccharide repeating unit defined by specific patterns of sulfation at the 0–2 and 0–4 positions. *J. Biol. Chem.* **1994**, *269*, 22113–22123.

96. Alves, A.P.; Mulloy, B.; Diniz, J.A.; Mourao, P.A. Sulfated polysaccharides from the egg jelly layer are species-specific

inducers of acrosomal reaction in sperms of sea urchins. *J. Biol. Chem.* **1997**, *272*, 6965–6971.

97. Vacquier, V.D.; Moy, G.W. The fucose sulfate polymer of egg jelly binds to sperm REJ and is the inducer of the sea urchin sperm acrosome reaction. *Dev. Biol.* **1997**, *192*, 125–135.

98. Alves, A.P.; Mulloy, B.; Moy, G.W.; Vacquier, V.D.; Mourao, P.A. Females of the sea urchin *Strongylocentrotus purpuratus* differ in the structures of their egg jelly sulfated fucans. *Glycobiology* **1998**, *8*, 939–946.

99. Vilela-Silva, A.C.; Castro, M.O.; Valente, A.P.; Biermann, C.H.; Mourao, P.A. Sulfated fucans from the egg jellies of the closely related sea urchins *Strongylocentrotus droebachiensis* and *Strongylocentrotus pallidus* ensure species-specific fertilization. *J. Biol. Chem.* **2002**, *277*, 379–387.

100. Endo, T.; Hoshi, M.; Endo, S.; Arata, Y.; Kobata, A. Structures of the sugar chains of a major glycoprotein present in the egg jelly coat of a starfish, *Asterias amurensis*. *Arch. Biochem. Biophys.* **1987**, *252*, 105–112.

101. Hoshi, M.; Nishigaki, T.; Ushiyama, A.; Okinaga, T.; Chiba, K.; Matsumoto, M. Egg-jelly signal molecules for triggering the acrosome reaction in starfish spermatozoa. *Int. J. Dev. Biol.* **1994**, *38*, 167–174.

102. Uno, Y.; Hoshi, M. Separation of the sperm agglutinin and the acrosome reaction-inducing substance in egg jelly of starfish. *Science* **1978**, *200*, 58–59.

103. Naruse, M.; Suetomo, H.; Matsubara, T.; Sato, T.; Yanagawa, H.; Hoshi, M.; Matsumoto, M. Acrosome reaction-related steroidal saponin, Co-ARIS, from the starfish induces structural changes in microdomains. *Dev. Biol.* **2010**, *347*, 147–153.

104. Koyota, S.; Wimalasiri, K.M.; Hoshi, M. Structure of the main saccharide chain in the acrosome reaction-inducing substance of the starfish, *Asterias amurensis*. *J. Biol. Chem.* **1997**, *272*, 10372–10376.

105. Gunaratne, H.M.; Yamagaki, T.; Matsumoto, M.; Hoshi, M. Biochemical characterization of inner sugar chains of acrosome reaction-inducing substance in jelly coat of starfish eggs. *Glycobiology* **2003**, *13*, 567–580.

106. Hoshi, M.; Moriyama, H.; Matsumoto, M. Structure of acrosome reaction-inducing substance in the jelly coat of starfish eggs: A

mini review. *Biochem. Biophys. Res. Commun.* **2012**, *425*, 595–598.

107. Suphamungmee, W.; Chansela, P.; Weerachatyanukul, W.; Poomtong, T.; Vanichviriyakit, R.; Sobhon, P. Ultrastructure, composition, and possible roles of the egg coats in *Haliotis asinin*. *J. Shellfish Res.* **2010**, *29*, 687–697.

108. Shigekawa, K.; Clark, W.H., Jr. Spermiogenesis in the marine shrimp, *Sicyonia ingentis*. *Dev. Growth Differ.* **1986**, *28*, 95–112.

109. Nagao, J.; Munehara, H. Annual cycle of testicular maturation in the helmet crab *Telmessus cheiragonus*. *Fish. Sci.* **2003**,*69*, 1200–1208.

110. Okumura, T.; Hara, M. Androgenic gland cell structure and spermatogenesis during the molt cycle and correlation to morphotypic differentiation in the giant freshwater prawn, *Macrobrachium rosenbergii*. *Zool. Sci.* **2004**, *21*, 621–628.

111. Garcia, T.M.; Silva, J.R.F. Testis and vas deferens morphology of the red-clawed mangrove tree crab (*Goniopsis cruentata*) (Latreille, 1803). *Braz. Arch. Biol. Technol.* **2006**, *49*, 339–345.

112. Jones, R.; Brown, C.R.; Lancaster, R.T. Carbohydrate-binding properties of boar sperm proacrosin and assessment of its role in sperm-egg recognition and adhesion during fertilization. *Dev. Genes Evol.* **1988**, *102*, 781–779.

113. Anakwe, O.O.; Gerton, G.L. Acrosome biogenesis begins during meiosis: Evidence from the synthesis and distribution of an acrosomal glycoprotein, acrogranin, during guinea pig spermatogenesis. *Biol. Reprod.* **1990**, *42*, 317–328.

114. Martinez-Menarguez, J.A.; Ballesta, J.; Aviles, M.; Castells, M.T.; Madrid, J.F. Cytochemical characterization of glycoproteins in the developing acrosome of rats. An ultrastructural study using lectin histochemistry, enzymes and chemical deglycosylation. *Histochemistry* **1992**, *97*, 439–449.

115. Pratt, S.A.; Scully, N.F.; Shur, B.D. Cell surface beta 1,4 galactosyltransferase on primary spermatocytes facilitates their initial adhesion to Sertoli cells *in vitro*. *Biol. Reprod.* **1993**, *49*, 470–482.

116. Raychoudhury, S.S.; Millette, C.F. Multiple fucosyltransferases and their carbohydrate ligands are involved in spermatogenic cell-Sertoli cell adhesion *in vitro* in rats. *Biol. Reprod.* **1997**, *56*, 1268–1273.

117. Schulz, R.W.; Miura, T. Spermatogenesis and its endocrine regulation. *Fish. Physiol. Biochem.* **2002**, *26*, 43–56.

118. Nobrega, R.H.; Batlouni, S.R.; Franca, L.R. An overview of functional and stereological evaluation of spermatogenesis and germ cell transplantation in fish. *Fish. Physiol. Biochem.* **2009**, *35*, 197–206.

119. Parenti, L.R.; Grier, H.J. Evolution and phylogeny of gonad morphology in bony fishes. *Integr. Comp. Biol.* **2004**, *44*, 333–348.

120. Desantis, S.; Zizza, S.; Garcia-Lopez, A.; Sciscioli, V.; Mananos, E.; de Metrio, V.G.; Sarasquete, C. Lectin-binding pattern of Senegalese sole *Solea senegalensis* (Kaup) testis. *Histol. Histopathol.* **2010**, *25*, 205–216.

121. Liguoro, A.; Prisco, M.; Mennella, C.; Ricchiari, L.; Angelini, F.; Andreuccetti, P. Distribution of terminal sugar residues in the testis of the spotted ray *Torpedo marmorata*. *Mol. Reprod. Dev.* **2004**, *68*, 524–530.

122. Tokalov, S.V.; Gutzeit, H.O. Lectin-binding pattern as tool to identify and enrich specific primary testis cells of the tilapia (*Oreochromis niloticus*) and medaka (*Oryzias latipes*). *J. Exp. Zool. Part A* **2007**, *308*, 127–138.

123. Varki, A.; Cummings, R.; Esko, J.; Freeze, H.; Hart, G.; Marth, J. Nuclear and Cytoplasmic Glycosylation. In *Essentials of Glycobiology*; Varki, A., Cummings, R., Esko, J., Freeze, H., Hart, G., Marth, J., Eds.; Cold Spring Harbor Laboratory Press: Cold Spring Harbor, NY, USA, 1999; pp. 171–183.

124. Martinez-Menarguez, J.A.; Aviles, M.; Madrid, J.F.; Castells, M.T.; Ballesta, J. Glycosylation in Golgi apparatus of early spermatids of rat. A high resolution lectin cytochemical study. *Eur. J. Cell Biol.* **1993**, *61*, 21–33.

125. Unuma, T.; Yamamoto, T.; Akiyama, T.; Shiraishi, M.; Ohta, H. Quantitative changes in yolk protein and other components in the ovary and testis of the sea urchin *Pseudocentrotus depressus*. *J. Exp. Biol.* **2003**, *206*, 365–372.

126. Verhey, C.A.; Moyer, F.H. Fine structural changes during sea urchin oogenesis. *J. Exp. Zool.* **1967**, *162*, 195–225.

127. Marsh, A.G.; Watts, S.A. Energy metabolism and gonad development. *Dev. Aquac. Fish. Sci.* **2001**, *32*, 27–42.

128. Miyata, S.; Sato, C.; Kitamura, S.; Toriyama, M.; Kitajima, K. A major flagellum sialoglycoprotein in sea urchin sperm contains a novel polysialic acid, an alpha2,9-linked poly-*N*-acetylneuraminic acid chain, capped by an 8-*O*-sulfated sialic acid residue. *Glycobiology* **2004**, *14*, 827–840.

129. Miyata, S.; Sato, C.; Kumita, H.; Toriyama, M.; Vacquier, V.D.; Kitajima, K. Flagellasialin: A novel sulfated alpha2,9-linked polysialic acid glycoprotein of sea urchin sperm flagella. *Glycobiology* **2006**, *16*, 1229–1241.

130. Kambara, Y.; Shiba, K.; Yoshida, M.; Sato, C.; Kitajima, K.; Shingyoji, C. Mechanism regulating Ca^{2+}-dependent mechanosensory behaviour in sea urchin spermatozoa. *Cell Struct. Funct.* **2011**, *36*, 69–82.

131. Ijuin, T.; Kitajima, K.; Song, Y.; Kitazume, S.; Inoue, S.; Haslam, S.M.; Morris, H.R.; Dell, A.; Inoue, Y. Isolation and identification of novel sulfated and nonsulfated oligosialyl glycosphingolipids from sea urchin sperm. *Glycoconj. J.***1996**, *13*, 401–413.

132. Miyata, S.; Yamakawa, N.; Toriyama, M.; Sato, C.; Kitajima, K. Co-expression of two distinct polysialic acids, alpha2,8- and alpha2,9-linked polymers of *N*-acetylneuraminic acid, in distinct glycoproteins and glycolipids in sea urchin sperm. *Glycobiology* **2011**, *21*, 1596–1605.

133. Desantis, S.; Labate, M.; Maiorano, P.; Tursi, A. An ultrastructural and histochemical study of the germinal cells contained in hemispermatophores of males of the *Aristaeomorpha foliacea* (Risso, 1827). *Hydrobiologia* **2006**, *557*, 41–49.

134. Subramonian, T. Spermatophore formation in two intertidal crabs *Albunea symnista* and *Emerita asiatica* (Decapoda: Anomura). *Biol. Bull.* **1984**, *166*, 78–95.

135. Radha, T.; Subramonian, T. Origin and nature of spermatophoric mass of spiny lobster *Panulirus homarus*. *Mar. Biol.***1985**, *86*, 13–19.

136. Yufeng, W.; Nanshan, D.; Wei, L. A histochemical study on the male reproductive system of *Macrobrachium rosenbergii*.*J. Northwest Atl. Fish. Sci.* **1997**, *4*, 13–16.

137. Sasikala, S.L.; Subramoniam, T. On the occurrence of acid mucopolysaccharides in the spermatophores of two marine prawns, *Penaeus indiens* (Milne-Edwards) and *Metapenaeus monoceros* (Fabricius) (Crustacea: Macrura). *J. Exp. Mar. Biol. Ecol.* **1987**, *113*, 145–153.

138. Subramonian, T. Chemical Composition of Spermatophores in Decapod Crustaceans. In *Crustacean Sexual Biology*; Bauer, R.T., Martin, J.A., Eds.; Columbia University Press: New York, NY, USA, 1991; pp. 308–321.

139. Desantis, S.; Labate, M.; Cirillo, F.; Labate, G.M. Testicular activity and sperm glycoproteins in giant red shrimp (*Aristaeomorpha foliacea*). *J. Northwest Atl. Fish. Sci.* **2003**, *31*, 205–212.

140. Yanagimachi, R. Fertility of mammalian spermatozoa: Its development and relativity. *Zygote* **1994**, *2*, 371–372.

141. Snell, W.J.; White, J.M. The molecules of mammalian fertilization. *Cell* **1996**, *85*, 629–637.

142. Vacquier, V.D. Evolution of gamete recognition proteins. *Science* **1998**, *281*, 1995–1998.

143. Tosti, E. Sperm activation in species with external fertilisation. *Zygote* **1994**, *2*, 359–361.

144. Wassarman, P.M.; Litscher, E.S. Towards the molecular basis of sperm and egg interaction during mammalian fertilization. *Cells Tissues Organs* **2001**, *168*, 36–45.

145. Monroy, A.; Rosati, F. A comparative analysis of sperm-egg interaction. *Gamete Res.* **1983**, *7*, 85–102.

146. Wassarman, P.M. The biology and chemistry of fertilization. *Science* **1987**, *235*, 553–560.

147. Moy, G.W.; Mendoza, L.M.; Schulz, J.R.; Swanson, W.J.; Glabe, C.G.; Vacquier, V.D. The sea urchin sperm receptor for egg jelly is a modular protein with extensive homology to the human polycystic kidney disease protein, PKD1. *J. Cell Biol.* **1996**, *133*, 809–817.

148. Mengerink, K.J.; Vacquier, V.D. Glycobiology of sperm-egg interactions in deuterostomes. *Glycobiology* **2001**, *11*, 37R–43R.

149. Vacquier, V.D.; Swanson, W.J.; Hellberg, M.E. What have we learned about sea urchin sperm bindin? *Dev. Growth Differ.* **1995**, *37*, 1–10.

150. Ohlendieck, K.; Partin, J.S.; Stears, R.L.; Lennarz, W.J. Developmental expression of the sea urchin egg receptor for sperm. *Dev. Biol.* **1994**, *165*, 53–62.

151. Dhume, S.T.; Lennarz, W.J. The involvement of *O*-linked oligosaccharide chains of the sea urchin egg receptor for sperm in fertilization. *Glycobiology* **1995**, *5*, 11–17.

152. Hirohashi, N.; Lennarz, W.J. Role of a vitelline layer-associated 350 kDa glycoprotein in controlling species-specific gamete interaction in the sea urchin. *Dev. Growth Differ.* **2001**, *43*, 247–255.

153. Hoshi, M.; Matsui, T.; Nishiyama, I.; Amano, T.; Okita, Y. Physiological inducers of the acrosome reaction. *Cell Differ. Dev.* **1988**, *25* Suppl., 19–24.

154. Longo, F.J.; Ushiyama, A.; Chiba, K.; Hoshi, M. Ultrastructural localization of acrosome reaction-inducing substance (ARIS) on sperm of the starfish *Asterias amurensis*. *Mol. Reprod. Dev.* **1995**, *41*, 91–99.

155. Ushiyama, A.; Araki, T.; Chiba, K.; Hoshi, M. Specific binding of acrosome-reaction-inducing substance to the head of starfish spermatozoa. *Zygote* **1993**, *1*, 121–127.

156. Hoshi, M.; Amano, T.; Okita, Y.; Okinaga, T.; Matsui, T. Egg signals for triggering the acrosome reaction in starfish spermatozoa. *J. Reprod. Fertil.* **1990**, *42*, 23–31.

157. Nishiyama, I.; Matsui, T.; Fujimoto, Y.; Ikekawa, N.; Hoshi, M. Correlation between the molecular structure and the biological activity of Co-ARIS, a cofactor for acrosome reaction-inducing substance. *Dev. Growth Differ.* **1987**, *29*, 171–176.

158. Nishigaki, T.; Chiba, K.; Miki, W.; Hoshi, M. Structure and function of asterosaps, sperm-activating peptides from the jelly coat of starfish eggs. *Zygote* **1996**, *4*, 237–245.

159. Nishigaki, T.; Chiba, K.; Hoshi, M. A 130-kDa membrane protein of sperm flagella is the receptor for asterosaps, sperm-activating peptides of starfish *Asterias amurensis*. *Dev. Biol.* **2000**, *219*, 154–162.

160. Matsumoto, M.; Hirata, J.; Hirohashi, N.; Hoshi, M. Sperm-egg binding mediated by sperm alpha-D-fucosidase in the ascidian, *Halocynthia roretzi*. *Zool. Sci.* **2002**, *19*, 43–48.

161. Matsumoto, M.; Kawase, O.; Islam, M.S.; Naruse, M.; Watanabe, S.N.; Ishikawa, R.; Hoshi, M. Regulation of the starfish sperm acrosome reaction by cGMP, pH, cAMP and Ca^{2+}. *Int. J. Dev. Biol.* **2008**, *52*, 523–526.

162. Swanson, W.J.; Vacquier, V.D. The abalone egg vitelline envelope receptor for sperm lysin is a giant multivalent molecule. *Proc. Natl. Acad. Sci. USA* **1997**, *94*, 6724–6729.

163. Vacquier, V.D.; Carner, K.R.; Stout, C.D. Species-specific sequences of abalone lysin, the sperm protein that creates a hole in the egg envelope. *Proc. Natl. Acad. Sci. USA* **1990**, *87*, 5792–5796.

164. Kresge, N.; Vacquier, V.D.; Stout, C.D. Abalone lysin: The dissolving and evolving sperm protein. *Bioessays* **2001**, *23*, 95–103.

165. Swanson, W.J.; Aquadro, C.F.; Vacquier, V.D. Polymorphism in abalone fertilization proteins is consistent with the neutral evolution of the egg's receptor for lysin (VERL) and positive darwinian selection of sperm lysin. *Mol. Biol. Evol.* **2001**, *18*, 376–383.

166. Hoshi, M.; de Santis, R.; Pinto, M.R.; Cotelli, F.; Rosati, F. Sperm glycosidases as mediators of sperm-egg binding in the ascidians. *Zool. Sci.* **1985**, *2*, 65–69.

167. Hoshi, M. Sperm glycosidase as a plausible mediator of sperm binding to the vitelline envelope in Ascidians. *Adv. Exp. Med. Biol.* **1986**, *207*, 251–260.

168. Godknecht, A.; Honegger, T.G. Isolation, characterization, and localization of a sperm-bound *N*-acetylglucosaminidase that is indispensable for fertilization in the ascidian, *Phallusia mammilla. Dev. Biol.* **1991**, *143*, 398–407.

169. Godknecht, A.J.; Honegger, T.G. Specific inhibition of sperm β-*N*-acetylglucosaminidase by the synthetic inhibitor *N*-acetylglucosaminono-1,5-lactone *O*-(phenylcarbamoyl)oxime inhibits fertilization in the ascidian, *Phallusia mammillata. Dev. Growth Differ.* **1995**, *37*, 183–189.

170. Xie, M.; Honegger, T.G. Ultrastructural investigations on sperm penetration and gamete fusion in the ascidians*Boltenia villosa* and *Phallusia mammillata. Mar. Biol. 1993 116*, 117–127.

171. Hoshi, M.; de Santis, R.; Pinto, M.R.; Cotelli, F.; Rosati, F. Is Sperm L-fucosidase Responsible for Sperm-Egg Binding in Ciona Intestinalis. In *The Sperm Cell*; André, J., Ed.; Martinus Nijhoff: Leiden, NL, USA, 1983; pp. 107–110.

172. Fukumoto, M. Fertilization in ascidians apical processes and gamete fusion in Ciona intestinalis spermatozoa. *J. Cell Sci.* **1988**, *89*, 189–196.

173. Fukumoto, M. Morphological aspects of ascidian fertilization: Acrosome reaction, apical processes and gamete fusion in *Ciona intestinalis*. *Invertebr. Reprod. Dev.* **1990**, *17*, 147–154.

174. Fukumoto, M.; Numakunai, T. Morphological aspects of fertilization in *Halocynthia roretzi* (Ascidiacea, Tunicata). *J. Struct. Biol.* **1995**, *114*, 157–166.

175. Sawada, H.; Sakai, N.; Abe, Y.; Tanaka, E.; Takahashi, Y.; Fujino, J.; Kodama, E.; Takizawa, S.; Yokosawa, H. Extracellular ubiquitination and proteasome-mediated degradation of the ascidian sperm receptor. *Proc. Natl. Acad. Sci. USA* **2002**, *99*, 1223–1228.

176. Sawada, H.; Yokosawa, H.; Ishii, S. Purification and characterization of two types of trypsin-like enzymes from sperm of the ascidian (Prochordata) *Halocynthia roretzi*. Evidence for the presence of spermosin, a novel acrosin-like enzyme.*J. Biol. Chem.* **1984**, *259*, 2900–2904.

177. Lambert, C.C.; Someno, T.; Sawada, H. Sperm surface proteases in ascidian fertilization. *J. Exp. Zool.* **2002**, *292*, 88–95.

178. Dupre, E.M.; Barros, C. *In vitro* fertilization of the rock shrimp, *Rhynchocinetes typus* (Decapoda, Caridea): A review.*Biol. Res.* **2011**, *44*, 125–133.

179. Barros, C.; Dupré, E.; Viveros, L. Sperm-egg interactions in the shrimp *Rhynchocinetes typus*. *Gamete Res.* **1986**, *14*, 171–180.

180. Rios, M.; Barros, C. Trypsin-like enzymes during fertilization in the shrimp *Rhynchocinetes typus*. *Mol. Reprod. Dev.***1997**, *46*, 581–586.

181. Bustamante, E.; Palomino, J.; Amoroso, A.; Moreno, R.D.; Barros, C. Purification and biochemical characterization of a trypsin-like enzyme present in the sperm of the rock shrimp, *Rhynchocinetes typus*. *Invertebr. Reprod. Dev.* **2001**, *39*, 175–181.

182. Gilbert, S.F. *Developmental Biology*, 5th ed; Sinauer Associates: Sunderland, MA, USA, 1997.

183. Foltz, K.R.; Shilling, F.M. Receptor-mediated signal transduction and egg activation. *Zygote* **1993**, *1*, 276–279.

184. Hirohashi, N.; Lennarz, W.J. Sperm-egg binding in the sea urchin: A high level of intracellular ATP stabilizes sperm attachment to the egg receptor. *Dev. Biol.* **1998**, *201*, 270–279.

185. Kamei, N.; Glabe, C.G. The species-specific egg receptor for sea urchin sperm adhesion is EBR1, a novel ADAMTS protein. *Genes Dev.* **2003**, *17*, 2502–2507.

186. Kamei, N.; Swanson, W.J.; Glabe, C.G. A rapidly diverging EGF protein regulates species-specific signal transduction in early sea urchin development. *Dev. Biol.* **2000**, *225*, 267–276.

187. Hagstrom, B.E. Further studies on cross fertilization in sea urchins. *Exp. Cell Res.* **1956**, *11*, 507–510.

188. Adelson, D.L.; Alliegro, M.C.; McClay, D.R. On the ultrastructure of hyalin, a cell adhesion protein of the sea urchin embryo extracellular matrix. *J. Cell Biol.* **1992**, *116*, 1283–1289.

189. Shapiro, B.M.; Somers, C.; Weidman, P.J. Extracellular Remodeling during Fertilization. In *Cell Biology of Fertilization*; Schatten, H., Schatten, G., Eds.; Academic Press: San Diego, CA, USA, 1989; pp. 251–276.

190. Somers, C.E.; Shapiro, B.M. Functional domains of proteoliaisin, the adhesive protein that orchestrates fertilization envelope assembly. *J. Biol. Chem.* **1991**, *266*, 16870–16875.

191. Nomura, K.; Suzuki, N. Sea urchin ovoperoxidase: Solubilization and isolation from the fertilization envelope, some structural and functional properties, and degradation by hatching enzyme. *Arch. Biochem. Biophys.* **1995**, *319*, 525–534.

192. Inoue, S.; Inoue, Y. Fertilization (activation)-induced 200- to 9-kDa depolymerization of polysialoglycoprotein, a distinct component of cortical alveoli of rainbow trout eggs. *J. Biol. Chem.* **1986**, *261*, 5256–5261.

193. Inoue, S.; Kitajima, K.; Inoue, Y.; Kudo, S. Localization of polysialoglycoprotein as a major glycoprotein component in cortical alveoli of the unfertilized eggs of *Salmo gairdneri*. *Dev. Biol.* **1987**, *123*, 442–454.

194. Laale, H.W. The perivitelline space and egg envelopes of bony fishes: A review. *Copeia* **1980**, *1980*, 210–226.

195. Rudy, P.P., Jr.; Potts, W.T. Sodium balance in the eggs of the Atlantic salmon, *Salmo salar*. *J. Exp. Biol.* **1969**, *50*, 239–246.

196. Eddy, F.B. Osmotic properties of the perivitelline fluid and some properties of the chorion of Atlantic salmon eggs (*Salmo salar*). *J. Zool.* **1974**, *174*, 237–243.

197. Peterson, R.H.; Martin-Robichaud, D.J. Rates of ionic diffusion across the egg chorion of Atlantic salmon (*Salmo salar*).*Physiol. Zool.* **1993**, *66*, 289–306.

198. Li, X.; Jenssen, E.; Fyhn, H.J. Effects of salinity on egg swelling in Atlantic salmon (*Salmo salar*). *Aquaculture* **1989**, *76*, 317–334.

199. Kudo, S. Fertilization, cortical reaction, polyspermy-preventing and anti-microbial mechanisms in fish eggs. *Bull. Inst. Zool. Acad. Sci. Monogr.* **1991**, *16*, 313–340.

200. Kobayashi, W.; Yamamoto, T.S. Factors inducing closure of the micropylar canal in the chum salmon egg. *J. Fish. Biol.***1993**, *42*, 385–394.

201. Kudo, S.; Teshima, C. Assembly *in vitro* of vitelline envelope components induced by a cortical alveolus sialoglycoprotein of eggs of the fish *Tribolodon hakonensis*. *Zygote* **1998**, *6*, 193–201.

202. Oppen-Berntsen, D.O.; Helvik, J.V.; Walther, B.T. The major structural proteins of cod (*Gadus morhua*) eggshells and protein crosslinking during teleost egg hardening. *Dev. Biol.* **1990**, *137*, 258–265.

203. Koyanagi, R.; Honegger, T.G. Molecular cloning and sequence analysis of an ascidian egg beta-*N*-acetylhexosaminidase with a potential role in fertilization. *Dev. Growth Differ.* **2003**, *45*, 209–218.

204. Lambert, C.C. Fertilization-induced modification of chorion *N*-acetylglucosamine groups blocks polyspermy in ascidian eggs. *Dev. Biol.* **1986**, *116*, 168–173.

205. Lambert, C.C. Ascidian eggs release glycosidase activity which aids in the block against polyspermy. *Development***1989**, *105*, 415–420.

206. Schneider, E.G.; Nguyen, H.T.; Lennarz, W.J. The effect of tunicamycin, an inhibitor of protein glycosylation, on embryonic development in the sea urchin. *J. Biol. Chem.* **1978**, *253*, 2348–2355.

207. Heifetz, A.; Lennarz, W.J. Biosynthesis of *N*-glycosidically linked glycoproteins during gastrulation of sea urchin embryos. *J. Biol. Chem.* **1979**, *254*, 6119–6127.

208. Lennarz, W.J. Glycoprotein synthesis and embryonic development. *CRC Crit. Rev. Biochem.* **1983**, *14*, 257–272.

209. Carson, D.D.; Lennarz, W.J. Inhibition of polyisoprenoid and glycoprotein biosynthesis causes abnormal embryonic development. *Proc. Natl. Acad. Sci. USA* **1979**, *76*, 5709–5713.

210. Carson, D.D.; Lennarz, W.J. Relationship of dolichol synthesis to glycoprotein synthesis during embryonic development. *J. Biol. Chem.* **1981**, *256*, 4679–4686.

211. Carson, D.D.; Farach, M.C.; Earles, D.S.; Decker, G.L.; Lennarz, W.J. A monoclonal antibody inhibits calcium accumulation and skeleton formation in cultured embryonic cells of the sea urchin. *Cell* **1985**, *41*, 639–648.

212. Farach, M.C.; Valdizan, M.; Park, H.R.; Decker, G.L.; Lennarz, W.J. Developmental expression of a cell-surface protein involved in calcium uptake and skeleton formation in sea urchin embryos. *Dev. Biol.* **1987**, *122*, 320–331.

213. Farach-Carson, M.C.; Carson, D.D.; Collier, J.L.; Lennarz, W.J.; Park, H.R.; Wright, G.C. A calcium-binding, asparagine-linked oligosaccharide is involved in skeleton formation in the sea urchin embryo. *J. Cell Biol.* **1989**, *109*, 1289–1299.

214. Decker, G.L.; Valdizan, M.C.; Wessel, G.M.; Lennarz, W.J. Developmental distribution of a cell surface glycoprotein in the sea urchin *Strongylocentrotus purpuratus*. *Dev. Biol.* **1988**, *129*, 339–349.

215. Anstrom, J.A.; Chin, J.E.; Leaf, D.S.; Parks, A.L.; Raff, R.A. Localization and expression of msp130, a primary mesenchyme lineage-specific cell surface protein in the sea urchin embryo. *Development* **1987**, *101*, 255–265.

216. McClay, D.R.; Matranga, V.; Wessel, G. Expression and Appearance of Germ-Layer Specific Antigens on the Surface of Embryonic Sea Urchin Cells. In *The Cellular and Molecular Biology of Invertebrate Development*; Sawyer, R.H., Showman, R.M., Eds.; University of South Carolina Press: Columbia, SC, USA, 1985; pp. 171–186.

217. Kabakoff, B.; Lennarz, W.J. Inhibition of glycoprotein processing blocks assembly of spicules during development of the sea urchin embryo. *J. Cell Biol.* **1990**, *111*, 391–400.

218. Hoshi, M.; Nagai, Y. Biochemistry of mucolipids of sea urchin gametes and embryos. 3. Mucolipids during early development. *Jpn. J. Exp. Med.* **1970**, *40*, 361–365.

219. Cheresh, D.A.; Klier, F.G. Disialoganglioside GD2 distributes preferentially into substrate-associated microprocesses on human melanoma cells during their attachment to fibronectin. *J. Cell Biol.* **1986**, *102*, 1887–1897.

220. Spiegel, S.; Yamada, K.M.; Hom, B.E.; Moss, J.; Fishman, P.H. Fluorescent gangliosides as probes for the retention and organization of fibronectin by ganglioside-deficient mouse cells. *J. Cell Biol.* **1985**, *100*, 721–726.

221. Wessel, G.M.; Marchase, R.B.; McClay, D.R. Ontogeny of the basal lamina in the sea urchin embryo. *Dev. Biol.* **1984**,*103*, 235–245.

222. Jackson, R.L.; Busch, S.J.; Cardin, A.D. Glycosaminoglycans: Molecular properties, protein interactions, and role in physiological processes. *Physiol. Rev.* **1991**, *71*, 481–539.

223. Vilela-Silva, A.C.; Werneck, C.C.; Valente, A.P.; Vacquier, V.D.; Mourao, P.A. Embryos of the sea urchin*Strongylocentrotus purpuratus* synthesize a dermatan sulfate enriched in 4-*O*- and 6-*O*-disulfated galactosamine units.*Glycobiology* **2001**, *11*, 433–440.

224. Oguri, K.; Yamagata, T. Appearance of a proteoglycan in developing sea urchin embryos. *Biochim. Biophys. Acta* **1978**,*541*, 385–393.

225. Kinoshita, S.; Yoshii, K. The role of proteoglycan synthesis in the development of sea urchins. II. The effect of administration of exogenous proteoglycan. *Exp. Cell Res.* **1979**, *124*, 361–369.

226. Akasaka, K.; Amemiya, S.; Terayama, H. Scanning electron microscopical study of the inside of sea urchin embryos (*Pseudocentotus depressus*). Effects of Aryl beta-xyloside, tunicamycin and deprivation of sulfate tions. *Exp. Cell Res.***1980**, *129*, 1–13.

227. Solursh, M.; Mitchell, S.L.; Katow, H. Inhibition of cell migration in sea urchin embryos by beta-D-xyloside. *Dev. Biol.***1986**, *118*, 325–332.

2

CHAPTER

Marine Non-Glycosaminoglycan Sulfated Glycans as Potential Pharmaceuticals

Vitor H. Pomin

Program of Glycobiology, Institute of Medical Biochemistry Leopoldo de Meis, University Hospital Clementino Fraga Filho, Federal University of Rio de Janeiro, Rio de Janeiro, RJ 21941-913, Brazil

ABSTRACT

Sulfated fucans (SFs) and sulfated galactans (SGs) are currently the marine non-glycosaminoglycan (GAG) sulfated glycans most studied in glycomics. These compounds exhibit therapeutic effects in several pathophysiological systems such as blood coagulation, thrombosis, neovascularization, cancer, inflammation, and microbial infections. As analogs of the largely employed GAGs and due to some limitations of the GAG-based therapies, SFs and SGs comprise new carbohydrate-based therapeutics available for clinical studies. Here, the principal structural features and the major mechanisms of action of the SFs and SGs in the above-mentioned pathophysiological systems are presented. Discussion is also given on the current challenges and the future perspectives in drug development of these marine glycans.

Keywords: angiogenesis; cancer; coagulation; inflammation; marine glycans; microbial infection; sulfated fucans; sulfated galactans; thrombosis

1. INTRODUCTION

Glycosaminoglycans (GAGs) are the most therapeutically explored carbohydrates of the pharmaceutical market [1,2]. They belong to a class of sugars named sulfated glycans. Among various types, the commonly used GAGs in medicine are heparin, chondroitin sulfate and keratan sulfate. Heparin is a potent anticoagulant and antithrombotic agent [3,4]. Chondroitin sulfated is used in cases of osteoarthritis, osteoarthrosis and sometimes osteoporosis [5,6,7]. Keratan sulfated is explored as a functional ingredient of eye-drops in therapies of corneal dysfunctions [8,9]. Despite the large application of these GAGs in medicine, they still present some downsides. For example, heparin-based treatments offer serious bleeding and hemorrhage risks [10,11,12]. Chondroitin sulfate-based formulation destined to oral administration was detected contaminated with another GAG type due to rough large-scale preparation methods [7]. Keratan sulfate is only clinically active in corneal dystrophies characterized by structural defects on keratan sulfate-containing proteoglycans [13]. This gives a restricted medical application to keratan sulfate.

However, non-GAG sulfated glycans endowed also with medical actions are becoming more and more available as the glycomics evolves. The marine sulfated fucans (SFs) and sulfated galactans (SGs) are examples of these glycans [14,15,16]. Two reasons to explain the growing interest of these molecules are as follows. (i) The chemical structures of the relatively new marine sulfated glycans are very unique and distinct if compared to GAG structures. Although SFs and SGs bear sulfation and are sometimes composed of disaccharide repeating units like GAGs, the marine sulfated glycans (especially those extracted from invertebrates and red algae) are much more homogeneous and regular in terms of backbone composition and sulfation patterns than GAGs [2,14,15,16]. (ii) Although resembling the mechanisms of action of the mammalian-derived GAGs of medicine, SFs and SGs can exhibit additional or slightly distinct effects. The serpin-independent anticoagulant action of some marine sulfated glycans is an example of additional effect since this mechanism is non-existent in heparin-based treatments [17,18,19]. The pronounced heparin cofactor II (HCII)-driven serpin-dependent anticoagulant activity of some marine SFs and SGs [20,21], as opposed to the main antithrombin (AT)-driven anticoagulant mechanism of action of heparin [21,22], is an example of slightly different effect since both sulfated glycans present serpin-dependent activities although acting on distinct primary serpin-targets (AT or HCII). These additional and slightly distinct effects of the marine sulfated glycans can be considered advantageous factors in the development of alternative anticoagulants, especially in cases in which heparin could be less effective or inactive. Examples of these cases are, respectively, heparin preparations with low concentration of the AT high-affinity pentasaccharide [23], and patients suffering from congenital or acquired disorders characterized by decreased levels of circulating serpins [21,24,25]. In these deficiencies, anticoagulant/antithrombotic therapies unrelated to serpin activities would be better in order to achieve the desired anticoagulant outcome. Moreover, as opposed to heparin which presents bleeding risks as already mentioned, certain SFs and SGs do not exhibit this side-effect [19,26].

This report aims therefore at discussing the prospects of the marine SFs and SGs as potential pharmaceuticals of the future medicine, not only in coagulation and thrombosis but also in other systems in which these molecules have been reported to be effective. The additional systems are angiogenesis, cancer, inflammation and microbial infections. Therefore, molecular details of major mechanisms of action of SFs and SGs in these pathophysiological systems are highlighted here in order to offer a comprehensive overview of their medical properties. Final discussion focuses on the current obstacles and expected challenges to implement these marine sulfated glycans in the next generation of potential carbohydrate-based therapeutics of the global market.

2. STRUCTURE

GAGs are composed of disaccharide repeating building blocks composed of alternating hexosamine and hexuronic acid or galactose (Gal) units (Figure 1). The hexosamine can be glucose-based (glucosamine, GlcN) as seen in heparin and keratan sulfate or galactose-based (galactosamine, GalN) as seen in chondroitin sulfate. The hexuronic acids can be glucuronic acid (GlcA) or its C5-epimer iduronic acid (IdoA). GlcA can be found in chondroitin sulfate and in heparin although at lower proportions since heparin is largely composed of IdoA. Besides variation in glycosidic linkage type, but never different from 3- or 4-positions, GAGs also vary in terms of substitutions at N- and/or O-positions (Figure 1). For instance, while heparin is largely N-sulfated, O-sulfated frequently at positions C6 and rarely at C3 of the composing GlcN units with additional sulfation at the C2 position of the IdoA units [27,28,29], chondroitin sulfate bears sulfation mostly at positions C4 and/or C6 of its N-acetyl GalN (GalNAc) units (Figure 1) [30,31]. In keratan sulfate, sulfation can occur at C6 positions of either unit (Gal or GlcNAc) but more often at the amino sugar (Figure 1) [7,8,13]. Although composed of disaccharide units and heavily sulfated, GAGs are very complex and heterogeneous in terms of structure. Sulfation is not absolutely regular in any GAG type. It varies within the disaccharide building blocks. The levels of epimerization in GlcA to IdoA in heparin also vary considerably throughout the composing disaccharides.

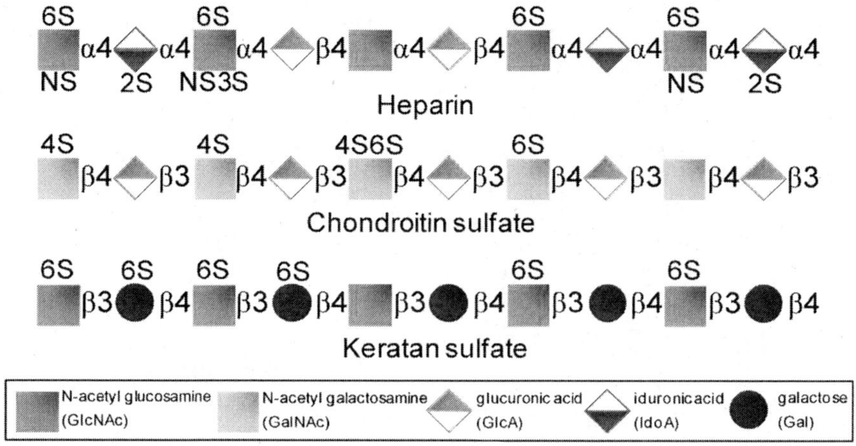

Figure 1. Structural representation of the most explored therapeutic glycosaminoglycans of the global pharmaceutical market.

On the other hand, certain SFs and SGs, especially those extracted from red algae and invertebrate animals, are fairly homogeneous in terms of monosaccharide composition/distribution and very regular in terms of sulfation patterns (Figure 2) if compared to the therapeutic GAGs (Figure 1). While red algae express only SGs composed of disaccharide repeating units (Figure 2) like GAG molecules (Figure 1), SFs and SGs of regular chemical structures are composed of oligosaccharide (mono to tetrasaccharides) repeating units. These molecules can be extracted from sea urchins, sea cucumbers and ascidians (Figure 2) [14,15,16]. While in sea urchins the SFs and SGs comprise extracellular matrix (ECM) components of the egg jelly coat in female gametes of this echinoderm, in sea cucumbers and ascidians they occur as ECM components of their body walls. While SFs are polymers solely composed of fucopyranose (Fuc) units, SGs are made up primarily of Gal units (Figure 2). In brown algae, only SFs have been reported, however within a more heterogeneous backbone composition [32,33,34,35]. These brown algal SFs are best known as fucoidans and other monosaccharide types have been detected [32,33,34,35]. SGs are the major sulfated glycan in green algae [14,19,36,37]. Although more homogeneous than brown algal fucoidans, green algal SGs are still more heterogeneous than the disaccharide-composed SGs of red algae [14,19,36,37]. Regardless of the phylogenetic classification, sulfation patterns are likely to vary among different species.

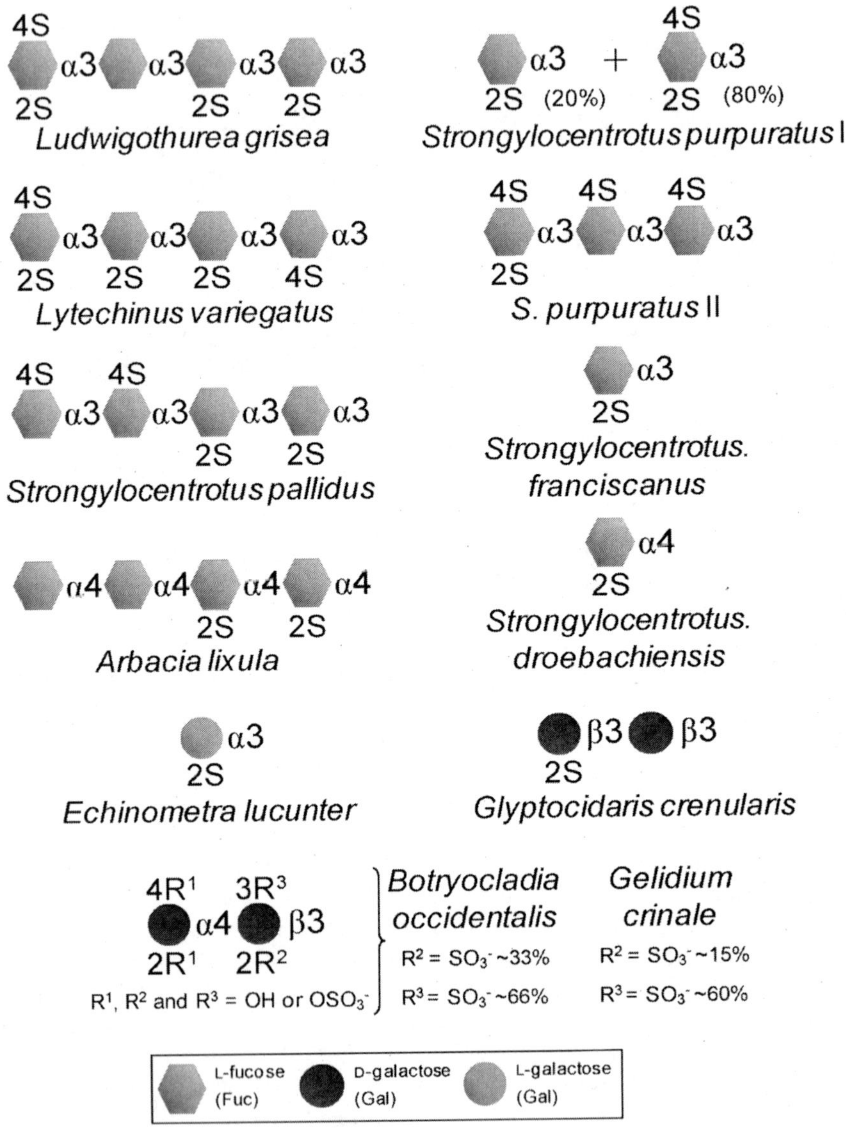

Figure 2. Structural representation of the marine, invertebrate and red algal, SFs and SGs of well-defined chemical structures. All species are sea urchins, except the first one which is a sea cucumber and the last two which are red algae.

In summary, SFs or SGs isolated from invertebrate animals and red algae are very homogeneous in terms of backbone constitution and sulfation patterns (Figure 2). This makes these particular marine sulfated glycans very useful molecular tools to be investigated in medicine specially for establishing advanced structure-activity relationship (SAR). These advanced SAR analyses are relevant in drug discovery since it helps to unveil (i) the underlying mechanisms of action of their medical properties in the different pathophysiological systems, and (ii) the best structural motifs responsible for the highest clinical responses. Although these two aspects can be safely investigated in GAG studies, it is much more difficult to design SAR analyses on GAGs than on the marine SFs and SGs of well-defined chemical structures. Table 1 highlights the major structural aspects of both sulfated glycan types discussed in this report. Outcomes, advantageous as well disadvantageous aspects regarding their medical potentials of each molecular type are also highlighted.

Table 1. Comparative structural and medical aspects of glycosaminoglycans (GAGs) and marine sulfated glycans of well-defined chemical structures.

Polysaccharide Type	Molecular Type	Structural Units	Overall Aspects	Medical Benefits
GAGs	Heparin	IdoA2S + GlcNS6S	As the most negatively charged biomacromolecule in nature, it interacts and regulates various protein types.	Potent anticoagulant and antithrombotic agent [3,4].
	Chondroitin sulfate	GlcA + GalNAc(4S and/or 6S)	The most abundant GAG of the body and of connective (cartilage) tissues.	Used in cases of osteoarthritis, osteoarthrosis and sometimes osteoporosis [5–7].
	Keratan sulfate	Gal + GlcNAc (6S at both units but more often at GlcNAc)	Highly abundant in cornea. Related with the proper visual functions.	Explored as functional ingredient in eye-drops [8,9].
Marine glycans	SFs	Fuc (2S and/or 4S)	Found in well-defined structures in sea urchins and sea cucumbers.	Exhibits potential effects in anticoagulation, antithrombosis, anti-angiogenesis, antitumor, anti-inflammation and antimicrobial infections [14–16].
	SGs	Gal (2S and/or 3S, and/or 4S)	Found in well-defined structures in red algae, sea urchins and ascidians.	

3. MEDICAL MECHANISMS OF ACTION

3.1. Anticoagulant/Antithrombotic Mechanisms of Action

3.1.1. The Serpin-Dependent Mechanisms

Under normal physiological conditions, GAGs on proteoglycans at the endothelial surface play a key role in haemostasis as accelerators of the serpins AT and HCII. These effects lead to enhanced AT and HCII activities of inhibition over certain procoagulant factors such as IIa (thrombin), IXa and Xa [25,38,39,40,41]. The GAG-catalyzed effects on AT and HCII over the blood serine-proteases result from two concomitant molecular events. (i) A template effect by which the sulfated glycans accelerates the formation of the serpin-protease-sulfated glycan ternary complex. In this molecular event the sulfated glycan serves as a 'molecular bridge' that will act bringing together both serpin and protease for a close contact. (ii) An allosteric mechanism by which the sulfated glycan induces a conformational change on the serpin in order to achieve a more active structure for interaction and complex formation.

Exogenous sulfated glycans, such as marine SFs and SGs endowed with the capacity of interacting with serpins and their blood targets, will work as synergic molecules in the two molecular events described above for the physiological GAGs. The contribution and quality of these events will depend, of course, on the dose of the exogenous SFs and SGs administered in the system as well as on the presence and concentration per molecule of the active motifs required for interaction with the blood co-(factors) [14,15,16,17,18,19,20,21]. Table 2 shows half maximal inhibitory concentration (IC_{50}) values from IIa and Xa inhibition assays via AT and HCII activities using purified enzymes and co-factors as well as the values determined in a coagulometer through the activated partial thromboplastin time (aPTT) method and expressed in units/mg. Values were obtained from curves of anticoagulation time as a function of increasing concentrations of the tested sulfated glycan. A parallel curve generated with commercially available Heparin Sodium sample was obtained as calibrator. In a comparative perspective taking together the anticoagulant potencies (Table 2) and their chemical structures (Figure 2) which enable confident SAR analyses, the structural requirements for the anticoagulant activities of the tested SFs and SGs were discovered. They are the 2-sulfated Gal units as seen in the sea urchin Echinometra lucunter SG and the 2,4-di-sulfated Fuc units seen in the SFs from the sea urchin Strongylocentrotus purpuratus as well as in the two SGs from the red algal speciesBotryocladia occidentalis and Gelidium crinale (Rhodophyta, Rhodymeniophycidae). These structural features have been proposed as the anticoagulant structural motifs of the marine sulfated glycans of well-defined chemical structures [42,43]. Although exhibiting lower anticoagulant activity than heparin, the active compounds do not show the high-bleeding risk of heparin as determined through in vivo assays of thrombosis [19,26].

Table 2. Anticoagulant activities of marine sulfated glycans of well-defined structures (Figure 2) measured by activated partial thromboplastin time (aPTT) and by IC_{50}values of thrombin (IIa) and factor Xa inhibition in the presence of antithrombin (AT) or heparin cofactor II (HCII).

Polysaccharide Type	Source	aPTT (units/mg) [a]	IC_{50} (μg/mL)		
			IIa/AT	IIa/HCII	Xa/AT
Invertebrate 3-linked α-L-SF	*S. purpuratus* I	76	0.3	0.3	2
	S. purpuratus II	10	0.9	2	nd [b]
	S. pallidus	18	>500	>500	>500
	L. variegatus I	3	>500	>500	>500
	S. franciscanus	~2	>500	>500	250
	L. grisea	<1	>500	>500	>500
Invertebrate 4-linked α-L-SF	*S. droebachiensis*	<1	nd	nd	nd
	A. lixula	~2	150	150	>500
Invertebrate α-L-SG	*E. lucunter*	20	3	6	20
	G. crenularis	<1	nd	nd	Nd
Red algal SGs	*B. occidentalis*	93	0.02	1.1	2.5
	G. crinale	65	0.02	25	1.5

[a] The activity was measured in units/mg through curves of coagulation time as a function of increasing concentrations of the polysaccharide. A parallel standard curve of an assayed unfractionated heparin sample (commercially available Heparin Sodium) was generated and the activity of this calibrator was measured as 193 units/mg. [b] not determined.

3.1.2. The Serpin-Independent Mechanism

In addition to its great serpin-dependent anticoagulant activity, it was recently discovered through assays using serpin-free plasma that the SG from the red algae B. occidentalis is also able to prolong coagulation time and delay generation course of factors IIa and Xa [17]. In contrast, heparin as the principal anticoagulant GAG of medicine does not exhibit such effects. Further investigation on this unusual anticoagulant effect using purified proteins of the blood coagulation system have led to conclusions that the SG from the red algae B. occidentalis is capable of impairing the proper molecular assembly in the intrinsic tenase and prothrombinase complexes. Figure 3 illustrates the physiological (panel A) and impaired (panel B) molecular assemblies. As depicted, these complexes are critical for activation of factors Xa and IIa, respectively.

Therefore, an anticoagulant outcome arises from the exogenous application of SG from B. occidentalis via an effect unrelated with AT and HCII activities [17].

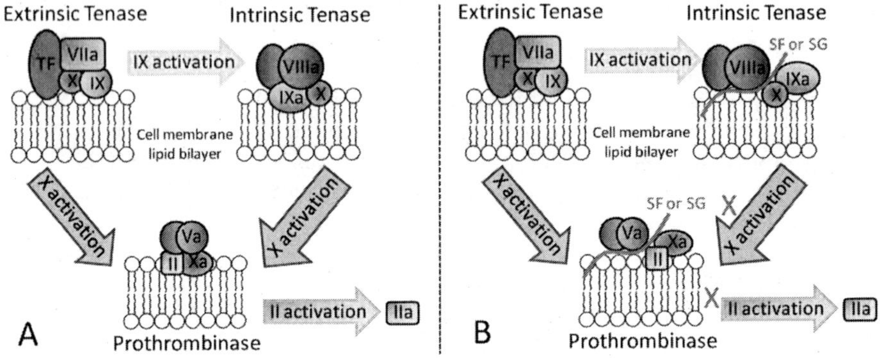

Figure 3. (**A**) Representation of the three key coagulation complexes: extrinsic tenase, intrinsic tenase and prothrombinase. (**B**) The serpin-independent anticoagulant mechanism of marine non-GAG sulfated glycans (SF or SG) relies on the assembly inhibition of the intrinsic tenase and protrombinase complexes.

Recent data have demonstrated that this serpin-independent anticoagulant activity of the SG from red algae can also be seen in in vivo models of thrombosis [19]. In fact, through both in vivo and in vitro assays, the serpin-independent effect comprises the main mechanism of action for the anticoagulant/antithrombotic outcomes of these molecules [18,19]. Results coming from a different research group also concerning the serpin-independent anticoagulant activity of marine sulfated glycans, but specifically on the holothurian GAG named fucosylated chondroitin sulfate, have led to the same conclusion [44]. Hence, although serpin-dependent anticoagulant and antithrombotic SFs and SGs can be measured and structural motifs responsible for such activities can be proposed [41,42], the serpin-unrelated mechanisms dominate in therapy. This conclusion is recent. Future investigations, especially those focused on the right biologically active structural motifs necessary to the serpin-independent mechanisms are worth being carried out.

3.2. Anti-Angiogenic and Anticancer Mechanisms of Action

In angiogenesis, the activity of angiogenic, mitogenic, chemotatic and growth-stimulating factors like vascular endothelial growth factor

(VEGF) and basic fibroblast growth factor (bFGF, also known as FGF-2) is relevant to the success of the event [45]. These growth factors have to interact and be activated by a resultant ternary complex made up with their canonical receptors and the endothelial surface proteoglycan GAGs [46,47,48,49,50,51,52,53]. Assuming that these factors are circulating free in the plasma, the attachment on surface GAGs is required for their activities in the specific sites of neovascularization. The actions of these growth factors are commonly seen in the cell differentiation event named mesenchymal-epithelial transition, a process which permits the formation of new endothelial cells from angioblasts, and these, in turn, form mesodermal cells [54]. Besides assisting the cell differentiation processes, growth factors also work on the molecular networks involved in the neovascularization [45]. Angiogenesis is a pivotal event in cancer growth (of the primary tumor) and metastasis [55,56]. Formation of new vessels intended to feed the tumor cells are needed for the development of this severe pathology [55,56].

Since binding and attachment of growth factors to endothelial GAGs is required for angiogenesis, administration of certain sulfated glycans in effective concentration will result in a competition process with the functional endothelial GAGs. This could therefore give rise to an anti-angiogenic outcome. Based on this premise, administration in the system of exogenous sulfated glycans with active structural features is likely able to decrease the rates of neovascularization [57,58]. Indeed marine sulfated glycans such as brown algal fucoidans display competitive structural and functional properties against endothelial GAGs and are considered great anti-angiogenic and anticancer sulfated polysaccharides. Both in vitroand in vivo experiments have been performed to evaluate the anti-angiogenic and anticancer potentials of these glycans [57,58,59].

Although useful in advanced SAR studies, data regarding the marine SFs and SGs of well-defined chemical structures (Figure 2) as anti-angiogenic and antitumor agents are virtually inexistent up to now. On the other hand, since the brown algal fucoidans have been preferred to the investigations related to these systems, some SAR information concerning fucoidans has been proposed. It seems that molecular size of fucoidans is an important feature in anti-angiogenesis [58]. Based on what has been highlighted by Prof. Boisson-Vidal [60] and supported by the recent analyses of Ustyuzhanina and associates [58], high-molecular weight fucoidans (usually above 30 kDa) bearing also high degrees of sulfation, are likely able to present anti-angiogenic affects. Conversely, low-molecular weight fucoidans (usually below 15 kDa) or fucoidan oligosaccharide fractions tend to promote angiogenesis. This molecular weight-dependent antagonic effects in angiogenesis of the brown algal fucoidan must be further investigated for anticancer therapy.

3.3. Anti-Inflammatory Mechanisms of Action

It is well-established that carbohydrates including sulfated glycans play key roles in events of cell–cell and cell–matrix communication. These molecules can either promote or inhibit the steps of inflammation. In addition, this category of molecules is attracting much attention nowadays in research programs involved with the molecular details and therapeutic agents of inflammation [61,62,63]. Amongst all glycans reported to be active in inflammation, surface GAGs of both circulating blood cells and endothelial proteoglycans are those with the highest number of roles. As illustrated in Figure 4, they are involved in (1) interactions with L-selectin during the initial steps of leukocyte recruitment and rolling; (2) interactions and storage of released chemokines from macrophages in the underlying tissue; (3) after translocation of the proteoglycan-retained chemokines facing the inner tissue, these peptides are presented to their respective leukocyte receptors in order to trigger leukocyte activation; (4) facilitation on the adequate extravasation of activated leukocytes through the tight spaces of the endothelial barrier; (5) after activated leukocytes have gained access into the inflamed inner tissue, the infiltrated leukocytes will interact with the yet-unbound chemokines available at that region via surface proteoglycan GAGs; and (6) in order to promote a pro-inflammatory upstream in the system, leukocytes that have entered into the inner tissue become able to release hydrolases capable of degrading proteoglycans and collagen fibers of the ECM close to the underlying basement membrane. This later process is important to facilitate leukocyte transmigration through the endothelial layer.

Since multiple roles can be assigned to GAGs in inflammation progress; it wouldn't be unexpected if exogenous sulfated glycans could be used as possible therapeutic anti-inflammatory agents. This outcome arises from the fact that when sulfated glycans are administered in effective concentrations in cases of acute or chronic inflammatory disorders, these molecules are likely able to compete with the physiological GAGs needed to keep the proper molecular workflow of this pathophysiological event. Hence, sulfated glycans from diverse sources and various structures such as the marine SFs and SGs are currently under investigation as potential anti-inflammatory agents [57,64,65,66,67,68]. It has been postulated that the anti-inflammatory activities of these marine sulfated glycans depend directly on the structure of the polysaccharide used for treatment [57,64,65,66,67,68].

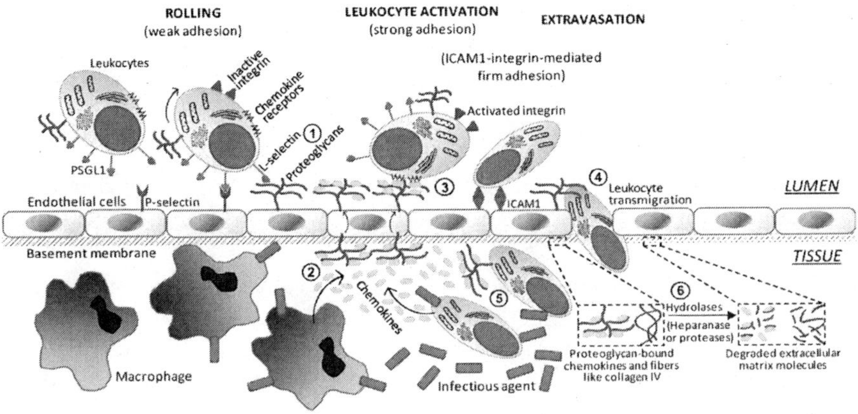

Figure 4. Simplified scheme regarding the major cells and molecular players involved in the progress of inflammation. The numbers indicate the sites of action of the anti-inflammatory sulfated glycans. Abbreviations used are TNF for tumor necrosis factor, ICAM for intercellular adhesion molecule and PSGL1 for P-selectin glycoprotein ligand-1.

3.4. Antimicrobial Mechanism of Action

As stated in the beginning of the previous section, GAGs are essential molecules to events involving cell–cell interactions. This can be easily exemplified by interactions of microbial pathogens with their target cells in host organisms as seen in the initial stages of microbial infectivity of mammalian cells. GAGs, as the main physiological sulfate glycan representatives of host cells, are one of the principal molecules involved in attachment, adhesion and entry of the various types of micropathogens (Figure 5A) including bacteria [69,70,71,72,73,74], virus [69,70,74,75,76,77,78,79], fungus [69,70,80], and protozoan parasites [69,70,81,82].

Based on the molecular mechanisms illustrated in Figure 5, sulfated glycans including commercially available GAGs and marine SFs and SGs are clearly able to result in an antimicrobial outcome when applied in events characterized by micropathogen infectivity [83,84,85,86,87,88,89]. The antimicrobial activity of sulfated glycans arises from their capacity to compete with the host cell GAGs necessary to the progress of the microbial recognition, attachment and invasion (Figure 5B). The levels of response for the antimicrobial treatment will

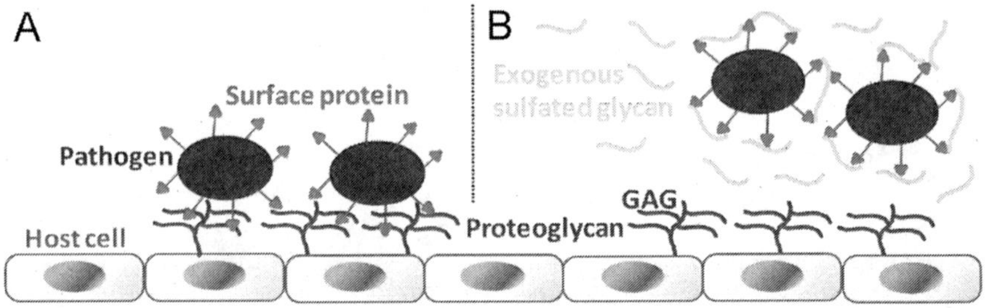

Figure 5. General schematic representation for the molecular mechanisms during (**A**) the microbial infection of pathogens such as bacteria, virus, fungus, and protozoan parasites; and during (**B**) the administration at certain concentrations of exogenous sulfated glycans endowed with antimicrobial activity. In (**A**), microbial infection is driven by molecular interactions between pathogen proteins and host cell GAGs. In (**B**), the antimicrobial sulfated glycan competes with host cell GAGs by binding to proteins displayed at the microbial surface, impairing or disrupting then the pathogen attachment onto host cells.

depend, of course, on the dose of the active glycan administered in the system as well as on the structural features of the chosen sulfated glycans. The marine sulfated glycans most investigated so far concerning their antimicrobial activities have been the brown algal fucoidans [86,87] and the red algal carrageenans [87,88,89], a type of SG composed of disaccharide repeating units and usually regular patterns of sulfation. No firm SAR conclusions about these red and brown algal sulfated glycans have been proposed so far as new antimicrobial agents.

4. CONCLUDING REMARKS AND EXISTING CHALLENGES IN DRUG DEVELOPMENT

Sulfated glycans are perhaps the carbohydrate types most studied in what concerns their medical activities. Mammalian-derived GAGs such as heparin, chondroitin sulfate and keratan sulfate are examples of these well-known medical sulfated glycans. Non-GAGs such as marine sulfated glycans are another type of sulfated glycans also endowed with medical properties. SFs and SGs are the main examples of these marine polysaccharides. They are available to be used as alternative or supplement agents in GAG-based therapies. Table 3 highlights in a comparative fashion the effects of the therapeutic GAGs and the marine

SFs and SGs in the series of pathophysiological systems discussed in this document.

Table 3. Medical effects of exogenous GAGs and marine sulfated glycans of well-defined chemical structures in different pathophysiological systems.

Medical System	GAGs	Marine Sulfated Glycans
Anticoagulation and antitrombosis	Serpin-dependent mechanism - Accelerate AT inhibition over factors IIa and Xa. - Accelarate HCII inhibition over factor IIa.	Serpin-dependent mechanism - Accelerate AT inhibition over factors IIa and Xa. - Accelarate HCII inhibition over factor IIa. Serpin-independent mechanism - Inhibit intrinsic tenase complex. - Inhibit prothrombinase complex.
Anti-angiogenesis and anticancer	- Inhibition activities on growth factors necessary for cell differentiation and neovascularization. - Inhibition activities on selectins necessary for cell migration, attachment and adhesion.	
Anti-inflammation	- Inhibit L-selectin during leukocyte recruitment and rolling. - Inhibit chemokine functions in leukocyte activation. - Decrease extravasation of activated leukocytes. - Inhibit binding of infiltrated leukocytes to free chemokines in the inflamed tissue. - Compete with hydrolase activity during ECM processing necessary to enhance leukocyte transmigration. Marine sulfated glycans maybe less active or inactive in this effect.	
Antimicrobial infections	- Compete with host cell GAGs during host cell-pathogen recognition, interaction and attachment during microbial infection.	

As detailed above, SFs and GSs of well-defined chemical structures are promising drug candidates because they exhibit a phenomenal number of medical benefits in multiple pathophysiological systems. Conversely, the real clinical application of these marine sulfated glycans is still some years away. Although studies of clinical trials using patients committed with chronic or acute disorders involving the pathologies which the marine sulfated glycans are known to be active and have already been approved and initiated [59,90], fucoidans seem to be the only representative molecules of marine sulfated glycans under current trials. As detailed here, these molecules are heterogeneous and accurate SAR conclusions regarding their structural features are hard to be proposed. In addition to this, some marine sulfated glycans can respond in a similar way to the oversulfated chondroitin sulfate in respect to the capacity of activating coagulation blood factor XII [19,91]. Factor XIIa can participate in inducing severe hypotension caused by kallikrein, a potent regulator of blood pressure via activation of bradykinin. This can lead to a harmful effect to health during a possible therapy of the marine sulfated glycans equipped with kallikrein-dependent effects. This downside of certain marine sulfated glycans is a serious challenge for the future implementation of these compounds in the next generation of potential carbohydrate-based drug candidates. Nonetheless, interest and efforts in research regarding the medicinal benefits of these marine sulfated glycans are growing considerably in the last few years across scientific groups of different countries. We expect some day that the amount of information concerning the beneficial and the harmful effects to health of the marine sulfated glycans reach a reliable point in order to make these molecules much more accessible to patients suffering from the pathologies they are active.

CONFLICTS OF INTEREST

Vitor H. Pomin declares no conflict of interest.

REFERENCES

1. Volpi, N. Therapeutic applications of glycosaminoglycans. Curr. Med. Chem. **2006**, 13, 1799–1810.
2. Pomin, V.H. A dilemma in the glycosaminoglycan-based therapy: Synthetic or naturally unique molecules? Med. Res. Rev. **2015**, 35, 1195–1219.
3. Gray, E.; Hogwood, J.; Mulloy, B. The anticoagulant and antithrombotic mechanisms of heparin. Handb. Exp. Pharmacol. **2012**, 207, 43–61.
4. Hirsh, J.; Anand, S.S.; Halperin, J.L.; Fuster, V. Guide to anticoagulant therapy: Heparin: A statement for healthcare professionals from the American Heart Association. Circulation **2001**, 103, 2994–3018.
5. Uebelhart, D. Clinical review of chondroitin sulfate in osteoarthritis. Osteoarthritis Cartilage **2008**, 16, S19–S21.

6. Conrozier, T. Anti-arthrosis treatments: Efficacy and tolerance of chondroitin sulfates (CS 4&6). Presse Med. **1998**, 27, 1862–1865.

7. Pomin, V.H.; Piquet, A.A.; Pereira, M.S.; Mourão, P.A. Residual keratan sulfate in chondroitin sulfate formulations for oral administration. Carbohydr. Polym. **2012**, 90, 839–846.

8. Pomin, V.H. Keratan sulfate: An up-to-date review. Int. J. Biol. Macromol. **2015**, 72, 282–289.

9. Maruyama, H.; Morikawa, K.; Tawada, A.; Miyauchi, S.; Yoshida, K.; Asari, A. Keratan Sulfate Oligosaccharide Fraction and Pharmaceutical Containing the Same. U.S. Patent 6,159,954 A, 12 December 2000.

10. Crowther, M.A.; Warkentin, T.E. Bleeding risk and the management of bleeding complications in patients undergoing anticoagulant therapy: focus on new anticoagulant agents. Blood **2008**, 111, 4871–4879.

11. Babikian, V.L.; Kase, C.S.; Pessin, M.S.; Norrving, B.; Gorelick, P.B. Intracerebral hemorrhage in stroke patients anticoagulated with heparin. Stroke **1989**, 20, 1500–1503.

12. Kelton, J.G.; Hirsh, J. Bleeding associated with antithrombotic therapy. Semin. Hematol. **1980**, 17, 259–379.

13. Funderburgh, J.L. Keratan sulfate: Structure, biosynthesis, and function. Glycobiology **2000**, 10, 951–958.

14. Pomin, V.H.; Mourão, P.A. Structure, biology, evolution, and medical importance of sulfated fucans and galactans.Glycobiology **2008**, 18, 1016–1027.

15. Pomin, V.H. Fucanomics and galactanomics: marine distribution, medicinal impact, conceptions, and challenges. Mar. Drugs **2012**, 10, 793–811.

16. Pomin, V.H. Fucanomics and galactanomics: Current status in drug discovery, mechanisms of action and role of the well-defined structures. Biochim. Biophys. Acta **2012**, 1820, 1971–1979.

17. Glauser, B.F.; Rezende, R.M.; Melo, F.R.; Pereira, M.S.; Francischetti, I.M.; Monteiro, R.Q.; Rezaie, A.R.; Mourão, P.A. Anticoagulant activity of a sulfated galactan: Serpin-independent effect and specific interaction with factor Xa.Thromb. Haemost. **2009**, 102, 1183–1193.

18. Glauser, B.F.; Mourão, P.A.; Pomin, V.H. Marine sulfated glycans with serpin-unrelated anticoagulant properties. Adv. Clin. Chem. **2013**, 62, 269–303.

19. Quinderé, A.L.; Santos, G.R.; Oliveira, S.N.; Glauser, B.F.; Fontes, B.P.; Queiroz, I.N.; Benevides, N.M.; Pomin, V.H.; Mourão, P.A. Is the antithrombotic effect of sulfated galactans independent of serpin? J. Thromb. Haemost. **2014**, 12, 43–53.

20. Pomin, V.H.; Pereira, M.S.; Valente, A.P.; Tollefsen, D.M.; Pavão, M.S.; Mourão, P.A. Selective cleavage and anticoagulant activity of a sulfated fucan: stereospecific removal of a 2-sulfate ester from the polysaccharide by mild acid hydrolysis, preparation of oligosaccharides, and heparin cofactor II-dependent anticoagulant activity.Glycobiology **2005**, 15, 369–381.

21. Mourão, P.A. Use of sulfated fucans as anticoagulant and antithrombotic agents: Future perspectives. Curr. Pharm. Des. **2004**, 10, 967–981.

22. Tollefsen, D.M.; Pestka, C.A.; Monafo, W.J. Activation of heparin cofactor II by dermatan sulfate. J. Biol. Chem. **1983**,258, 6713–6716.

23. Santos, G.R.; Tovar, A.M.; Capillé, N.V.; Pereira, M.S.; Pomin, V.H.; Mourão, P.A. Structural and functional analyses of bovine and porcine intestinal heparins confirm they are different drugs. Drug Discov. Today **2014**, 19, 1801–1807.

24. Weitz, J.I. Low-molecular-weight heparins. N. Engl. J. Med. **1992**, 337, 688–698.

25. Patnaik, M.M.; Moll, S. Inherited antithrombin deficiency: A review. Haemophilia **2008**, 14, 1229–1239.

26. Fonseca, R.J.; Oliveira, S.N.; Melo, F.R.; Pereira, M.G.; Benevides, N.M.; Mourão, P.A. Slight differences in sulfation of algal galactans account for differences in their anticoagulant and venous antithrombotic activities. Thromb. Haemost.**2008**, 99, 539–545.

27. Sasisekharan, R.; Venkataraman, G. Heparin and heparan sulfate: Biosynthesis, structure and function. Curr. Opin. Chem. Biol. **2000**, 4, 626–631.

28. Rabestein, D.L. Heparin and heparan sulfate: Structure and function. Nat. Prod. Rep. **2002**, 19, 312–331.

29. Pomin, V.H.; Mulloy, B. Current structural biology of the heparin interactome. Curr. Opin. Struct. Biol. **2015**, 34, 17–25.

30. Pomin, V.H.; Park, Y.; Huang, R.; Heiss, C.; Sharp, J.S.; Azadi, P.; Prestegard, J.H. Exploiting enzyme specificities in digestions of chondroitin sulfates A and C: Production of well-defined hexasaccharides. Glycobiology **2012**, 22, 826–838.

31. Mizumoto, S.; Yamada, S.; Sugahara, K. Molecular interactions between chondroitin-dermatan sulfate and growth factors/receptors/matrix proteins. Curr. Opin. Struct. Biol. **2015**, 34, 35–42.

32. Li, B.; Lu, F.; Wei, X.; Zhao, R. Fucoidan: Structure and bioactivity. Molecules **2008**, 13, 1671–1695.

33. Fitton, J.H. Therapies from fucoidan; multifunctional marine polymers. Mar. Drugs **2011**, 9, 1731–1760.

34. Wijesinghe, W.A.J.P.; Jeon, Y.-J. Biological activities and potential industrial applications of fucose rich sulfated polysaccharides and fucoidans isolated from brown seaweeds: A review. Carbohydr. Pol. **2012**, 88, 13–20.

35. Fitton, J.H.; Stringer, D.N.; Karpiniec, S.S. Therapies from fucoidan: An update. Mar. Drugs **2015**, 13, 5920–5946.

36. Farias, E.H.; Pomin, V.H.; Valente, A.P.; Nader, H.B.; Rocha, H.A.; Mourão, P.A. A preponderantly 4-sulfated, 3-linked galactan from the green alga Codium isthmocladum. Glycobiology **2008**, 18, 250–259.

37. Pomin, V.H. Structural and functional insights into sulfated galactans: A systematic review. Glycoconj. J. **2010**, 27, 1–12.

38. Danielsson, A.; Björk, I. Properties of antithrombin-thrombin complex formed in the presence and in the absence of heparin. Biochem. J. **1983**, 213, 345–353.

39. Liu, L.; Dewar, L.; Song, Y.; Kulczycky, M.; Blajchman, M.A.; Fenton, J.W., 2nd; Andrew, M.; Delorme, M.; Ginsberg, J.; Preissner, K.T.; et al. Inhibition of thrombin by antithrombin III and heparin cofactor II in vivo. Thromb. Haemost. **1995**,73, 405–412.

40. Maaroufi, R.M.; Jozefowicz, M.; Tapon-Bretaudière, J.; Jozefonvicz, J.; Fischer, A.M. Mechanism of thrombin inhibition by heparin cofactor II in the presence of dermatan sulphates, native or oversulphated, and a heparin-like dextran derivative. Biomaterials **1997**, 18, 359–366.

41. Tollefsen, D.M. Vascular dermatan sulfate and heparin cofactor II. Prog. Mol. Biol. Transl. Sci. **2010**, 93, 351–372.

42. Pomin, V.H. Anticoagulant motifs of marine sulfated glycans. Glycoconj. J. **2014**, 31, 341–344.

43. Pomin, V.H.; Mourão, P.A. Specific sulfation and glycosylation—a structural combination for the anticoagulation of marine carbohydrates. Front. Cell. Infect. Microbiol. **2014**, 4, 33.

44. Zhao, L.; Wu, M.; Xiao, C.; Yang, L.; Zhou, L.; Gao, N.; Li, Z.; Chen, J.; Chen, J.; Liu, J.; Qian, H.; Zhao, J. Discovery of an intrinsic tenase complex inhibitor: Pure nonasaccharide from fucosylated glycosaminoglycan. Proc. Natl. Acad. Sci. USA **2015**, 112, 8284–8289.

45. Otrock, Z.K.; Mahfouz, R.A.; Makarem, J.A.; Shamseddine, A.I. Understanding the biology of angiogenesis: Review of the most important molecular mechanisms. Blood Cells Mol. Dis. **2007**, 39, 212–220.

46. Faham, S.; Hileman, R.E.; Fromm, J.R.; Linhardt, R.J.; Rees, D.C. Heparin structure and interactions with basic fibroblast growth factor. Science **1996**, 271, 1116–1120.

47. Pellegrini, L.; Burke, D.F.; von Delft, F.; Mulloy, B.; Blundell, T.L. Crystal structure of fibroblast growth factor receptor ectodomain bound to ligand and heparin. Nature **2000**, 407, 1029–1034.

48. Schlessinger, J.; Plotnikov, A.N.; Ibrahimi, O.A.; Eliseenkova, A.V.; Yeh, B.K.; Yayon, A.; Linhardt, R.J.; Mohammadi, M. Crystal structure of a ternary FGF-FGFR-heparin complex reveals a dual role for heparin in FGFR binding and dimerization. Mol. Cell. **2000**, 6, 743–750.

49. Goodger, S.J.; Robinson, C.J.; Murphy, K.J.; Gasiunas, N.; Harmer, N.J.; Blundell, T.L.; Pye, D.A.; Gallagher, J.T. Evidence that heparin saccharides promote FGF2 mitogenesis through two distinct mechanisms. J. Biol. Chem. **2008**,283, 13001–13008.

50. Robinson, C.J.; Harmer, N.J.; Goodger, S.J.; Blundell, T.L.; Gallagher, J.T. Cooperative dimerization of fibroblast growth factor 1 (FGF1) upon a single heparin saccharide may drive the formation of 2:2:1 FGF1.FGFR2c.heparin ternary complexes. J. Biol. Chem. **2005**, 280, 42274–42282.

51. Guerrini, M.; Hricovíni, M.; Torri, G. Ineraction of heparins with fibroblast growth factors: conformational aspects.Curr. Pharm. Des. **2007**, 13, 2045–2056.

52. Guglier, S.; Hricovíni, M.; Raman, R.; Polito, L.; Torri, G.; Casu, B.; Sasisekharan, R.; Guerrini, M. Minimum FGF2 binding structural

requirements of heparin and heparan sulfate oligosaccharides as determined by NMR spectroscopy. Biochemistry **2008**, 47, 13862–13869.

53. Teran, M.; Nugent, M.A. Synergistic binding of vascular endothelial growth factor-A and its receptors to heparin selectively modulates complex affinity. J. Biol. Chem. **2015**, 290, 16451–16462.

54. Yao, D.; Dai, C.; Peng, S. Mechanism of the mesenchymal-epithelial transition and its relationship with metastatic tumor formation. Mol. Cancer Res. **2011**, 9, 1608–1620.

55. Khosravi Shahi, P.; Soria Lovelle, A.; Pérez Manga, G. Tumoral angiogenesis and breast cancer. Clin. Transl. Oncol.**2009**, 11, 138–142.

56. Nishida, N.; Yano, H.; Nishida, T.; Kamura, T.; Kojiro, M. Angiogenesis in cancer. Vasc. Health Risk Manag. **2006**, 2, 213–219.

57. Cumashi, A.; Ushakova, N.A.; Preobrazhenskaya, M.E.; D'Incecco, A.; Piccoli, A.; Totani, L.; Tinari, N.; Morozevich, G.E.; Berman, A.E.; Bilan, M.I.; et al. A comparative study of the anti-inflammatory, anticoagulant, antiangiogenic, and antiadhesive activities of nine different fucoidans from brown seaweeds. Glycobiology **2007**, 17, 541–552.

58. Ustyuzhanina, N.E.; Bilan, M.I.; Ushakova, N.A.; Usov, A.I.; Kiselevskiy, M.V.; Nifantiev, N.E. Fucoidans: Pro- or antiangiogenic agents? Glycobiology **2014**, 24, 1265–1274.

59. Kwak, J.Y. Fucoidan as a Marine anticancer agent in preclinical development. Mar. Drugs **2014**, 12, 851–870.

60. Zemani, F.; Benisvy, D.; Galy-Fauroux, I.; Lokajczyk, A.; Colliec-Jouault, S.; Uzan, G.; Fischer, A.M.; Boisson-Vidal, C. Low-molecular-weight fucoidan enhances the proangiogenic phenotype of endothelial progenitor cells. Biochem. Pharmacol. **2005**, 15, 1167–1175.

61. Dube, D.H.; Bertozzi, C.R. Glycans in cancer and inflammation—potential for therapeutics and diagnostics. Nat. Rev. Drug Discov. **2005**, 4, 477–488.

62. Lowe, J.B. Glycan-dependent leukocyte adhesion and recruitment in inflammation. Curr. Opin. Cell Biol. **2003**, 15, 531–538.

63. McEver, R.P. Selectin-carbohydrate interactions during inflammation and metastasis. Glycoconj. J. **1997**, 14, 585–591.

64. Pomin, V.H. Sulfated glycans in inflammation. Eur. J. Med. Chem. **2015**, 92, 353–369.

65. Albuquerque, I.R.; Cordeiro, S.L.; Gomes, D.L.; Dreyfuss, J.L.; Filgueira, L.G.; Leite, E.L.; Nader, H.B.; Rocha, H.A. Evaluation of anti-nociceptive and anti-inflammatory activities of a heterofucan from Dictyota menstrualis. Mar. Drugs**2013**, 11, 2722–2740.

66. De Sousa, A.A.; Benevides, N.M.; de Freitas Pires, A.; Fiúza, F.P.; Queiroz, M.G.; Morais, T.M.; Pereira, M.G.; Assreuy, A.M. A report of a galactan from marine alga Gelidium crinale with in vivo anti-inflammatory and antinociceptive effects. Fundam. Clin. Pharmacol. **2013**, 27, 173–180.

67. Li, X.J.; Ye, Q.F. Fucoidan reduces inflammatory response in a rat model of hepatic ischemia-reperfusion injury. Can. J. Physiol. Pharmacol. **2015**, 93, 999–1005.

68. Lean, Q.Y.; Eri, R.D.; Fitton, J.H.; Patel, R.P.; Gueven, N. Fucoidan extracts ameliorate acute colitis. PLoS ONE **2015**, 10, e0128453.

69. Jinno, A.; Park, P.W. Role of glycosaminoglycans in infectious disease. Methods Mol. Biol. **2015**, 1229, 567–585.

70. Sawitzky, D. Protein-glycosaminoglycan interactions: Infectiological aspects. Med. Microbiol. Immunol. **1996**, 184, 155–161.

71. Baron, M.J.; Wong, S.L.; Nybakken, K.; Carey, V.J.; Madoff, L.C. Host glycosaminoglycan confers susceptibility to bacterial infection in Drosophila melanogaster. Infect. Immun. **2009**, 77, 860–866.

72. Chang, Y.C.; Wang, Z.; Flax, L.A.; Xu, D.; Esko, J.D.; Nizet, V.; Baron, M.J. Glycosaminoglycan binding facilitates entry of a bacterial pathogen into central nervous systems. PLoS Pathog. **2011**, 7, e1002082.

73. Sava, I.G.; Zhang, F.; Toma, I.; Theilacker, C.; Li, B.; Baumert, T.F.; Holst, O.; Linhardt, R.J.; Huebner, J. Novel interactions of glycosaminoglycans and bacterial glycolipids mediate binding of enterococci to human cells. J. Biol. Chem. **2009**, 284, 18194–18201.

74. Wadström, T.; Ljungh, A. Glycosaminoglycan-binding microbial proteins in tissue adhesion and invasion: key events in microbial pathogenicity. J. Med. Microbiol. **1999**, 48, 223–233.

75. Leistner, C.M.; Gruen-Bernhard, S.; Glebe, D. Role of glycosaminoglycans for binding and infection of hepatitis B virus. Cell. Microbiol. **2008**, 10, 122–133.

76. Hallak, L.K.; Spillmann, D.; Collins, P.L.; Peeples, M.E. Glycosaminoglycan sulfation requirements for respiratory syncytial virus infection. J. Virol. **2000**, 74, 10508–10513.

77. Bose, S.; Banerjee, A.K. Role of heparan sulfate in human parainfluenza virus type 3 infection. Virology **2002**, 298, 73–83.

78. Vivès, R.R.; Imberty, A.; Sattentau, Q.J.; Lortat-Jacob, H. Heparan sulfate targets the HIV-1 envelope glycoprotein gp120 coreceptor binding site. J. Biol. Chem. **2005**, 280, 21353–21357.

79. Patel, M.; Yanagishita, M.; Roderiquez, G.; Bou-Habib, D.C.; Oravecz, T.; Hascall, V.C.; Norcross, M.A. Cell-surface heparan sulfate proteoglycan mediates HIV-1 infection of T-cell lines. AIDS Res. Hum. Retroviruses **1993**, 9, 167–174.

80. Srinoulprasert, Y.; Kongtawelert, P.; Chaiyaroj, S.C. Chondroitin sulfate B and heparin mediate adhesion of Penicillium marneffei conidia to host extracellular matrices. Microb. Pathog. **2006**, 40, 126–132.

81. Muller, H.M.; Reckmann, I.; Hollingdale, M.R.; Bujard, H.; Robson, K.J.; Crisanti, A. Thrombospondin-related anonymous protein (TRAP) of Plasmodium falciparum binds specifically to sulfated glycoconjugates and to HepG2 hepatoma cells, suggesting a role for this molecule in sporozoite invasion of hepatocytes. EMBO J. **1993**, 12, 2881–2889.

82. Yokoyama, N.; Okamura, M.; Igarashi, I. Erythrocyte invasion by Babesia parasites: Current advances in the elucidation of the molecular interactions between the protozoan ligands and host receptors in the invasion stage. Vet. Parasitol. **2006**, 138, 22–32.

83. Pomin, V.H. Antimicrobial sulfated glycans: Structure and function. Curr. Top. Med. Chem. **2015**.

84. Rosett, W.; Hodges, G.R. Antimicrobial Activity of Heparin. J. Clin. Microbiol. **1980**, 11, 30–34.

85. Yamada, S.; Sugahara, K. Potential therapeutic application of chondroitin sulfate/dermatan sulfate. Curr. Drug Discov. Technol. **2008**, 5, 289–301.

86. Schaeffer, D.J.; Krylov, V.S. Anti-HIV activity of extracts and compounds from algae and cyanobacteria. Ecotoxicol. Environ. Saf. **2000**, 45, 208–227.

87. Wang, W.; Wang, S.X.; Guan, H.S. The antiviral activities and mechanisms of marine polysaccharides: An overview.Mar. Drugs **2012**, 10, 2795–2816.

88. Pangestuti, R.; Kim, S.K. Biological activities of carrageenan. Adv. Food Nutr. Res. **2014**, 72, 113–124.

89. Ahmed, A.B.; Adel, M.; Karimi, P.; Peidayesh, M. Pharmaceutical, cosmeceutical, and traditional applications of marine carbohydrates. Adv. Food Nutr. Res. **2014**, 73, 197–220.

90. Irhimeh, M.R.; Fitton, J.H.; Lowenthal, R.M. Pilot clinical study to evaluate the anticoagulant activity of fucoidan.Blood Coagul. Fibrinolysis **2009**, 20, 607–610.

91. Fonseca, R.J.; Oliveira, S.N.; Pomin, V.H.; Mecawi, A.S.; Araujo, I.G.; Mourão, P.A. Effects of oversulfated and fucosylated chondroitin sulfates on coagulation. Challenges for the study of anticoagulant polysaccharides. Thromb. Haemost. **2010**, 103, 994–1004.

3

CHAPTER

Extraction, Isolation, Structural Characterization and Anti-Tumor Properties of an Apigalacturonan-Rich Polysaccharide from the Sea Grass Zostera caespitosa Miki

Youjing Lv [1], Xindi Shan [1], Xia Zhao [1,2], Chao Cai [1,2], Xiaoliang Zhao [1], Yinzhi Lang [1], He Zhu [1] and Guangli Yu [1,2]

[1]*Key Laboratory of Marine Drugs, Ministry of Education, School of Medicine and Pharmacy, Ocean University of China, Qingdao 266003, China*

[2]*Shandong Provincial Key Laboratory of Glycoscience and Glycotechnology, Ocean University of China, Qingdao 266003, China*

ABSTRACT

An apigalacturonan (AGA)-rich polysaccharide, ZCMP, was isolated from the sea grass Zostera caespitosa Miki. The depolymerized fragments derived from ZCMP were obtained by either acidic degradation or pectinase degradation, and their structures were characterized by electrospray ionization collision-induced-dissociation mass spectrometry (ESI-CID-MS^2) and nuclear magnetic resonance (NMR) spectroscopy. The average molecular weight of ZCMP was 77.2 kD and it consisted of galacturonic acid (GalA), apiosefuranose (Api), galactose (Gal), rhamnose (Rha), arabinose (Ara), xylose (Xyl), and mannose (Man), at a molar ratio of 51.4?15.5?6.0?11.8?4.2?4.4?4.2. There were two regions of AGA (70%) and rhamnogalacturonan-I (RG-I, 30%) in ZCMP. AGA was composed of an α-1,4-D-galactopyranosyluronan backbone mainly substituted at the O-3 position by single Api residues. RG-I possessed a backbone of repeating disaccharide units of →4GalAα1,2Rhaα1→, with a few α-L-arabinose and β-D-galactose residues as side chains. The anti-angiogenesis assay showed that ZCMP inhibited the migratory activity of human umbilical vein endothelial cell (HUVECs), with no influence on endothelial cells growth. ZCMP also promoted macrophage phagocytosis. These findings of the present study demonstrated the potential anti-tumor activity of ZCMP through anti-angiogenic and immunoregulatory pathways.

Keywords: Zostera caespitosa Miki; apigalacturonan; oligosaccharides; ESI-CID-MS^2; anti-angiogenesis; immunoregulation

1. INTRODUCTION

Angiogenesis plays a critical role in tumor growth and metastasis [1]. Previous reports have shown that several polysaccharides can inhibit angiogenesis via different signaling pathways [2,3,4,5]. Plant polysaccharides are ideal candidates as immunomodulators in anti-tumor therapy because of their macrophage modulatory effects and relative non-toxicity [6]. Alga-derived polysaccharides exhibit a wide range of bioactivities and it is feasible to find potential anti-tumor drugs from marine polysaccharides.

Apigalacturonan (AGA) is a kind of Apiose-rich pectin that exclusively occurs in a small number of aquatic monocots. Two types of AGA, namely, lemnan and zosterin, have been extracted from the duckweed, Lemna minor [7,8,9] and the marine phanerogam, Zostera marina [10,11,12], respectively. Both of them possess a backbone comprising α-1,4-D-galactopyranosyluronan. The structure of lemnan consists of a hairy region composed of β-1,3′-Api_f, terminal and α-1,5-linked Ara_f, terminal, β-1,3- and β-1,4-linked Gal_p, terminal and β-1,4-linked Xyl_p [7]. The structure of zostein has been extensively investigated in the 1960s and 1970s [11,12,13]; however, the specific linkage between side Ara_f residues remained unknown until 2010, when Gloaguen reported that the side chains were composed of 1,2-linked Api_f oligosaccharides [10].

Lemnan and zosterin exhibit a wide range of physiological activities. Lemnan imparts a positive effect on the immune system by activating the phagocytosis [8] and the inflammatory response [14]. On the other hand, zosterin strongly suppresses the proliferation, migration and invasion of A431 human epidermoid carcinoma cells by inhibiting the expression of metalloproteases [10]. Zosterin also possesses high metal-binding activity [15] and disrupts protein-synthesis in mouse liver cells [16].

Zostera caespitosa Miki (Z. caespitosa Miki) is a marine phanerogam and widely distributed in the coastal area of Liaoning, China, the southern coast of Japan, and the eastern coast of North Korea. It is one of most important species of Zostera; however, information on polysaccharides from Z. caespitosa Miki has not been reported. In the present study, an AGA-rich polysaccharide, ZCMP, was extracted and purified from Z. caespitosa Miki and its structure was determined. The anti-tumor activity of ZCMP was also evaluated by using anti-angiogenesis and macrophage phagocytosis assays.

2. RESULTS AND DISCUSSION

2.1. Extraction, Purification and General Analysis of ZCMP

Ammonium oxalate is a calcium-chelating agent that is commonly used to increase pectin solubility. The yield of ZCMP extracted from Z. caespitosa Miki using 2% ammonium oxalate solution was 10.8% (w/w). ZCMP contained low levels of protein (4.3%) and sulfate (1.7%) and showed an average molecular weight of 77.2 kD. A single and symmetric peak on the Q-Sepharose Fast Flow (Figure 1a) and the Shodex OHpak SB-804 HQ column (Figure 1b) indicated that the extracted ZCMP was of high purity. Monosaccharide composition analysis demonstrated that ZCMP was composed of galacturonic acid (GalA), apiose (Api), galactose (Gal), rhamnose (Rha), arabinose (Ara), xylose (Xyl) and mannose (Man) at a molar ratio of 51.4:15.5:6.0:11.8:4.3:4.4:4.2 (Table 1), which was similar to that of lemnan and zosterin [7,10].

2.2. Preparation of ZCMP-Derived Oligosaccharides

2.2.1. Degradation of ZCMP

ZCMP was shown to be sensitive under acidic conditions such as 0.1 mol/L HCl and H_2SO_4, and the Api residues were rapidly released as monosaccharides in our model experiment. A three-level acid solution (0.1 mol/L CH_3COOH, 0.2 mol/L HCl, and 0.5 mol/L HCl) was added to degrade the polysaccharide progressively, and pectinase was also used to generate oligosaccharides with methyl esters and acetyl groups. The depolymerized oligosaccharides were collected by precipitation using different concentrations of ethanol. The degradation process of ZCMP is shown in Figure 2.

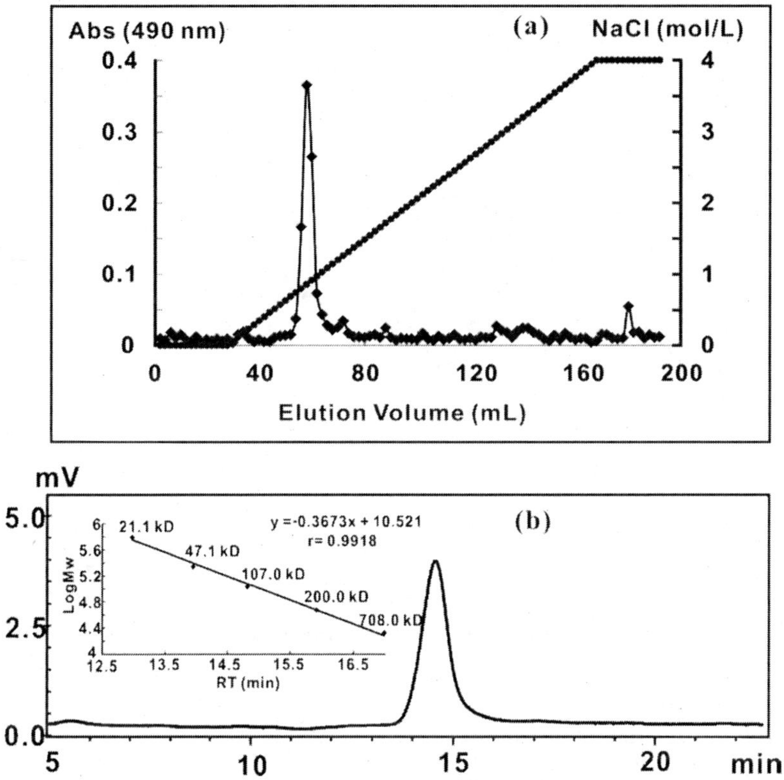

Figure 1. Separation and purification of ZCMP from Z. caespitosa Miki. (**a**) Elution profiles of ZCMP on a Q-Sepharose Fast Flow ion-exchange chromatography column; (**b**) The average molecular weight of ZCMP was determined using the High Performance Gel Permeation Chromatography (HPGPC) method on a Shodex OHpak SB804 column.

Monosaccharide composition analysis (Table 1) showed that the fractions ZCMP-P, ZCMP-PS, and ZCMP-SS contained different molar ratios of monosaccharides, indicating that they were derived from different regions of ZCMP. Api was the major monosaccharides released in the first step of hydrolysis, and only little Api in ZCMP-PS and ZCMP-P was detected, thus promoting us to speculate that the Api residues were at the terminal position or side chains. Meanwhile, Gal, Xyl, and Ara were mainly detected in ZCMP-SS and ZCMP-PS, which suggested that the three residues existed in the branches. ZCMP-PS mainly contained GalA and Rha with 7%–10% Gal and Ara residues, suggesting that the major component of ZCMP-PS was the fragment of RG-I. The content of GalA increased with the enhancement of acid strength and it was almost the only monosaccharide in ZCMP-PP, which indicated that GalA was present in the backbone of ZCMP structure.

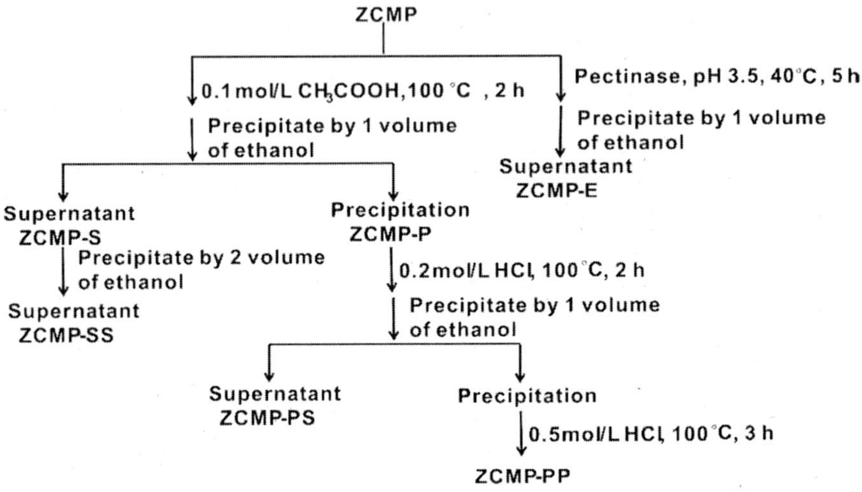

Figure 2. Flow chart of the degradation process of ZCMP.

Table 1. Molecular weight and monosaccharide analysis of ZCMP and its oligosaccharides.

	Molecular Weight	Monosaccharides (%)							
	(kD)	Man	Rha	GlcA	GalA	Api	Gal	Xyl	Ara
ZCMP	77.2	4.2	11.8	-	51.4	15.5	6.0	4.4	4.3
ZCMP-SS	-	2.3	6.7	3.9	5.7	52.1	10.7	5.8	12.2
ZCMP-PS	-	3.0	27.0	5.0	39.9	2.5	10.1	4.7	7.8
ZCMP-P	23.4	3.9	5.8	0.9	79.9	2.9	3.1	1.2	2.2
ZCMP-PP	-	3.4	3.0	0.4	93.2	-	-	-	-

2.2.2. Purification of ZCMP-Derived Oligosaccharides

The mixtures of ZCMP-derived oligosaccharides were fractionated by gel filtration chromatography (Figure 3). The proposed structural composition (abundance >10% in the ESI-MS spectrum) of oligosaccharides is presented in Table 2, which was based on the monosaccharide composition and ESI-MS analysis in the negative mode. Oligosaccharides with similar molecular mass and charge coexisted as one broad peak during gel-permeation chromatography, which was mainly due to the heterogeneous structure of ZCMP. Most of the Api coexisted in the salt peak and only minor Api-oligosaccharides were detected in ZCMP-S2-4 with a low polymerization degree (<5; Figure 3a).

Table 2. Components (abundance of >10% in the ESI-MS spectrum) of the oligosaccharide fractions degraded from ZCMP.

Fraction	Ions	Mw (H Form)	Dp	Composition
E1	272.05 (z = 2)	546.10	3	$GalA_3$
	338.07 (z = 2)	678.14	4	$GalA_3Api$
E2	448.08 (z = 2)	898.16	5	$GalA_5$
	360.06 (z = 2)	722.12	4	$GalA_4$
	342.40 (z = 3)	1030.10	6	$GalA_5Api$
	386.41 (z = 3)	1162.23	7	$GalA_5Api_2$
E3	536.09 (z = 2)	1074.18	6	$GalA_6$
	401.07 (z = 3)	1206.21	7	$GalA_6Api$
	445.09 (z = 3)	1338.27	8	$GalA_6Api_2$
E4	415.74 (z = 3)	1250.22	7	$GalA_7$
	355.56 (z = 4)	1426.24	8	$GalA_8$
	459.75 (z = 3)	1382.25	8	$GalA_7Api$
	503.76 (z = 3)	1514.28	9	$GalA_7Api_2$
	547.78 (z = 3)	1646.34	10	$GalA_7Api_3$
	591.79 (z = 3)	1778.37	11	$GalA_7Api_4$
S1	149.05 (z = 1)	150.05	1	Api
S2	281.10 (z = 1)	282.10	2	Api_2
S3	413.14 (z = 1)	414.14	3	Api_3
S4	545.18 (z = 1)	546.18	4	Api_4
PS1	339.09 (z = 1)	340.09	2	GalARha
	369.10 (z = 1)	370.10	2	$GalA_2$
	545.10 (z = 1)	546.10	3	$GalA_3$
PS2	661.17 (z = 1)	662.17	4	$GalA_2Rha_2$
	721.12 (z = 1)	722.12	4	$GalA_4$
PP1	193.08 (z = 1)	194.08	1	GalA
PP2	369.10 (z = 1)	370.10	2	$GalA_2$
	339.09 (z = 1)	340.09	2	GalARha
PP3	545.10 (z = 1)	546.10	3	$GalA_3$
PP4	721.12 (z = 1)	722.12	4	$GalA_4$
PP5	448.08 (z = 2)	898.16	5	$GalA_5$
PP6	536.09 (z = 2)	1074.18	6	$GalA_6$
PP7	624.10 (z = 2)	1250.20	7	$GalA_7$
PP8	474.41 (z = 3)	1426.24	8	$GalA_8$

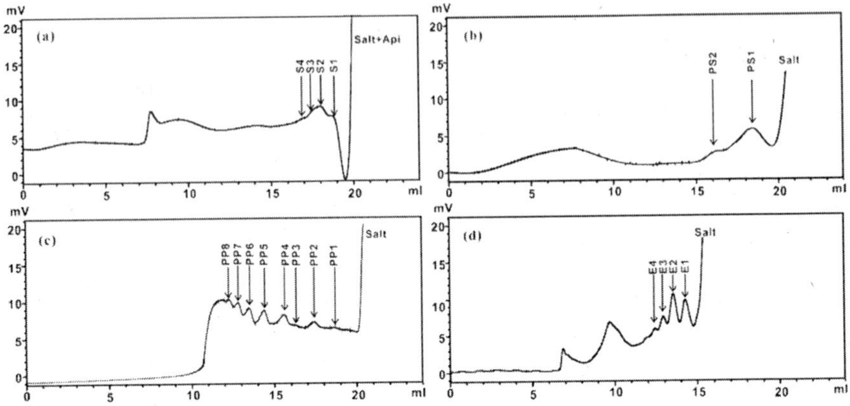

Figure 3. Low pressure gel-permeation chromatography of ZCMP-derived oligosaccharides from Z. caespitosa Miki. (**a**) ZCMP-S; (**b**) ZCMP-PS; (**c**) ZCMP-PP; (**d**) ZCMP-E.

2.3. ESI-CID-MS^2 Analysis of the Oligosaccharides Derived from ZCMP

Several reports have summarized the major contributions of mass spectrometry to the structural elucidation of carbohydrates [17,18,19]. The formation of $^{0,2}X$ and $^{0,2}A$ ions requires hydrogen on C3-OH and occurs at the 4-linked monosaccharide residue [20,21]. $^{1,3}A$-type cleavage usually arises with 2-linked residues [22,23]. Reduction of the hemiacetal to alditol is a common method to determine the reducing terminal of oligosaccharides. A reducing terminal ion will have a 2 Da increment after reduction by sodium borohydride [24]. Therefore, the multistage mass spectrum facilitates in determining the linkages and sequences of oligosaccharides.

2.3.1. ESI-CID-MS^2 Analysis of the Sequences of Oligosaccharides from ZCMP-S

A series of Api-oligosaccharides was released from ZCMP after CH_3COOH hydrolysis and the ESI-CID-MS^2 spectra of di-, tri- and hexa-saccharides were detected. The results demonstrated that all of them possessed the same fragment ion pattern. Taking the product-ion spectrum of Api_4 (m/z 545) as an example (Figure 4), a series of ions of glycosidic bond cleavage at m/z 263 (B_2/Y_2), 281 (C_2/Z_2), 395 (B_3/Y_3) and 413 (C_3/Z_3) indicated a linear chain. In addition, a series of notably $^{2,3}A$ type ions (m/z 191, 323, 455) were generated by cross-cleavage of the C2-C3 and C3-C4 bonds of Api, which in turn led to the loss of $C_3H_6O_3$ of m/z 90.

The ion at m/z 485 was deduced as $^{1,3}A_4$ or $^{0,2}A_4$ cleavage. Similarly, the ion at m/z 353 was deduced as $^{1,3}A_3$ or $^{0,2}A_3$ cleavage. Guo et al. [25] also determined a series of linear oligo-galatofuranoses by negative-ion ESI-CID-MS^2. Cross-ring fragment ions of $^{3,4}A$ and $^{0,3}A$-type fragment ions were observed as well and used in the identification of linkages between the β-D-(1→5)-linked Gal_f oligosaccharides. Based on the proposed similar ion pattern of fragments, the linkage between Api residues was deduced to be 3′-linked. After reduction (Figure 4b), four glycosidic ions of tetrasaccharide Api_4 shifted to m/z 415 (Z_3), 397 (Y_3), 283 (Z_2) and 265 (Y_2) from m/z 413, 395, 281 and 263, respectively. No cross-ring cleavage ions shifted after reduction, suggesting that all of these were produced from the non-reducing end.

The structures of lemnan and zosterin are restricted to algal species. Lemnan has a side chain of 3′-linked Api residues [8], whereas zosterin has a side chain of 2-linked Api residues [10]. However, the AGA obtained from Z. caespitosa Miki showed a similar side chain as that observed in lemnan.

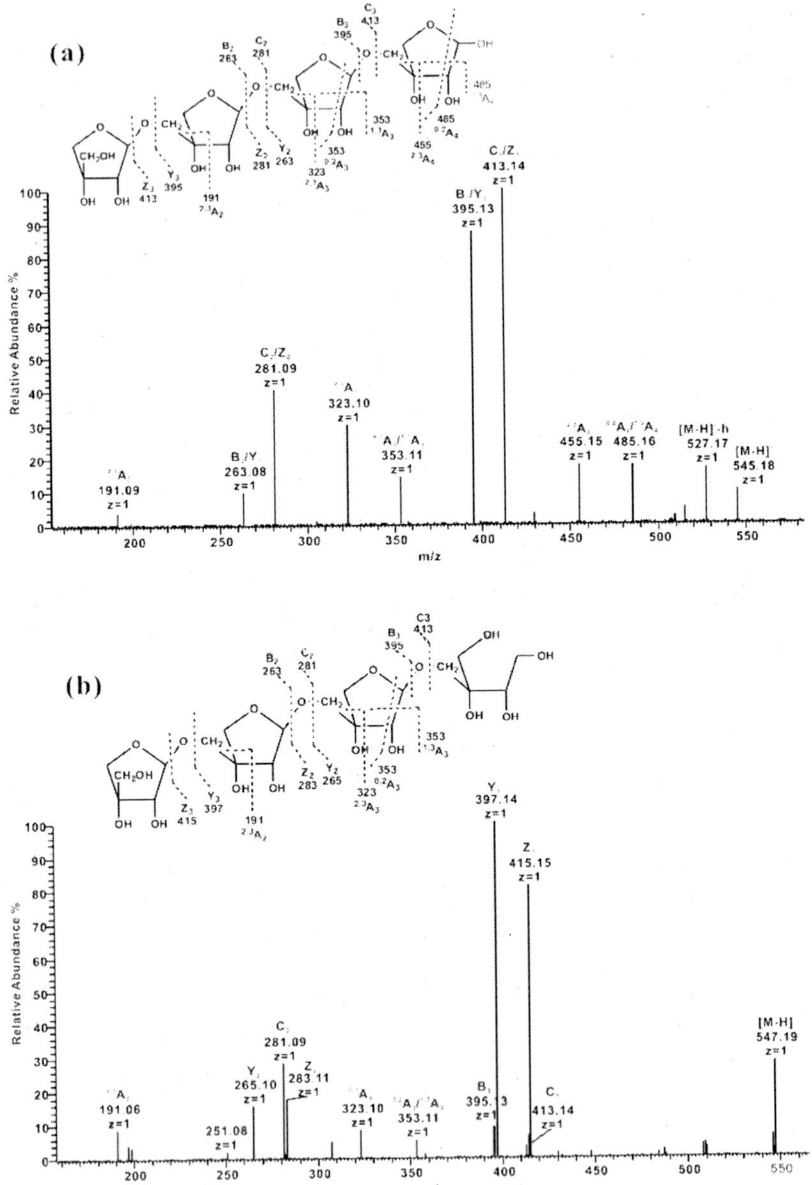

Figure 4. Negative-ion ESI-CID-MS^2 product-ion spectra of the tetrasaccharide Api$_4$from ZCMP-S. (**a**) Sequence analysis of the tetrasaccharide Api$_4$ at m/z 545; (**b**) Sequence analysis of the tetrasaccharide alditol at m/z 547.

2.3.2. ESI-CID-MS^2 Analysis of Oligosaccharides from ZCMP-PS

Even-numbered oligosaccharides with equal amounts of Rha and GalA were identified in the fractions of ZCMP-PS as GalA-Rha and $GalA_2Rha_2$, which indicated the presence of a repeating disaccharide unit. Its product ion spectra were acquired by ESI-CID-MS^2.

The ESI-CID-MS^2 spectrum (Figure 5) of the tetrasaccharide $GalA_2Rha_2$ (m/z 661) showed a series of ions of glycosidic bond cleavage at m/z 321 (B_2/Z_2), 339 (C_2/Y_2), 485 (Z_3), 497 (B_3), and 515 (C_3), indicating a linear chain with repeating linkages of GalA and Rha. Comparison of the spectra of $GalA_2Rha_2$ (m/z 661) with its alditol (m/z 663) after reduction showed that the two glycosidic ions shifted to m/z 323 (Z_2) and 487 (Y_3) from m/z 321 and 485, respectively, thus revealing that Rha was at the reducing terminal. The $^{1,3}A_4$ ion (m/z 557) from the reduced Rha and the $^{1,3}A_2$ ion (m/z 235) from the internal Rha revealed the presence of 2-linked Rha. The $^{0,2}A_3$ (m/z 455) and $^{2,4}X_3$ (m/z 601) ions were the characteristic evidence for 4-linked GalA. A similar fragment ion pattern was observed in the ESI-CID-MS^2 spectrum of the disaccharide GalA-Rha (Supplementary Figure 1). In the present study, GalA and Rha residues in the backbone of pectin were determined to be in the α-configurations, which was similar to the findings of previous NMR results [7,8,10,26,27]. Therefore, the structure of the main oligosaccharides in ZCMP-PS were identified as -[4)-α-GalA-(1→2)-α-Rha-(1]$_n$-, which was assigned as the backbone of RG-I and was in agreement with the findings of previous reports [26,27].

2.3.3. ESI-CID-MS^2 Analysis of Oligosaccharides from ZCMP-PP

The backbone of ZCMP was completely degraded into the oligosaccharides ZCMP-PP by 0.5 mol/L HCl after removing the branches and the RG-I region. ZCMP-PP comprised a series of GalA oligosaccharides, except for a few GalA-Rha disaccharides (Table 2). The ESI-CID-MS^2 spectra were obtained from disaccharides to octasaccharides, and all of these showed a similar fragment ions pattern. For instance, in the negative-ion production-ion spectrum of $GalA_7$ at m/z 624.10 (Figure 6), a linear sequence was deduced from the major fragment ions m/z 175/193, 351/369, 527/545, 703/721, 879/897 and 1055/1073, which all had arisen from glycosidic bond cleavages. As described in previous reports, the formation of $^{0,2}X$ and $^{0,2}A$ ions requires hydrogen on C3-OH and only occurs at the 4-linked monosaccharide residue [17,28]. The observation of continuous cross-ring cleavage of $^{0,2}A$ ions suggests that GalA oligomers in ZCMP-PP were homogenous 4-linked. All $^{0,2}A$ ions were accompanied by ions derived from dehydration, e.g., $^{0,2}A_3$, m/z 485/467 (weak); $^{0,2}A_4$, m/z 661/643; $^{0,2}A_5$, m/z837/819; $^{0,2}A_6$, m/z 1013/995; and $^{0,2}A_7$, m/z 594/585 (double charged).

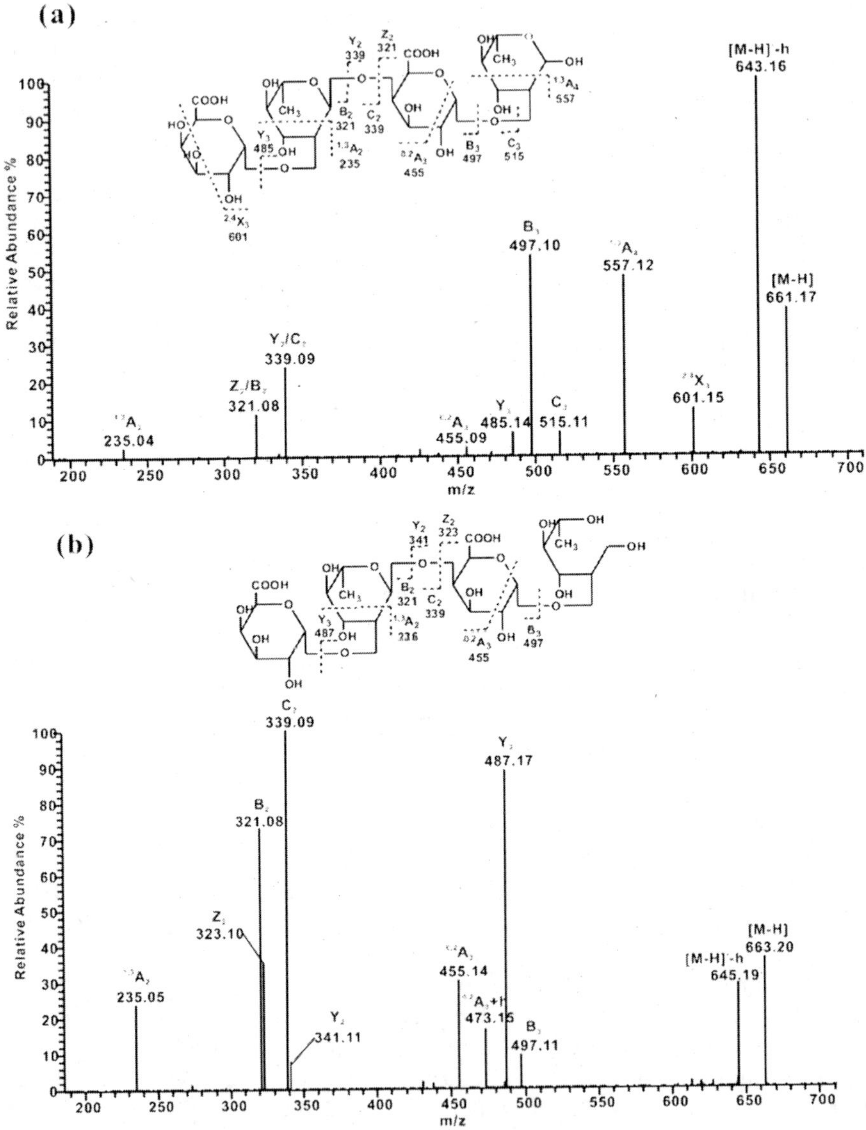

Figure 5. Negative-ion ESI-CID-MS^2 product-ion spectra of the tetrasaccharide (GalA-Rha)$_2$from ZCMP-PS. (**a**) Sequence analysis of (GalA-Rha)$_2$ at m/z 661; (**b**) Sequence analysis of tetrasaccharide alditol at m/z 663.

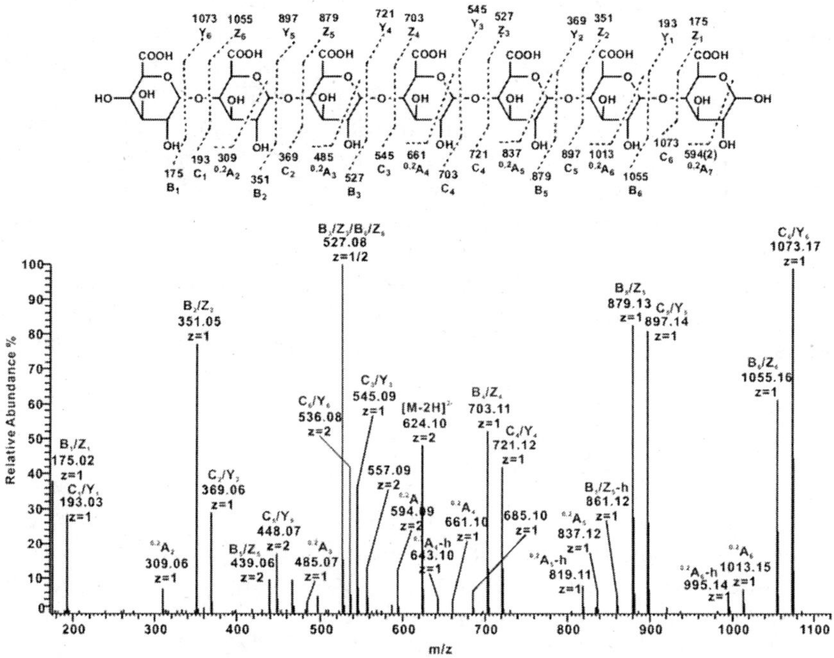

Figure 6. Negative-ion ESI-CID-MS² product-ion spectrum of the heptasaccharide $GalA_7$ derived from ZCMP-PP.

2.3.4. ESI-CID-MS² Analysis of Oligosaccharides from ZCMP-E

Pectinase can specifically cleave the glycosidic bond between GalA residues. ESI-MS analysis of pectinase hydrolysate ZCMP-E1-4 showed that all fractions were mixtures of different oligosaccharides. For example, $GalA_7$, $GalA_7Api_1$, $GalA_7Api_2$, $GalA_7Api_3$, and $GalA_7Api_4$ were observed in ZCMP-E4 (Figure 7a). The GalA residues are usually methyl esterified or acetylated partially at the O-2 and/or O-3 positions in pectin [29,30]. Weak fragment ions at m/z 508.44, 515.77, 523.10 and 537.76 (triple charged), assigned to $[M-3H]^{3-}$, $[M-4H+Na]^{3-}$, $[M-5H+2Na]^{3-}$ and $[M-7H+4Na]^{3-}$ of $GalA_7Api_2Me$, respectively, were found after magnifying the spectrum by five-fold (Figure 7b). Low abundance (<4%) of these peaks suggested that a small number of GalA residues were methyl esterified. No acetylated oligosaccharides were detected in E4.

To investigate the linkages between GalA residues and Api residues, ion at 503.76 ($GalA_7Api_2$, triple charged) from E4 was selected as precursor ion to get an ESI-CID-MS² spectrum (Figure 7c). Ion at m/z 690 (double charged, $Y_{7\alpha}/Y_{7\beta}$) confirmed that the Api residues were on the side chains. Ion detected at m/z 809 (C_3) was assigned as $GalA_3Api_2$, indicating that there were four unsubstituted GalA residues on the terminal and the two Api residues were distributed on the other three GalA residues. The appearance of Z_6 at m/z 602 (double charged) demonstrated that the disaccharide Api-GalA was on the terminal and the B_2 ion at m/z 483 indicated that the trisaccharide Api-GalA-GalA was on the terminal. The deduced sequence of this nonasaccharide is shown in Figure 7c. The Api residues were randomly distributed on different GalA residues in the form of monosaccharides rather than oligosaccharides, indicating that the level of the Api oligosaccharides in ZCMP was relatively low, and most of Api residues were monosaccharides. The linkage between GalA and Api residues was not deduced due to the absence of cross-ring cleavages.

2.4. Methylation Analysis of ZCMP

The linkage between GalA and Api residues was also confirmed by carboxyl-reduction and methylation analysis (Table 3), in which 1,3,4,5-Ac_4-2,6-Me_2-D-galactitol$_{red}$ was detected, suggesting that GalA residues were substituted at the O-3 position by Api. Large amounts of 1,4-Ac_2-2,3,3′-Me_3-Api$_f$ and lower amounts of 1,3′,4-Ac_3-2,3-Me_2-Api$_f$ were detected, indicating that there was a high level of terminal Api and a low level of 3′-linked Api residues in ZCMP. In addition, 1,4,5-Ac_3-2,3-Me_2-D-arabinitol, 1,4,5-Ac_3-2,3-Me_2-D-xylitol, 1,2,5-Ac_3-6-deoxy-3,4-Me_2-D-mannitol, 1,4,5-Ac_3-2,3,6-Me_3-D-galactitol, and 1,3,4,5-Ac_4-2-Me_2-D-arabinitol were detected, which originated from the 5-substituted Ara$_f$, 4-substituted Xyl, 2-substituted Rha, 4-substituted Gal, and 3,5-substituted Ara residues, respectively.

2.5. NMR Analysis of ZCMP-SS

The structural features of ZCMP-SS were also characterized by using a combination of one-dimensional ^1H NMR, ^{13}C NMR, and Distortionless Enhancement by Polarization Transfer (DEPT) experiments (Figure 8), as well as heteronuclear two-dimensional ^1H-^{13}C Heteronuclear Multiple Quantum Coherence (HMQC) experiment. The proton-carbon correlation was assigned based on the HMQC spectrum (Figure 9), and seven cross peaks corresponding to the anomeric signals were clearly detected. Correlations between H1 at 5.24 ppm and C1 at 104.70 ppm, H1 at 5.33 ppm and C1 at 97.60 ppm, H1 at 5.54 ppm and C1 at 99.18 ppm, and H1

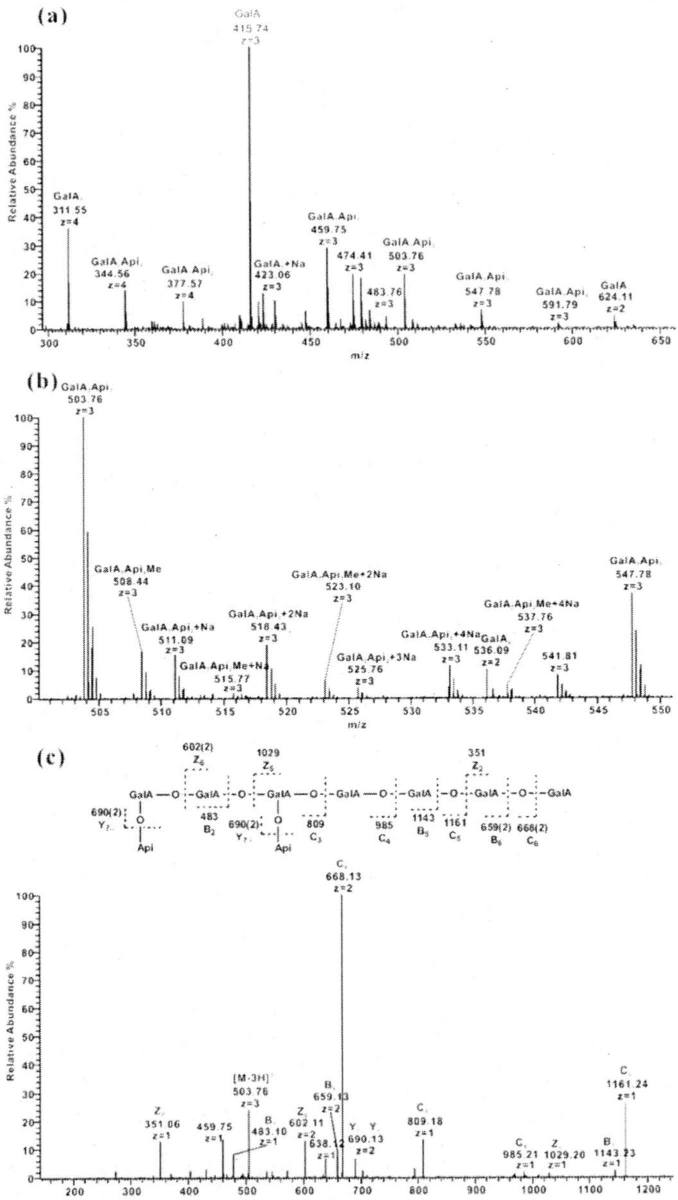

Figure 7. Negative-ion ESI-MS product-ion spectra of fractions from ZCMP-E. (**a**) Negative-ion ESI-MS spectrum of ZCMP-E4; (**b**) Five-fold magnification of the region between 500 and 550 m/z in the negative-ion ESI-MS spectrum of ZCMP-E; (**c**) Sequence analysis of nonasaccharide $GalA_7Api_2$ at m/z 503.76 (triple charged).

Table 3. Methylation analysis of ZCMP.

Permethylated Alditol Acetate	Primary Mass Fragments (m/z)	Linkages	Molar Ratio
2,3,3′-Me$_3$-Api$_f$	118, 132, 161	Api$_f$-(1→	12.81
2,3-Me$_2$-Ara$_f$ 2,3-Me$_2$-Xyl	118, 129, 189	→5)-Ara$_f$-(1→ →4)-Xyl-(1→	14.26
3,4-Me$_2$-Rha	131, 190, 234, 304	→2)-Rha-(1→	17.75
2,3,6-Me$_3$-Gal	118, 131, 173, 233	→4)-Gal-(1→	6.55
2,3,6-Me$_3$-Gal$_{red}$	118, 174, 234	→4)-GalA-(1→	32.40
2,6-Me$_2$-Gal$_{red}$	118, 130, 186, 306	→3,4)-GalA-(1→	8.35
2-Me-Ara$_f$	118, 159, 201, 261	→3,5)-Ara$_f$-(1→	3.07
2,3-Me$_2$-Api$_f$	118, 129, 189, 234	→3′)-Api$_f$-(1→	4.82

Gal$_{red}$: Gal residues generated from GalA residues by reduction with $NaBD_4$.

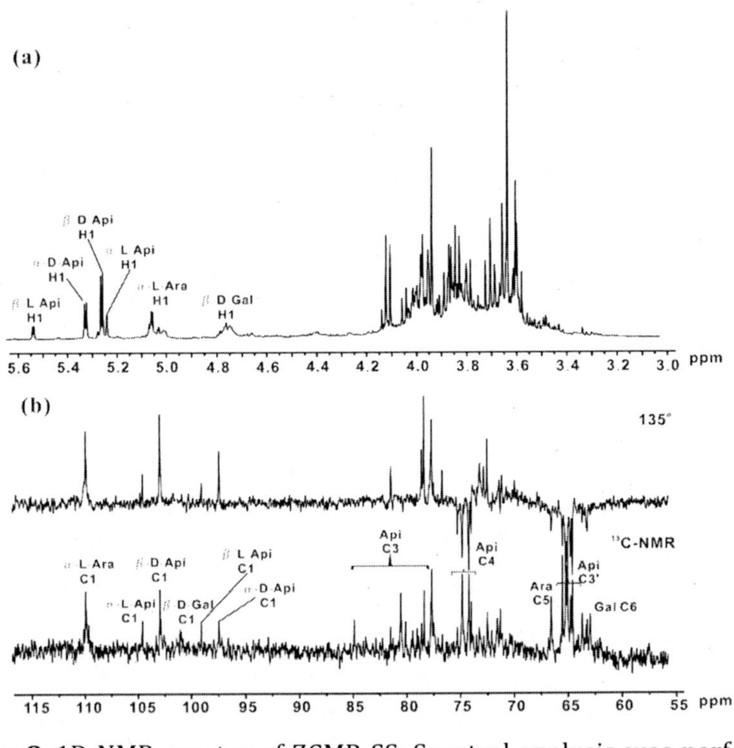

Figure 8. 1D NMR spectra of ZCMP-SS. Spectral analysis was performed at 25 °C on a JEOL ECP 600 MHz spectrometer using acetone as internal standard. (**a**) ^1H NMR spectrum and (**b**) ^{13}C NMR and DEPT spectra.

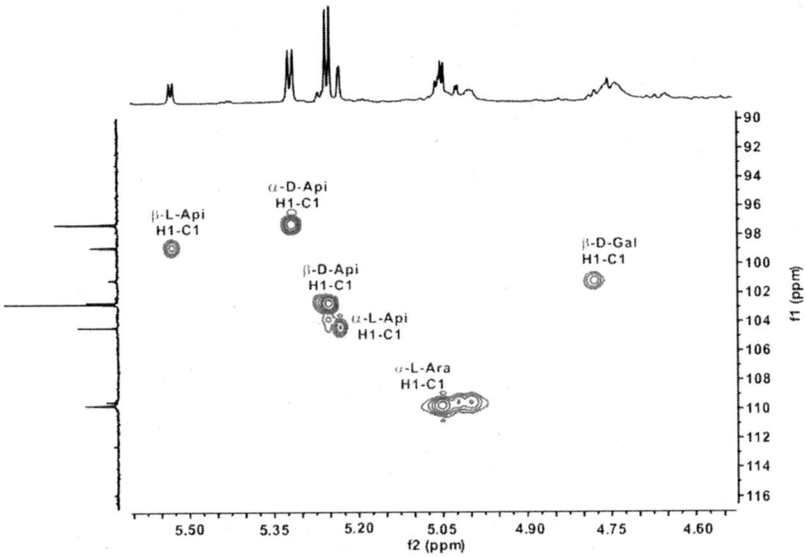

Figure 9. The ^{1}H-^{13}C HSQC spectrum of ZCMP-SS. Spectral analysis was performed at 25 °C on a JEOL ECP 600 MHz spectrometer using acetone as internal standard.

at 5.27 ppm and C1 at 103.08 ppm were assigned to α-L-Api$_f$, α-D-Api$_f$, β-L-Api$_f$, and β-D-Api$_f$ respectively [10,31]. Its molar ratio was determined to be 1.0:3.0:4.0:1.3, based on the integral area ratio detected in ^{1}H-NMR. No correlations of Api oligosaccharides in the side chains were detected due to its instability under acidic conditions. The anomeric proton signals of linked α-L-Ara and terminal α-L-Ara residues at 5.05 ppm and 5.02 ppm were correlated to the anomeric carbon signals at 110.1 ppm and 109.80 ppm, respectively. The correlation of H1 at 4.76 ppm with C1 at 101.44 ppm was assigned to β-D-Gal residues.

According to the proposed general structural model for lemnan [7], zosterin [10], pectin [32], and the results obtained in the present study, we propose the following structural model for Z. caespitosa Miki polysaccharide ZCMP (Figure 10). ZCMP is composed of AGA and RG-I regions. AGA has a backbone of α-1,4-D-galactopyranosyluronan with an extremely low degree of etherification, whereas the side chains were linked to the O-3 of GalA by most of the single Api residues and minor short (1→3′)-linked β-D-Api oligosaccharides with different degree of polymerization (<5). RG-I has a backbone of repeating 4-linked GalA and 2-linked Rha with minor 5-linked α-L-Ara residues and 4-linked β-D-Gal residues as side chains.

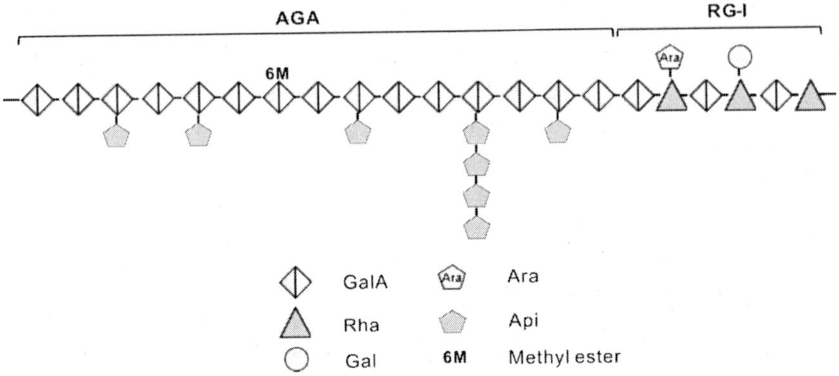

Figure 10. Proposed structural model of ZCMP.

2.6. ZCMP Inhibited the Migration of HUVECs

To assess the anti-angiogenic properties of ZCMP in vitro, its inhibitory effects on the chemotactic motility of human umbilical vein endothelial cell (HUVECs) were investigated using the wound-healing migration assay. As shown in Figure 11a, untreated HUVECs migrated into the wounded area of the cell monolayer, whereas ZCMP treatment significantly inhibited the HUVEC migration in a dose-dependent manner (Figure 11a,b).

HUVEC viability was tested to determine whether the migration inhibitory effect was the result of the inhibition of HUVEC proliferation after treatment with various concentrations of ZCMP for 24 h. As shown in Figure 11c, ZCMP had no significant effect on the viability of HUVECs.

Angiogenesis plays an important role in providing nutrients and oxygen to the growing tumor, whereas endothelial cell migration is essential for angiogenesis [33]. ZCMP inhibited angiogenesis by suppressing migration of endothelial cells.

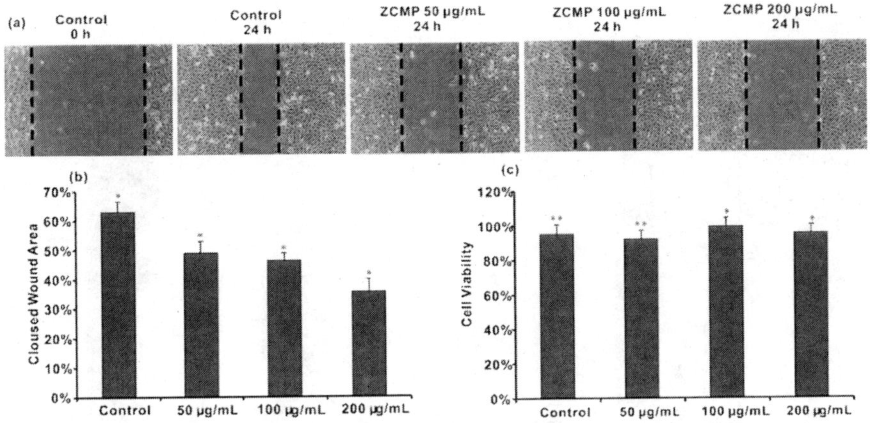

Figure 11. ZCMP inhibited the migration of HUVECs (**a**) HUVEC monolayer was scraped to generate a wound (0 h), and the cells were incubated with different concentrations of ZCMP (50 μg/mL, 100 μg/mL, 200 μg/mL) or vehicle (Control). After 24 h, the cells were imaged at 40× magnification. The wound areas at 0 and 24 h are indicated by dotted lines; (**b**) Quantification of the effect of ZCMP on HUVEC migration in the wound healing assay; (**c**) HUVEC viability was determined by using the MTT assay after incubating with different concentrations of ZCMP (50 μg/mL, 100 μg/mL, 200 μg/mL) or vehicle (Control). Each experiment was performed at least 3 times, and the values represent the mean ± S.D. * $P < 0.05$; ** $P < 0.01$, as determined by unpaired student's t-test.

2.7. ZCMP Enhanced Macrophage Phagocytosis

The effects of ZCMP treatment on macrophage phagocytosis were examined by using Grifola polysaccharide (50 μg/mL) as positive control (Figure 12). Grifola polysaccharide is a glucan that consists of a backbone of β-1,3 glucosidic bond with β-1,6 side chains, and it has been used clinically for tumor immunotherapy in several countries [34]. The promotion of macrophage phagocytosis was enhanced after increasing the ZCMP dose from 50 μg/mL to 200 μg/mL.

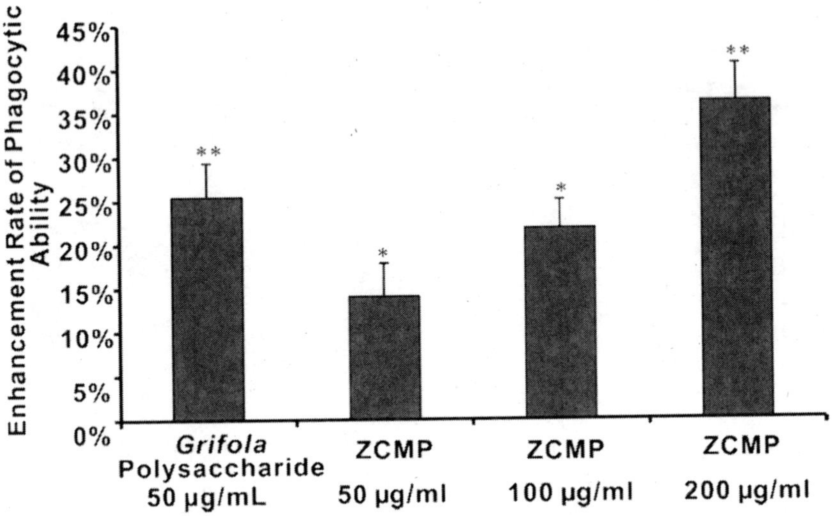

Figure 12. Effects of different doses of ZCMP on the phagocytic ability of the mouse macrophage cell line Raw 264.7. Results are expressed as means ± S.D. * $P < 0.05$; ** $P < 0.01$, as determined by unpaired student's t-test.

3. EXPERIMENTAL SECTION

3.1. Samples and Materials

The seagrass Z. caespitosa Miki was collected from Bohai Gulf, China. Monosaccharide standards (Man, Glc, Gal, Xyl, Fuc, GlcA, GalA and Api), sodium borohydride ($NaBH_4$), deuterium sodium borohydride ($NaBD_4$), 1-ethyl-3-(3-dimethyaminopropyl) carbodimide (EDC), 1-phenyl-3-methyl-5-pyrazolone (PMP), and 5-diphenyl tetrazolium bromide (MTT) were purchased from Sigma-Aldrich (St. Louis, MO, USA). Superdex™ Peptide 10/300 GL column (1.0 × 30 cm) was procured from GE Healthcare (Uppsala, Sweden). Shodex OHpak SB-804 HQ column (8.0 × 300 mm) was obtained from Showa Denko (Tokyo, Japan). Amberlite IR120 resin was purchased from Sigma (St. Louis, MO, USA). All other reagents were of analytical grade.

3.2. Extraction, Isolation, and Purification of ZCMP

Z. caespitosa Miki was pulverized and passed through a 60 mesh sieve, extracted with 80% ethanol for 3 h at 80 °C (3 times) to remove lipids, and the residue was dried. Approximately 50 g of the residue was pretreated with 75 mL of 0.5% HCl at 50 °C for 3 h to break the cell wall and release the polysaccharides. The residue was washed with water and further extracted with 2% ammonium oxalate solution at 70 °C for 3 h (3 times). Acid polysaccharide was precipitated by adding 3 mol/L HCl until a pH level of 1.0 was attained. The precipitated polysaccharide was dialyzed (7 kD MWCO) against water for 2 days and then freeze-dried.

For purification, the crude polysaccharide (200 mg) was applied to a Q-Sepharose Fast Flow column connected to an ÄKTA-Fast Protein Liquid Chromatography (FPLC) system (General Electric Company, Fairfield, USA), and washed with water, followed by a linear gradient elution with an NaCl solution (from 0 to 4 mol/L) at a flow rate of 120 mL/h. The absorbance of each fraction (10 mL) at 490 nm was tested by using the phenol-sulfuric acid method. The fraction containing polysaccharide was collected, dialyzed, and concentrated. The purified fraction (ZCMP) was dialyzed against distilled water and lyophilized.

3.3. General Analysis of ZCMP

Crude protein content was determined by Lowry method [35]. Sulfate content was determined by $BaCl_2$-Gelatin method [36]. Purity and relative molecular weight (Mw) were determined by gel filtration chromatography on a Shodex OHpak SB-804 HQ (Showa Denko, Tokyo, Japan) eluted with 0.1 mol/L Na_2SO_4 at a flow rate of 0.5 mL/min at 35 °C. The column was calibrated with dextran standards, and the corrected regression equation was log (Mw) = −0.3673T_R (retention time) + 10.521 (r = 0.9918). Monosaccharide composition was determined by a 1-phenyl-3-methyl-5-pyrazolone (PMP)-High Performance Liquid Chromatography (HPLC) method as described by Chen et al. [37]. The Fourier Transform Infrared (FTIR) spectra of the polysaccharides ZCMP prepared as a KBr pellet was recorded with a Nicolet Nexus 470 Thermo instrument (Thermo Fisher Scientific, Waltham, MA, USA).

3.4. Carboxyl-Group Reductions and Methylation Analysis of ZCMP

ZCMP was first reduced to convert the carboxyl into hydroxyl groups by using the method described by Taylor and Conrad [38]. Methylation of the reduced polysaccharides was performed according to the method of Hakomori [39]. The partially methylated alditol acetates were analyzed by GC-MS equipped with a DB-225MS fused-silica capillary column. Mass spectra of the derivatives were analyzed by using the Complex Carbohydrate Structural Database (CCSD) of the Complex Carbohydrate Research Centre (http://www.ccrc.uga.edu/).

3.5. Preparation and Purification of Oligosaccharides of ZCMP

ZCMP (10 mg/mL) was hydrolyzed by adding 0.1 mol/L CH_3COOH at 100 °C for 2 h, and precipitated with an equivalent volume of ethanol. After centrifugation, the precipitation was named ZCMP-P, and the supernatant was named ZCMP-S. The supernatant was added ethanol to 70% and then precipitated. After centrifugation, the supernatant (ZCMP-SP) and the precipitate (ZCMP-SS) were lyophilized.

ZCMP-P (10 mg/mL) was hydrolyzed by adding 0.2 mol/L HCl at 100 °C for 2 h and precipitated with an equivalent volume of ethanol. After centrifugation, the supernatant (ZCMP-PS) and the precipitation were lyophilized. The precipitate (10 mg/mL) was hydrolyzed by 0.5 mol/L HCl at 100 °C for 3 h, neutralized, and finally lyophilized (ZCMP-PP).

ZCMP (10 mg/mL) was depolymerized with 1% pectinase (w/v) in a buffer solution (0.1 mol/L citric acid: 0.1 mol/L sodium citric acid = 15.5:4.5, v/v, pH 3.5) at 40 °C for 5 h. An equivalent volume of ethanol was added to remove the residual fraction. After centrifugation, the supernatant was lyophilized (ZCMP-E).

ZCMP-S, PS, PP and E were respectively fractionated on a Superdex Peptide column (GE Healthcare, Uppsala, Sweden) and eluted with 0.1 mol/L NH_4HCO_3 at a flow rate of 0.1 mL/min, and monitored by a refractive index detector. All fractions were collected and freeze-dried.

3.6. Oligosaccharide Reduction

Oligosaccharides reduction was conducted as described by Yu et al. [24]. Briefly, 20 μL of a $NaBH_4$ reagent (0.05 mol/L $NaBH_4$ in 0.01 mol/L NaOH) was added to the oligosaccharide (typically 20 μg). After overnight reduction at 4 °C, CH_3COOH was added to destroy the borohydride. The mixture was then passed through a mini-column of Amberlite IR120 resin (Sigma, St. Louis, MO, USA), and the boric acid in the eluent was removed by repeated co-evaporation with methyl alcohol.

3.7. MS Analysis of Oligosaccharides Derived from ZCMP

ESI-MS of ZCMP oligosaccharides was performed on a LTQ-Orbitrap XL instrument (Thermo Fisher Scientific, Waltham, MA, USA). Samples were dissolved in CH_3CN/H_2O (1:1, v/v) at a concentration of 10 pmol/μL and 5 μL was injected. Solvent volatilization temperature and capillary temperatures were 275 °C, and the sheath flow gas flow rate was 8 arb. The flow rate was 8 μL/min in the ESI-MS analysis and 3–5 μL/min in the ESI-CID-MS^2 analysis. Helium was used as collision gas with a collision energy of 20–25 eV.

3.8. NMR Analysis of ZCMP-SS

ZCMP-SS (7 mg) was co-evaporated with D_2O (99.96%) three times by lyophilization before it was finally dissolved in 500 μL of D_2O. Acetone was used as internal standard (2.225 ppm for ^{1}H-NMR and 30.83 ppm for ^{13}C-NMR). The results of the^{1}H-NMR, ^{13}C-NMR, DEPT, and HMQC experiments were recorded on JEOL JNM-ECP 600 spectrometer (JEOL, Tokyo, Japan) at 25 °C.

3.9. Cell Lines and Culture Conditions

HUVECs and the mouse macrophage cell line Raw 264.7 were purchased from the Cell Bank of the Type Culture Collection Center of the Chinese Academy of Sciences in Shanghai, China. These cells were maintained in DMEM culture medium (Gino Biological Medicine Technology Co., Ltd., Hangzhou, China) supplemented with 10% fetal bovine serum (FBS) (v/v), 100 U/mL penicillin and 100 μg/mL streptomycin, and cultured in an incubator at 37 °C under a humidified atmosphere containing 5% CO_2.

3.10. HUVEC Proliferation (MTT) Assays

Cell proliferation was measured by using the MTT tetrazolium assay. Briefly, HUVECs (cell density: 1×10^4 cells/well) were seeded into 96-well tissue culture plates and cultured with or without ZCMP (50 μg/mL, 100 μg/mL, and 200 μg/mL). After 44 h, the MTT solution was added and the cells were incubated at 37 °C for another 4 h. The insoluble violet formazan product was solubilized by adding 150 μL of DMSO. The color absorbance was recorded at a wavelength 490 nm using a Bio-Tek Elx 808 micro plate reader (BioTek China Shanghai Office, Shanghai, China). The effect of ZCMP on cell viability was calculated in terms of percentage of control, which was arbitrarily assigned a value of 100% viability.

3.11. HUVEC Migration Assays

To assess the effect of ZCMP on the mobility of HUVEC cells, a cell migration assay was performed. A total of 1 × 10^5HUVEC cells were seeded into each well of 24-well plates and incubated in DMEM medium for 24 h. An artificial line was then created, and the cells were washed and supplied with fresh culture medium and various concentrations of ZCMP (50 μg/mL, 100 μg/mL, and 200 μg/mL). The migration of cells through the line area was examined after 24 h. Images of the migrated cells were captured using a microscope (Olympus, CKX41, Tokyo, Japan).

3.12. Macrophage Phagocytosis Assays

Cell suspensions (200 μL), containing 2 × 10^4 mouse macrophage cells from the cell line Raw 264.7, were added into each well of 96-well plates. After a 6 h incubation to allow the cells to attach to the plate bottom, the cells were cultured with different ZCMP concentrations for another 24 h. Following that, the supernatant was discarded, and 0.075% of a neutral red dye was added to each well (200 μL per well). The plates were incubated for another 30 min. Then, the plates were washed three times with PBS solution (pH 7.2) to remove the redundant neutral red dye. Finally, 200 μL of a lysis solution (acetic acid and ethanol in the ratio of 1:1) was pipetted into each well. The mixtures were thoroughly mixed and evaluated at a wavelength of 540 nm on a Bio-Rad microplate reader (Bio-Rad Laboratories, Hercules, CA, USA).

4. CONCLUSIONS

The polysaccharide ZCMP extracted from Z. caespitosa Miki is composed of AGA and RG-I regions. The backbone of the AGA region consists of (1→4)-α-D-GalA residues with an extremely low degree of etherification, whereas the side chains predominantly contained single Api and a few (1→3′)-linked β-D-Api oligosaccharides linked to the O-3 position of GalA. RG-I contains a backbone of repeating disaccharide units composed of 4-linked GalA and 2-linked Rha, with minor 5-linked α-L-Ara residues, and 4-linked β-D-Gal residues were attached as side chains. ZCMP showed anti-angiogenesis activity by inhibiting migration of HUVECs, and immunoregulation activity by enhancing the phagocytosis of macrophages. Further research studies on the inhibition of tumor metastasis of ZCMP and its structure-activity relationship are warranted.

ACKNOWLEDGMENTS

This work was supported in part by NSFC-Shandong Joint Fund for Marine Science Research Centers (U1406402) and National Science & Technology Support Program of China (2013BAB01B02). The authors thank Sumei Ren and Ni Song, Ocean University of China, for providing technical support of MS and Xiuli Zhang and Cong Wang, Ocean University of China, for providing technical support of NMR.

AUTHOR CONTRIBUTIONS

Youjing Lv, Xia Zhao and Guangli Yu conceived and designed the experiments. Youjing Lv performed the extraction, purification and structural analysis experiments and drafted the manuscript. Xindi Shan performed the bioactivity experiments. Chao Cai, Xiaoliang Zhao, He Zhu and Yinzhi Lang participated in writing the paper. Guangli Yu comprehensively revised the manuscript and approved the final version for submission to the target journal.

CONFLICTS OF INTEREST

The authors declare no conflict of interest.

REFERENCES

1. Ichihara, E.; Kiura, K.; Tanimoto, M. Targeting angiogenesis in cancer therapy. Acta Med. Okayama **2011**, 65, 353–362.
2. Huang, X.; Zhang, Q.; Jiang, Q.; Kang, X.; Zhao, L. Polysaccharides derived from Lycium barbarum suppress IGF-1-induced angiogenesis via PI3K/HIF-1α/VEGF signalling pathways in MCF-7 cells. Food Chem. **2012**, 131, 1479–1484.
3. Zong, A.; Zhao, T.; Zhang, Y.; Song, X.; Shi, Y.; Cao, H.; Liu, C.; Cheng, Y.; Qu, X.; Cao, J.; et al. Anti-metastatic and anti-angiogenic activities of sulfated polysaccharide of Sepiella maindroni ink. Carbohydr. Polym. **2013**, 91, 403–409.
4. Rujitanaroj, P.; Aid-Launais, R.; Chew, S.Y.; le Visage, C. Polysaccharide electrospun fibers with sulfated poly (fucose)

promote endothelial cell migration and VEGF-mediated angiogenesis. Biomater. Sci. **2014**, 2, 843–852.

5. Wang, P.; Zhang, L.; Yao, J.; Shi, Y.; Li, P.; Ding, K. An arabinogalactan from flowers of Panax notoginseng inhibits angiogenesis by BMP2/Smad/Id1 signaling. Carbohydr. Polym. **2015**, 121, 328–335.

6. Schepetkin, I.A.; Quinn, M.T. Botanical polysaccharides: Macrophage immunomodulation and therapeutic potential.Int. Immunopharmacol. **2006**, 6, 317–333.

7. Golovchenko, V.V.; Ovodova, R.G.; Shashkov, A.S.; Ovodov, Y.S. Structural studies of the pectic polysaccharide from duckweed Lemna minor L. Phytochemistry **2002**, 60, 89–97.

8. Ovodova, R.G.; Golovchenko, V.V.; Shashkov, A.S.; Popov, S.V.; Ovodov, Y.S. Structural studies and physiological activity of lemnan, a pectin from Lemna minor L. Russ. J. Bioorganic Chem. **2000**, 26, 743–751.

9. Hart, D.A.; Kindel, P.K. Isolation and partial characterization of apiogalacturonans from the cell wall of Lemna minor.Biochem. J. **1970**, 116, 569–579.

10. Gloaguen, V.; Brudieux, V.; Closs, B.; Barbat, A.; Krausz, P.; Sainte-Catherine, O.; Kraemer, M.; Maes, E.; Guerardel, Y. Structural Characterization and Cytotoxic Properties of an Apiose-Rich Pectic Polysaccharide Obtained from the Cell Wall of the Marine Phanerogam Zostera marina. J. Nat. Prod. **2010**, 73, 1087–1092.

11. Ovodov, Y.S.; Ovodova, R.G.; Bondarenko, O.D.; Krasikova, I.N. The pectic substances of zosteraceae: Part IV. Pectinase digestion of zosterine. Carbohydr. Res. **1971**, 18, 311–318.

12. Ovodov, Y.S.; Mikheyskaya, L.V.; Ovodova, R.G.; Krasikova, I.N. The pectic substances of Zosteraceae: Part V. Smith degradation of zosterine. Carbohydr. Res **1971**, 18, 319–322.

13. Ovodova, R.G.; Vaskovsky, V.E.; Ovodov, Y.S. The pectic substances of Zosferaceae. Carbohydr. Res. **1968**, 6, 328–332.

14. Popov, S.V.; Ovodova, R.G.; Ovodov, Y.S. Effect of lemnan, pectin from Lemna minor L., and its fragments on inflammatory reaction. Phytother. Res. **2006**, 20, 403–407.

15. Sergushchenko, I.S.; Kovalev, V.V.; Bednyak, V.E.; Khotimchenko, Y.S. A comparative evaluation of the metal-binding activity of low-esterified pectin from the seagrass Zostera marine and other sorbents. Russ. J. Mar. Biol. **2004**, 30, 70–72.

16. Sgrebneva, M.N.; Tsygankov, V.I.; Anisimov, A.P.; Khasina, E.I. Effect of zosterin on protein-synthesizing activity of hepatocytes. Bull. Exp. Biol. Med. **2005**, 140, 425–427.

17. Chai, W.; Piskarev, V.; Lawson, A.M. Negative-ion electrospray mass spectrometry of neutral underivatized oligosaccharides. Anal. Chem. **2001**, 73, 651–657.

18. Zaia, J. Mass spectrum of oligosaccharides. Mass Spectrom. Rev. **2004**, 23, 161–227.

19. Lang, Y.; Zhao, X.; Liu, L.; Yu, G. Applications of mass spectrometry to structural analysis of marine oligosaccharides.Mar. Drugs **2014**, 12, 4005–4030.

20. Saad, O.M.; Leary, J.A. Delineating mechanisms of dissociation for isomeric heparin disaccharides using isotope labeling and ion trap tandem mass spectrometry. J. Am. Soc. Mass Spectrom. **2004**, 15, 1274–1286.

21. Wang, P.; Zhao, X.; Lv, Y.; Liu, Y.; Lang, Y.; Wu, J.; Liu, X.; Li, M.; Yu, G. Analysis of structural heterogeneity of fucoidan from Hizikia fusiforme by ES-CID-MS/MS. Carbohydr. Polym. **2012**, 90, 602–607.

22. Quemener, B.; Vigouroux, J.; Rathahao, E.; Tabet, J.C.; Dimitrijevic, A.; Lahaye, M. Negative electrospray ionization mass spectrometry: A method for sequencing and determining linkage position in oligosaccharides from branched hemicelluloses. J. Mass Spectrom. **2015**, 50, 247–264.

23. Everest-Dass, A.V.; Abrahams, J.L.; Kolarich, D.; Packer, N.H.; Campbell, M.P. Structural feature ions for distinguishing N- and O-linked glycan isomers by LC-ESI-IT MS/MS. J. Am. Soc. Mass Spectrom. **2013**, 24, 895–906.

24. Yu, G.; Zhao, X.; Yang, B.; Ren, S.; Guan, H.; Zhang, Y.; Lawson, A.M.; Chai, W. Sequence determination of sulfated carrageenan-derived oligosaccharides by high-sensitivity negative-ion electrospray tandem mass spectrometry. Anal. Chem. **2006**, 78, 8499–8505.

25. Guo, S.; Mao, W.; Yan, M.; Zhao, C.; Li, N.; Shan, J.; Lin, C.; Liu, X.; Guo, T.; Guo, T.; et al. Galactomannan with novel structure produced by the coral endophytic fungus Aspergillus ochraceus. Carbohydr. Polym. **2014**, 105, 325–333.

26. Renard, C.M.G.C.; Peau, M.C.; Thibault, J. Structure of the repeating units in the rhamnogalacturonic backbone of apple, beet and citrus pectin. Carbohydr. Res. **1995**, 275, 155–165.

27. Coenen, G.J.; Bakx, E.J.; Verhoef, R.P.; Schols, H.A.; Voragen, A.G.J. Identification of the connecting linkage between hono- or xylogalacturonan and rhamnogalacturonan type I. Carbohydr. Polym. **2007**, 70, 224–235.

28. Zhang, Z.; Yu, G.; Zhao, X.; Liu, H.; Guan, H.; Lawson, A.M.; Chai, W. Sequence analysis of alginate-derived oligosaccharides by negative-ion electrospray tandem mass spectrometry. J. Am. Soc. Mass Spectrom. **2006**, 17, 621–630.

29. Quéméner, B.; Cabrera Pino, J.C.; Ralet, M.C.; Bonnin, E.; Thibault, J.F. Assignment of acetyl groups to O-2 and/or O-3 of pectic oligogalacturonides using negative electrospray ionization ion trap mass spectrometry. J. Mass Spectrom.**2003**, 38, 641–648.

30. Bauer, S. Mass spectrometry for characterizing plant cell wall polysaccharides. Front. Plant Sci. **2012**, 3, 45.

31. Snyder, J.R.; Serianni, A.S. DL-apiose substituted with stable isotopes: Synthesis, NMR-spectral analysis, and furanose anomerization. Carbohydr. Res. **1987**, 166, 85–99.

32. Yapo, B.M. Pectic substances: From simple pectic polysaccharides to complex pectins—A new hypothetical model.Carbohydr. Polym. **2011**, 86, 373–385.

33. Huang, Z.; Lin, H.; Wang, Y.; Cao, Z.; Lin, W.; Chen, Q. Studies on the anti-angiogenic effect of Marsdenia tenacissimaextract in vitro and in vivo. Oncol. Lett. **2013**, 5, 917–922.

34. Masuda, Y.; Inoue, H.; Ohta, H.; Miyake, A.; Konishi, M.; Nanba, H. Oral administration of soluble beta-glucans extracted from Grifola frondosa induces systemic antitumor immune response and decreases immunosuppression in tumor-bearing mice. Int. J. Cancer **2013**, 133, 108–119.

35. Ledoux, M.; Lamy, F. Determination of proteins and sulfobetaine with the folin-phenol reagent. Anal. Biochem. **1986**,157, 28–31.

36. Dodgson, K.S.; Price, R.G. A note on the determination of the ester sulphate content of sulphated polysaccharides.Biochem. J. **1962**, 84, 106.

37. Chen, S.; Xu, J.; Xue, C.; Dong, P.; Sheng, W.; Yu, G.; Chai, W. Sequence determination of a non-sulfated glycosaminoglycan-like polysaccharide from melanin-free ink of the

squid Ommastrephes bartrami by negative-ion electrospray tandem mass spectrometry and NMR spectroscopy. Glycoconj. J. **2008**, 25, 481–492.

38. Taylor, R.L.; Conrad, H.E. Stoichiometric depolymerization of polyuronides and glycosaminoglycuronans to monosaccharides following reduction of their carbodiimide-activated carboxyl group. Biochemistry **1972**, 11, 1383–1388.

39. Hakomori, S. A rapid permethylation of glycolipid, and polysaccharide catalyzed by methylsulfinyl carbanion in dimethyl sulfoxide. J. Biochem. **1964**, 55, 205–208.

4

CHAPTER

Sialyltransferases of marine bacteria efficiently utilize glycosphingolipid substrates

Yasunori Kushi[1,2], Hisashi Kamimiya[2], Hiroko Hiratsuka[2], Hirofumi Nozaki[2], Hiroshi Fukui[2], Mayumi Yanagida[2], Makoto Hashimoto[3], Kimihide Nakamura[4], Shinobu Watarai[5], Takeshi Kasama[6], Hitomi Kajiwara[7] and Takeshi Yamamoto[7]

[2]*Department of Agriculture and Life Science, Obihiro University of Agriculture and Veterinary Medicine, Inada-cho, Obihiro, Hokkaido, 080-8555*

[3]*Graduate School of Agriculture, Hokkaido University; Kita 9, Nishi 9, Kita-ku, Sapporo 060-8589*

[4]*Health Care Administration Center, Obihiro University of Agriculture and Veterinary Medicine, Inada-cho, Obihiro, Hokkaido, 080-8555*

[5]*Laboratory of Veterinary Immunology, Division of Veterinary Science, Graduate School of Life and Environmental Sciences, Osaka Prefecture University, Izumisano, Osaka, 598-8531*

[6]*Instrumental Analysis Research Center, Tokyo Medical Dental University, 1-5-45, Yushima, Bunkyo-ku, Tokyo, 113-8510*

[7]*Glycotechnology Business Unit, Plant Innovation Center, Japan Tabacco Inc., 700 Higashibara, Iwata, Shizuoka, 483-0802, Japan*

ABSTRACT

Bacterial sialyltransferases (STs) from marine sources were characterized using glycosphingolipids (GSLs). Bacterial STs were found to be β-galacotoside STs. There were two types of STs: (1) ST obtained from strains such as ishi-224, 05JTC1 (#1), ishi-467, 05JTD2 (#2), and faj-16, 05JTE1 (#3), which form α2-3 sialic acid (Sia) linkages, named α2-3ST, (2) ST obtained from strains such as ISH-224, N1C0 (#4), pda-rec, 05JTB2 (#5), and pda-0160, 05JTA2 (#6), which form α2-6 Sia linkages, named α2-6ST. All STs showed affinity to neolacto- and lacto-series GSLs, particularly in neolactotetraosyl ceramide (nLc$_4$Cer). No large differences were observed in the pH and temperature profiles of enzyme activities. Kinetic parameters obtained by Lineweaver–Burk plot analysis showed that #3 and #4 STs had practical synthetic activity and thus it became easily possible to achieve large-scale ganglioside synthesis (100–300 μM) using these recombinant enzymes. Gangliosides synthesized from nLc$_4$Cer by α2-3 and α2-6STs were structurally characterized by several analytical and immunological methods, and they were identified as IV^{3}αNeuAc-nLc$_4$Cer(S2-3PG) and IV^{6}αNeuAc-nLc$_4$Cer (S2-6PG), respectively. Further characterization of these STs using lactotetraosylceramide (Lc$_4$Cer), neolactohexaosylceramide (i antigen), and IV^{6}kladoLc$_8$Cer (I antigen) showed the synthesis of corresponding gangliosides as well. Synthesized gangliosides showed binding activity to the influenza A virus [A/panama/2007/99 (H3N2)] at a similar level to purified S2-3PG and S2-6PG from mammalian sources. The above evidence suggests that these STs have unique features, including substrate specificities restricted to lacto- and neolactoseries GSLs, as well as catalytic potentials for ganglioside synthesis. This demonstrates that efficient in vitro ganglioside synthesis could be a valuable tool for selectively synthesizing Sias modifications, thereby permitting the exploration of unknown functions.

Key words bacterial sialyltransferase, gangliosides synthesis, Ii antigens, in vitro synthesis, sialylparagloboside, Previous SectionNext Section

INTRODUCTION

Gangliosides are glycosphingolipids (GSLs) containing one or more sialic acid (Sia) residues and are found predominantly in the plasma membranes of all vertebrates (Svennerholm et al. 1980). Their biological significance remains vague, although they have been implicated in a variety of crucial phenomena, including cell–cell recognition, neurite outgrowth, synaptogenesis, transmembrane signaling, cell growth and differentiation, and oncogenic transformation (Hakomori and Kannagi 1983; Yogeewaran1983; Hakomori 1984, 2002; Varki 1993; Ariga and Yu 2005).

Sias are present in a variety of mammalian glycoproteins and GSLs, usually at the non-reducing terminal of carbohydrate chain, in which four linkage patterns, NeuAcα2-3Gal, NeuAcα2-6Gal, NeuAcα2-6GalNAc, and NeuAcα2-8NeuAc are known to be found. These Sias play critical roles in recognition- and adherence-related biological and physiological processes

(Schauer 1982). The enzymes that transfer the Sia moiety from cytidinemonophospho 5′-*N*-acetylneuraminic acid (CMP-NeuAc) to the terminal positions of these key glycoconjugates are known as sialyltransferases (STs) (Paulson 1989).

Influenza A and B viruses infect host cells through the binding of viral hemagglutinins to sialylglycoproteins or sialylGSLs that serve as receptors on the host cell surface. The viruses recognize not only Sia in the receptors, but also Sia containing sugar chain structures such as sialyllacto-series type I (Siaα2-3/6Galβ1-4GlcNAcβ1-) and type II (Siaα2-3/6Galβ1-3GlcNAcβ1-). It has been reported that influenza viruses differ in their recognition of two types of *N*-acetylneuraminic acid (NeuAc) and*N*-glycolylneuraminic acid (NeuGc) and linkages (α2-3 or α2-6) of Sia residues (Suzuki et al. 1992, 2000). Avian and equine viruses preferentially bind Sia α2,3-galactose, while human influenza viruses preferentially bind Sia α2,6-galactose linkage (Connor et al. 1994). In addition, many biological phenomena, including the distinctive expression of NeuGc in human tissues by a potential feeding and recycling process, have been reported. (Varki 2001; Tangvoranuntakul et al. 2003; Bardor et al. 2005). Thus, evidence is accumulating from experimental studies relevant to Sia biology (Gagneux and Varki 1999; Schauer 2004). To further investigate the detailed mechanisms of Sia biology, more efficient purification and synthetic techniques for sialocompounds are required.

In addition to ST-mediated transfer in mammals, some microbial cells have also been reported to synthesize Sia in lipo-oligosaccharide (LOS) outer core structures (Gilbert et al. 2000). Molecular mimicry between Sia-containing LOSs and gangliosides has been suggested to act as a trigger of autoimmune responses in the development of Guillain–Barré syndrome (GBS) (Yuki et al. 1992; Gilbert et al. 2000). Despite their important biological roles, little is understood about the mechanisms of action or the molecular structures of these STs.

One of the authors of this study previously cloned STs of marine bacteria and enzymatically characterized them using fluorescence-labeled oligosaccharides, and the bacterial ST from marine sources is found to be β-galactoside α2-3 or α2-6ST which catalyzes the incorporation of NeuAc from CMP-NeuAc into the terminal galactose residue in glycoconjugates at position 3 or 6 (Yamamoto et al. 2006).

In this paper, we tested GSL substrates for ganglioside synthesis using bacterial STs from marine sources. Lacto-series gangliosides were effectively synthesized by these STs. Synthesized structures were confirmed by several analytical techniques and with mAbs. Furthermore, using these in vitro synthesized gangliosides, the binding activity of S2-3PG and S2-6PG to influenza viruses was investigated by measuring cytopathic effects in virus-infected Madine-Darby canine kidney (MDCK) cells.

RESULTS

TLC of ST products

Enzyme reactions were performed under the conditions described in the text. TLC analysis results for the reaction products are shown in Figure 1. STs showed affinity to substrates although some quantitative differences were observed between STs, despite using the same units used for sialyllactose synthesis, suggesting differences in optimal assay conditions between oligosaccharides and gangliosides. All bands appeared newly on TLC plates after ST enzyme reactions were reacted with resorcinol-HCl reagent (data not shown).

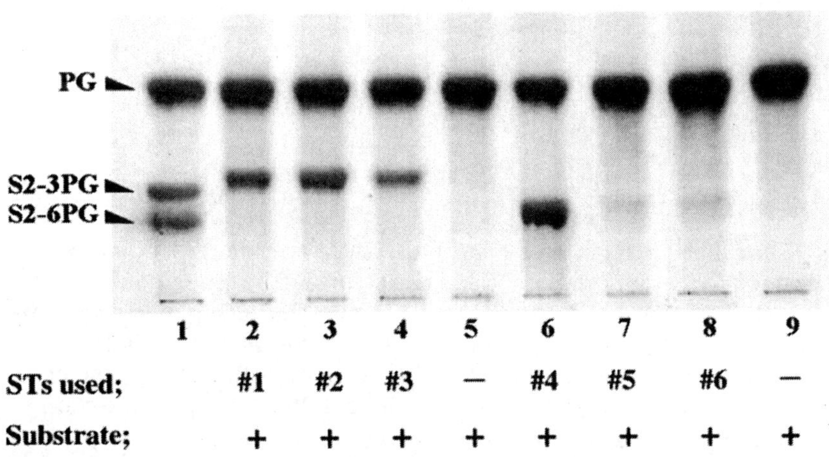

Fig. 1 Sialylation of PG by marine STs. After incubation at 37°C for 3 h, lipid fractions were prepared from enzyme assay mixtures (30 μL), as described in the text. One-fifth of the lipid fractions was applied to TLC, developed with solvent system I and visualized by orcinol/H_2SO_4. Lane 1 contains the standard GSLs: PG, S2–3PG, and S2–6PG from the top. In lanes 2–4, #1, #2, and #3 α2–3STs were used, respectively, and #4, #5, and #6 α2–6STs were used in lanes 6–8. Lanes 5 and 9 represent TLC without each STs. STs were added to the assay mixtures (2 μL of 0.1 U ST).

All gangliosides synthesized by α2-3STs migrated almost to the same position as standard S2-3PG (Figure 1, lanes 2–4), and gangliosides synthesized by α2-6STs migrated almost to the same position as S2-6PG (Figure 1, lanes 6–8). No other bands except the predicted gangliosides and un-reacted substrates were detected in all lanes. No corresponding bands were detected without both STs (Figure 1, lanes 5 and 9). After neuraminidase treatment of the synthesized gangliosides, all Rf values shifted back to the original position (PG) on TLC, confirming the presence of Sia residues in the molecules (data not shown). This is the first demonstration of in vitro ganglioside biosynthesis using STs from marine bacteria.

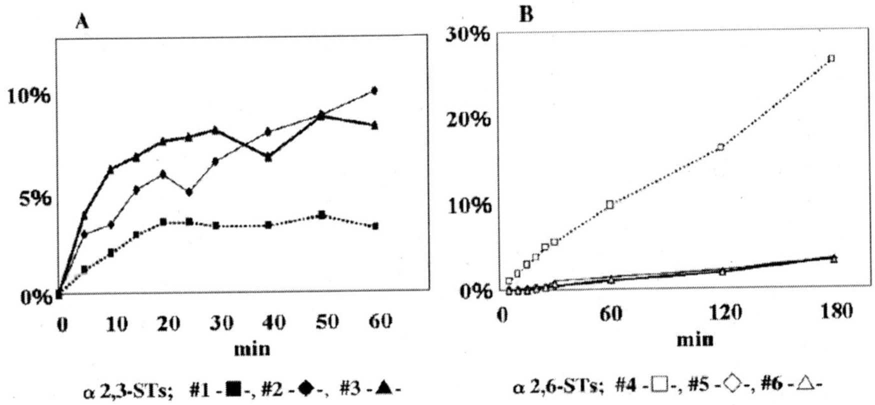

Fig. 2 Incubation time course profiles. After enzyme reactions using the indicated incubation times, lipid fractions purified by Sep-Pak C18 cartridge were chromatographed on TLC. The synthesized gangliosides were quantitated using NIH image software (Becker et al. 1996) and calibrated with plots of S2–3PG and S2–6PG. The incubation times used for enzyme reactions are shown in each plot in the figure. Panel **A** shows the profiles of ganglioside products from α2,3-STs #1 (-▪-), #2 (-♦-), and #3 (-▴-), whereas in panel **B** the profiles are from α2,6-STs #4 (-□-), #5 (-◊-), and #6 (-Δ-). The vertical plots are the percentages of synthesized gangliosides in the reaction mixtures.

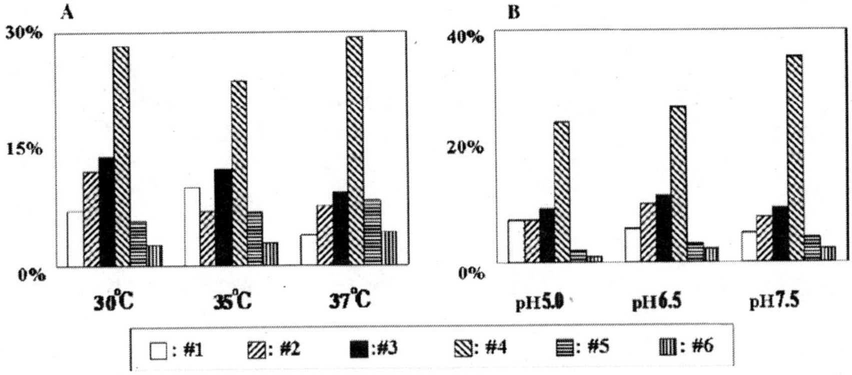

Fig. 3 Optimal temperature and pH profiles on ganglioside synthesis. Optimal temperature conditions were investigated at 30, 35, and 37°C. Other conditions were as described in the text. In order to obtain optimal pH profiles, enzyme assays were performed at pH 5.0, 6.5, and 7.5. Panel **A** shows optimal temperature and panel **B** shows optimal pH. Incubation times were 30 min for α2–3STs (#1–3) and 180 min for α2–6STs (#4–6). The vertical plots are the percentages of synthesized gangliosides in the reaction mixtures.

Enzyme characteristics

In order to compare the individual properties of STs, incubation time, pH, and temperature profiles were investigated. The results for different incubation times are shown in Figure 2. The reaction products synthesized by #1, #2, and #3 STs that form α2-3 Sia linkages were proportional to incubation times up to 30 min and reached a plateau; however, at 60 min, there was a slight reduction in S2–3PG synthesis (Figure 2A). In contrast, the products synthesized by #4, #5, and #6 STs, which form α2-6 Sia, gradually increased with incubation times up to 3 h (Figure 2B). Incubation times over 3 h were not examined. On the other hand, no large differences were found in temperature and pH profiles between STs in accordance with findings previously reported for oligosaccharide synthesis (Figure 3). However, at 30°C and 35°C, STs #2 and #3 showed slightly higher enzyme activities than at 37°C. From the above results, the enzyme assay was performed under the following conditions: 30°C and pH 5.0. The incubation time differed with STs (30 min for α2-3STs and 3 h for α2-6STs).

Enzyme kinetics

To further investigate the mechanisms of these STs, Line weaver–Burk plots were generated and individual Km and $V_{\max}$ values were obtained. The results are shown in Table I. Among STs #1 to #3, *Km* values ranged from 0.43 to 0.94 mM, and #1 had the lowest *Km* value. ST #1 also showed the lowest V_{max}, whereas #3 had the highest V_{max}/Km value and showed more efficient S2-3PG synthesis. In contrast, #4, which forms α2-6 Sia linkages, exhibited the lowest *Km* and highest V_{max} value among #4 to #6. These findings suggest that STs #3 and #4 are most suitable for synthesizing Sia linkages.

Table I Kinetic parameters of various STs for the acceptor substrate (PG)

Enzyme	K_m (mM)	V_{max}	V_{max}/K_m
#1	0.430	2.334	5.428
#2	0.944	5.049	5.349
#3	0.870	6.247	7.180
#4	0.252	1.416	5.619
#5	0.353	0.683	1.935
#6	1.153	0.626	0.543

Reactivity of newly synthesized gangliosides to mAbs specific for S2-3PG and S2-6PG

TLC/immunostaining was carried out using mAbs, as shown in Figure 4. Figure 4B shows the results obtained with an antibody SPS-20, which is directed against terminal NeuAc/NeuGcα2-3Gal carbohydrate chains. Bands of newly synthesized gangliosides showed strong reactivity with mAb SPS-20 in lanes 2–4. Further TLC/immunostaining using a mAb specific for NeuAcα2-6Gal (MFH-1) also confirmed gangliosides synthesized by STs #4 to #6 with NeuAcα2-6Gal epitopes (data not shown). These observations clearly indicate that gangliosides synthesized by STs #1 to #3 have the NeuAcα2-3Gal linkages (Figure 4, lanes 2–4), while gangliosides synthesized by STs #4 to #6 have NeuAcα2-6Gal linkages. The mobilities of the newly synthesized gangliosides on TLC were slightly upper than standard gangliosides S2-3PG and S2-6PG. This is probably due to a small amount of detergent Triton X-100 and salts remaining in ST assay products (Figures 1 and 4).

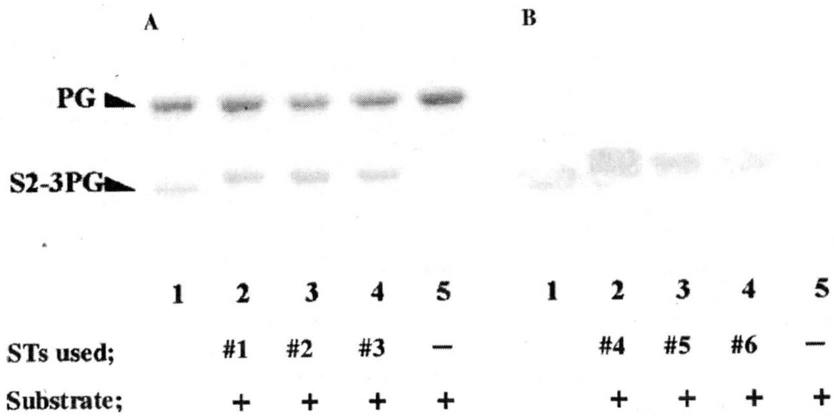

Fig. 4 TLC/immunostaining. After TLC development, TLC/immunostaining was performed as described with SPS-20 mAb (Watarai et al.1995). Other conditions including solvent system I were identical to those described in Figure 1. In **A**, the standard GSLs, PG, and S2–3PG are shown in lane 1. In lanes 2–4, #1, #2, and #3 α2-3STs were used, respectively. Lane 5 is an assay without enzymes. In **B**, the same TLC procedure used in **A** was performed and was followed by TLC/immunostaining.

Structural elucidation by SIMS spectrometry after TLC blotting, ¹H-NMR, and permethylation analysis of the synthesized gangliosides X and Y

Figures 1–4 showed that STs #1–3 catalyzed the formation of α2-3-linked Sia, while STs #4–6 catalyzed the formation of α2-6-linked Sia; however, different ST activities were detected between individual clones even when the same assay conditions were used in all reaction mixtures (Figure 1). Therefore, two typical gangliosides synthesized by ST #3 and ST #4 were temporally designated as X and Y, respectively, and structurally analyzed more details since STs #3 and #4 have comparatively strong activities and can more easily synthesize new gangliosides on a large scale. First, gangliosides X and Y were analyzed by in situ PVDF membrane transfer, followed by SIMS; the results are shown in Figure 5. Although PVDF membranes from different spots were used (Figure 5A), in the negative ion SIMS spectra of gangliosides X and Y, the same major pseudo-molecular ion $[M-H]^-$ was detected at m/z 1628, which was consistent with values calculated for the proposed structures with C24:0- and C24:1-C18 sphingenine (Figure 5C and D). This finding indicates the presence of sugars with the same composition in X and Y. Furthermore, the ions responsible for successive elimination of sugar moieties with ceramide portions were detected at *m/z* 1337 for [M-H-NeuAc]⁻, at *m/z* 1175 for [M-H-NeuAc-Gal]⁻, at *m/z* 972 for [M-H-NeuAc-Gal-GlcNAc]⁻, at *m/z*810 for [M-H-NeuAc-Gal-GlcNAc-Gal]⁻, and at *m/z* 648 for [M-H-NeuAc-Gal-GlcNAc-Gal-Glc]⁻, as well as for pseudomolecular ions. In contrast, the spectrum of substrate PG gave a pseudo-molecular ion at *m/z* 1337 for [M-H]⁻ (Figure 5B). From the difference in mass units between the two pseudo-molecular ions, *m/z* 1628–1337, the attached sugar residue at the non-reducing terminal of the molecules corresponded to the mass units of NeuAc, 291 (Figure 5B–D). This confirmed that NeuAc from CMP-NeuAc is transferred to the substrate PG via ST activity.

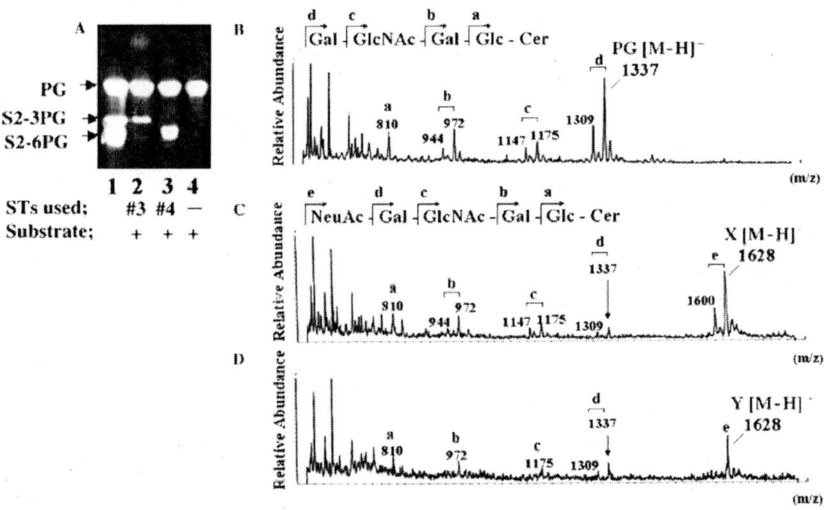

Fig. 5 TLC and mass spectrometric analyses by in situ PVDF membrane transfer of gangliosides synthesized by STs. Enzyme reactions were performed as described in the text. #3 and #4 STs with strong enzyme activities were chosen for structural analyses. After gangliosides synthesized were developed with TLC and sprayed with primurin reagent (Skipski 1975), PVDF membranes were blotted and GSLs were transferred to the PVDF membrane, as described by Ishikawa and Taki (Taki et al. 1995; Ishikawa and Taki 2000). Then, the blotted PVDF membranes were analyzed by negative detection of SIMS after cutting out the spots representing newly synthesized gangliosides (X and Y) and the remaining un-reacted substrate (PG). (**A**) A blotted PVDF membrane detected with primurin reagent. Lane 1, standard GSLs; lane 2, synthesized ganglioside by #3 α2-3ST; lane 3, synthesized ganglioside by α2-6ST; lane 4, the result without STs; (**B**) mass spectrum of substrate GSL (PG); (**C**) mass spectrum of synthesized ganglioside X; (**D**) mass spectrum of synthesized ganglioside Y.

These findings were further supported by ^{1}H-NMR spectroscopy, as shown in Figure 6. In the anomeric proton region of ganglioside X (Figure 5B), four signals can be seen at 4.18 ppm (J = 8.0 Hz) (IV), 4.24 ppm (J = 7.4 Hz) (III), 4.28 ppm (J = 6.7 Hz) (I), and 4.79 ppm (J = 8.0 Hz) (II), which were identified as anomeric protons of β-glucose (I), β-galactose (II), β-galactose (III), and β-*N*-acetylglucosamine (IV), respectively. In the anomeric proton regions of ganglioside Y (Figure 6C), four anomeric signals with the same chemical shift as that of ganglioside X were identified. Furthermore, the appearance of an equatorial 3-H signal for NeuAc at 2.59 ppm for α2-6 linkages and 2.73 ppm for α2-3 linkages indicates that NeuAc is linked to the 3/6 position of galactose. *N*-Acetyl signals at 1.8–1.9 ppm indicate that NeuAc and *N*-acetylglucosamine were present at an approximate molar ratio of 1:1 (Figure 6B and C). Additionally, amide proton signals of NeuAc from X and Y were clearly detected at 8.0–8.1 ppm (data not shown). These data show that newly synthesized gangliosides X and Y contain NeuAc with α-anomeric protons.

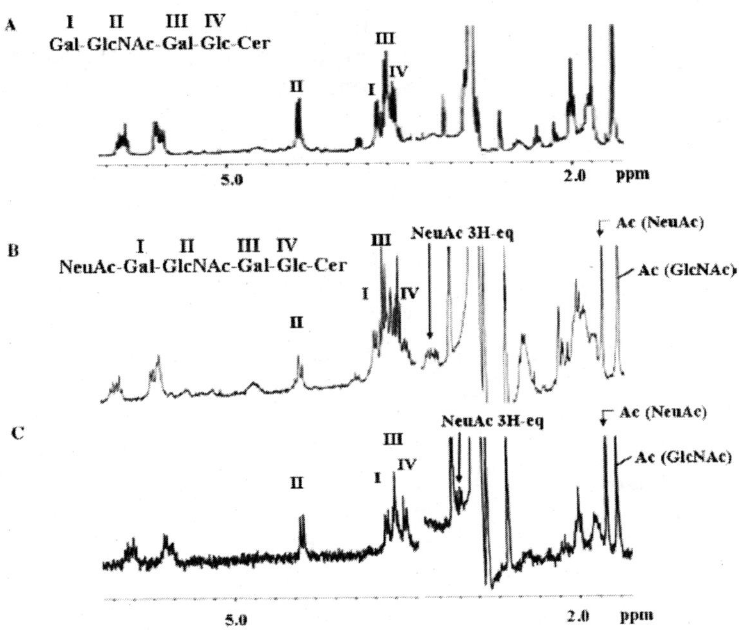

Fig. 6 ^{1}H-NMR spectroscopy of the synthesized X and Y gangliosides. Gangliosides synthesized by #3 and 4 STs were further purified by Sep-pak C18 cartridge and DEAE-Sephadex A25 (acetate form) column. Purified X and Y were exchanged for DMSO-$d_6$containing 2% D_2O and analyzed by ^{1}H-NMR spectroscopy. Tetramethylsilane was used as an internal standard for chemical shifts. (**A**) Substrate PG; (**B**) synthesized ganglioside X; (**C**) synthesized ganglioside Y.

On permethylation analysis using GC/MS, ganglioside X synthesized by α2-3ST #3 was demonstrated to contain 2,3,6-tri-*O*-methyl-1,4,5-tri-*O*-acetylglucitol (-4Glc1-), 2,4,6-tri-*O*-methyl-1,3,5-tri-*O*-acetylgalactitol (-3Gal1-), and 3,6-di-*O*-methyl-1,4,5-tri-*O*-acetyl-2-deoxy-2-*N*-methylacetoamidoglucitol (-4GlcNAc1-) at a molar ratio of 1.00:1.78:0.81. Ganglioside Y synthesized by α2-6ST #4 was demonstrated to contain 2,3,6-tri-*O*-methyl-1,4,5-tri-*O*-acetylglucitol (-4Glc1-), 2,4,6-tri-*O*-methyl-1,3,5-tri-*O*-acetylgalactitol (-3Gal1-), 2,3,4-tri-*O*-methyl-1,5,6-tri-*O*-acetylgalactitol (-6Gal1-), and 3,6-di-*O*-methyl-1,4,5-tri-*O*-acetyl-2-deoxy-2-*N*-methylacetoamidoglucitol (-4GlcNAc1-) at a molar ratio of 1.00:1.08:0.81:0.75. Further permethylation analysis of desialylated gangliosides X and Y additionally yielded 2,3,4,6-tetra-*O*-methyl-1,5-di-*O*-acetylgalactitol, showing that Sia was linked to C-3 and C-6 positions of the terminal galactose, respectively. No other peaks of alditol acetate derivatives were observed with GC/MS analysis. These results indicate that ganglioside X synthesized by α2-3ST #3 is S2–3PG, whereas ganglioside Y synthesized by α2-6ST #4 is S2–6PG. Identical results were also obtained on permethylation analysis of gangliosides from #1 and #2 and from #5 and #6 (data not shown). Taken together, the ganglioside structures generated by #1, #2, and #3 had the same structures and were identified as ganglioside X, while those generated by STs #4, #5, and #6 had the same structures as ganglioside Y.

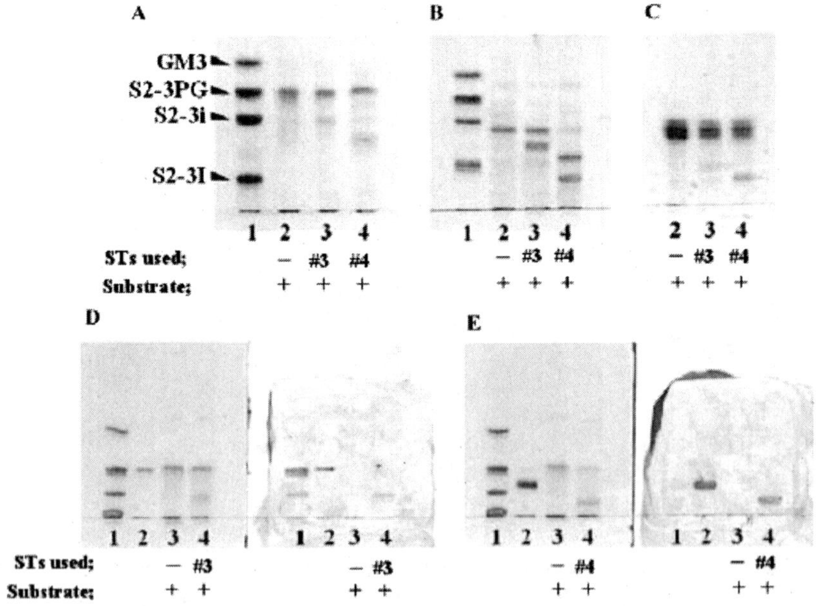

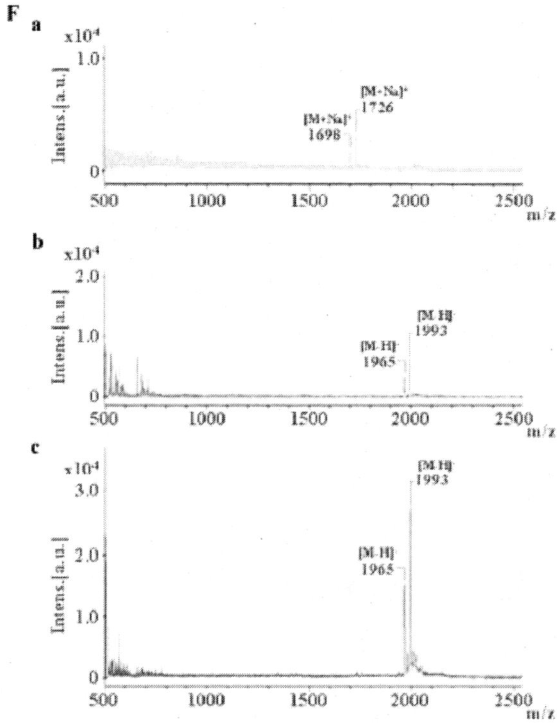

Fig. 7 Sialylation of homologus GSL substrates by marine STs. Instead of PG, i, I antigen GSLs, and type I lactosamine-containing Lc_4Cer were used as substrates in **A–C**, respectively. Other conditions and TLC were as described in the text. The solvent system II was used in this figure. TLCs of **A–D** (left panel) and **E** (left panel) were visualized with orcinol/H_2SO_4 reagent. Lane 1 of **A**, **B**, **D**, and **E**shows the standard gangliosides {GM3, S2–3PG, S-i (S2–3i), and S-I (S2–3I) GSLs} from the top of the TLC. Lane 2 of **A–C** shows the results without enzymes, while lanes 3 and 4 of **A–C** are the results with #3 α2,3-ST and # 4 α2,6-ST, respectively. In **D** and**E**, i antigen GSL was used as the substrate and TLC and TLC/immunostaining (right panels of **D** and **E**) were performed to identify the sialylated products by the respective mAbs specific for S2–3PG (SPS-20) and S2–6PG (MFH-1). Lane 2 of **D** is S2–3PG, whereas lane 2 of **E** is S2–6PG. Lane 3 of **D** and **E** shows the results without enzymes. Lane 4 of **D** and **E** shows results with #3 α2,3-ST and # 4 α2,6-ST, respectively. In **F**, the results of MALDI/TOF-MS analyses of gangliosides synthesized by STs are shown. (a) substrate i antigen; (b) synthesized ganglioside by #3 α2,3-ST; (c) synthesized ganglioside by # 4 α2,6-ST.

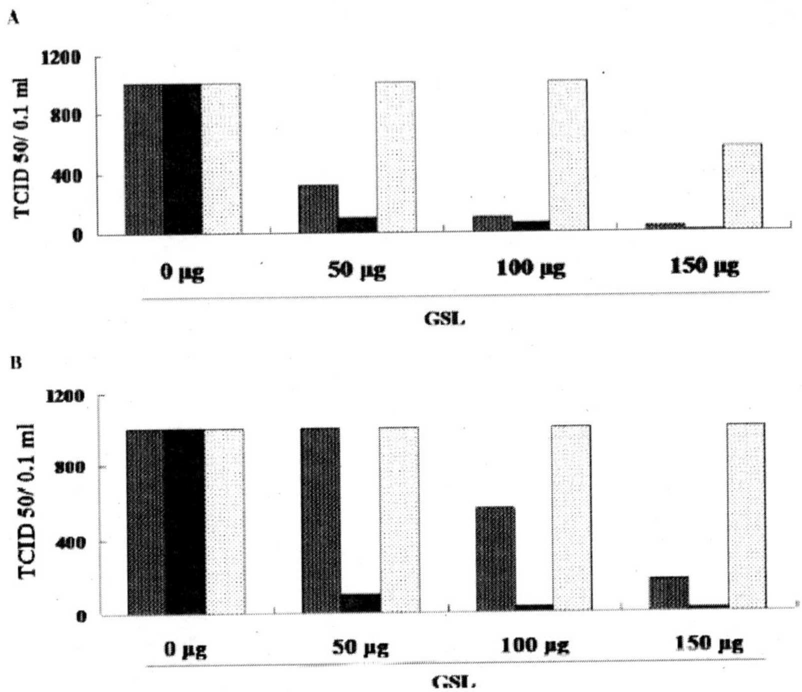

Fig. 8 Binding activity of in vitro synthesized gangliosides to influenza A virus. One hundred microliters of 1000 TCID50 of the virus {A/panama/2007/99 (H3N2)} was incubated at 37°C for 1 h with liposomes containing purified and in vitro synthesized gangliosides (S2–3PG and S2–6PG). After incubation, the virions not absorbed in the liposomes were isolated, diluted, and re-infected into MDCK cells for 3 days. Decreases in TCID50 were investigated for different ganglioside concentrations. In **A**, standard gangliosides (S2–3PG and S2–6PG) were used, while in **B**the results using in vitro synthesized gangliosides (S2–3PG and S2–6PG) are shown. PG, S2–3PG, and S2–6PG were prepared by scraping from the TLC plate and extracting with a solvent after the enzyme reaction. PG was prepared from the TLC plate without STs. Each assay was carried out in duplicate. (■), (■), and( )are S2–3PG, S2–6PG, and PG, respectively.

Synthesis of sialyl-i and -I antigen GSLs (S-i and S-I)

To investigate characteristic features of these unique STs, enzyme reactions were further performed using i and I antigen GSLs to synthesize S-i and S-I;

the results are described in Figure 7. Gangliosides newly synthesized by α2-3ST #3 migrated to the same position as S-i (Figure7A, lane 3). Gangliosides synthesized by α2-6ST #4 migrated between S-i and S-I (Figure 7A, lane 4). All newly synthesized bands were positive to resorcinol as well as orcinol reagent. In order to confirm the identity of the synthesized gangliosides immunologically, SPS-20 and MFH-1, which are mAbs specific to S2–3PG and S2–6PG, respectively, were used. The results are shown in Figure 7D and E. The synthesized ganglioside was reactive to SPS-20 mAb (right panel of Figure 7D, lane 4), while the ganglioside synthesized by α2-6ST #4 was positive for MFH-1 mAb (right panel of Figure 7E, lane 4). Further analysis of the synthesized gangliosides by MALDI/TOF-MS confirmed the Sia addition to the substrate i antigen GSL as well as the results of Figure 5 (Figure 7Fa, b, and c).

In the case of S-I synthesis from I antigen GSL, two new bands appeared after ST reaction (Figure 7B, lanes 3 and 4). The upper band migrated between S-i and S-I, corresponding to a monosialoganglioside attached by NeuAc to either nonreducing terminal of the substrate (I antigen GSL). The lower band (very low intensity) migrated to almost the same Rf value as S-I and corresponded to a disialoganglioside with two NeuAc molecules (VIII^3, $VI^3NeuAc\alpha_2$-$IV^6kladoLc_8Cer$) (Figure 7B, lane 3). Interestingly, the band of the substrate (I antigen GSL) was almost eliminated by the enzyme reactions and, correspondingly, the intensity of product disialoganglioside band was synthesized by α2-6ST increased (Figure 7B, lane 4). No further detailed chemical elucidation about S-I was performed due to an insufficient substrate amount. Thus, it was clearly shown that in synthesized gangliosides, one and two molecules of NeuAc are covalently linked to substrate I antigen GSL, respectively.

Synthesis of type I lactosamine-containing gangliosides

To demonstrate further that the potential synthesis of not only type II chain lactosamines (Galβ 1-4GlcNAc), but also type I chain lactosamine (Galβ1-3GlcNAc) can be performed using these STs, the same in vitro enzyme reactions were performed using Lc_4Cer. Synthesized gangliosides were observed after enzyme reactions using α2-3ST #3 and α2-6ST #4 as well (Figure 7C, lanes 2 and 3). Thus, unique STs are able to catalyze the substrate, despite different lactosamine units, to form gangliosides as reaction products.

From the findings obtained from Figure 7, STs from marine bacteria can react well not only with a PG substrate, such as short chain lactosamine-containing GSL, but also with type I lactosamine-containing GSL (Lc_4Cer), as well as i and I antigen GSLs with oligolactosamine chains.

Binding activity of in vitro synthesized gangliosides to influenza virus A

To confirm the biological activities of in vitro synthesized gangliosides, binding to the influenza A virus (H3N2) was compared between synthesized and purified gangliosides using liposome methods. As expected, viral tissue-culture infectious dose 50 (TCID50) was decreased by standard S2–3PG and S2–6PG in a dose-dependent manner (Figure 8A). In contrast, PG showed no decrease, indicating that no binding occurred. In vitro synthesized S2–3PG and S2–6PG showed a nearly identical decrease in the TCID50 value as purified gangliosides (Figure 8B). These data clearly show that in vitro synthesized gangliosides have affinity to bind to the influenza virus as well as natural gangliosides purified from mammalian sources.

DISCUSSION

There have been many reports about the molecular structures and chemical features of STs from mammalian sources that are involved in the biosynthesis of various sialoglycoconjugates (Taniguchi et al. 2002). These STs have very restricted substrate specificities that are controlled genetically; therefore, it is likely that sialoglycoconjugates are involved in many biological phenomena on cell surfaces (Schauer 1982; Paulson 1989; Angata and Varki 2002; Varki 2007).

Recently, Sia molecules were discovered and identified chemically in prokaryote species (Gilbert et al. 2002; Guerry et al. 2002; Chiu et al.2004; Fox et al. 2006; Godschalk et al. 2007), but their biological functions remain obscure. We have already reported the purification and cloning of STs from marine bacteria and some applications in glycoconjugation (Takakura et al. 2007; Tsukamoto et al. 2007; Yamamoto et al. 2007). In this study, we have extended the application to ganglioside synthesis and demonstrated the following: (1) 100 μM in vitro ganglioside synthesis using recombinant STs from marine sources became possible; (2) synthesis of sialylparaglobosides such as S2–3PG and S2–6PG as well as extended lacto- and neolacto-series gangliosides such as sialy-Lc_4Cer, S-i, and S-I was also possible; (3) in vitro synthesized and purified mammalian gangliosides showed equivalent influenza A virus binding activity. These results indicate the potential for mass production of gangliosides by using recombinant STs and in turn could lead to new developments in Sia biology. In the standard ST assay, the concentration of the 300 μM acceptor substrate (PG) was used through this study. As about 30% of the substrate is converted to the products by these ST reactions, 100 μM level in vitro ganglioside synthesis using these recombinant STs became possible. And almost the same recovery percent of the gangliosides products was obtained, even when the volume of the ST assay was scaled up to roughly 15-fold. Thus, these ST assays were very reliable and useful for large-scale ganglioside preparations.

In addition to the sialylparaglobosides (S2–3PG and S2–6PG) synthesis, these STs were successfully further applied to the synthesis of sialyl-Lc_4Cer, S-i, and S-I, though detailed chemical analyses for the synthesized gangliosides were limited due to insufficient substrates. As STs from marine bacteria form specific Sia linkages (Figures 1–6), S-i and S-I synthesized by α2,3-ST (Figure 7A–C, lane 3) and S-i and S-I synthesized by α2,6-ST (shown in Figure 7A–C, lane 4) likely have NeuAcα2-3Gal and NeuAcα2-6Gal linkages, respectively.

To confirm the structures, we used a combination of in situ PVDF membrane transfer analysis followed by SIMS and binding assay, such as mAbs. The basic method introduced by Ishikawa and Taki (Taki et al.1995; Ishikawa and Taki 2000) is a simple and attractive method to demonstrate that the Sia molecule is attached to the substrate molecule using simultaneous chemical and binding analyses.

Interestingly, the SPS-20 mAb (Watarai et al. 1995), and recently established MFH-1 clone specific to S2–6PG, showed negative reactions on TLC/immunostaining for both sialylated products, S-I synthesized by α2-3ST and α2-6ST. This result suggests that branched lacto-series containing gangliosides are not able to recognize these two mAbs in spite of the possible binding to extended straight lactosamine-containing gangliosides (S-i) (Figure 7D and E, lanes 3 and 4). In addition to PG and Lc_4Cer, several other substrates were used in this enzyme reaction. Almost none of the GSLs were good substrates in this study except for two or three GSLs. The concentration of detergent (Triton X-100) may be important for solubilization of the substrates, and another detergent may therefore be more useful. It is interesting that large differences in substrate consumption were clearly observed even though the same reaction conditions (pH, buffers, etc.) were used (Figure 7B, lanes 3 and 4). α2,6-ST #4 always exhibited strong activity, as shown in Figures 1 and 7, but we are unable to explain the reasons for the above findings. It has been reported that α2,3-STs (#1–3) have both neuraminidase and ST activities (Takakura et al. 2007; Tsukamoto et al. 2007); however, in our assay system, such neuraminidase activity was not detected during ganglioside synthesis. GSL structures, especially lipid portions, may markedly affect the neuraminidase activity.

Campylobacter jejuni is one of the most common causes of bacterial diarrhea worldwide and also frequently associated with Guillain-Barré syndrome (GBS), a post-infectious polyneuropathy. GBS is considered an autoimmune disease with the immune system mistakenly attacking myelin or axons, the nerve conduits for signal to and from the brain (Yu et al.2006). The association of *C. jejuni* and the development GBS is thought to result from the molecular mimicry between outer core structures of bacterial lipo-oligosaccharide (LOS) and gangliosides. A specific ST gene (cst-II) is closely involved with the biosynthesis of ganglioside-like LOS. Thus, for example, the core of different isolates of *C. jejuni* has shown to mimic GM1, GM2, GD3, GD1a, and GQ1b (Aspinall et al. 1994; Salloway et al. 1996; Moran and Prendergast 1998; Prendergast et al. 1998). Yuki et al. also reported that all the class of *C. jejuni* expressed both GM1-like and GD1a-like LOS, which could induce the production of autoantibodies not only to GM1 and GD1a but also to GM1a/GD1a complex, and might increase the risk of the development (Yuki 2005).

All strains from marine bacteria in this study were involved in gram-negative bacteria and homologically much conservative in the ST gene to those from *C. jejuni* (Yamamoto et al. 2006, 2008). Accordingly, we think that it is important to elucidate the precise LOS structures to know the evidence of the Sia linkages and resembles of linkages of LOS from bacteria used. The reason that SPG was preferentially and effectively synthesized in the current study will be more obvious from these detailed LOS core structures (Wakarchuk et al. 1998). In addition to the in vitro efficient lactoseries-ganglioside synthesis, we applied the synthesized SPGs to the biological assay of the influenza A virus binding. Synthesized S2–3PG and S2–6PG showed a nearly identical decrease in the TCID50 value as purified gangliosides in the liposome assay. As the binding specificity of influenza virus A/panama/2007/99 (H3N2) was also investigated using the inhibitory effect of standard SPGs on haemagglutination (data not shown), the results obtained here will reflect the binding specificity of A/panama/2007/99 (H3N2). However, the reason that only synthesized S2–3PG did not exhibit a remarkable reduction of the TCID50 value in spite of the same amount of gangliosides used was unclear. It may be influenced by the extracts including with synthesized S2–3PG since the preparations were performed by scraping the TLC plate. As glycopolymers containing multivalent sialyloligosaccharides with poly(L-glutamic acid) backbone were chemoenzymatically synthesized as inhibitors of infection by human influenza viruses (Totani et al. 2003), the STs assays by immobilized enzymes from marine bacteria will probably become more useful to the synthesis of sialoglycoconjugates with many biological functions.

During the past decade, the importance of GSLs in cell activities, such as cellular differentiation and oncogenic transformation, has been increasingly recognized (Hakomori 1984, 2002). Furthermore, exogenously added gangliosides have been found to evoke cell growth arrest (Bremer et al. 1986; Nojiri et al. 1986, 1991), cell differentiation (Nojiri et al. 1986), and neurite extension (Tsuji et al. 1983). Such gangliosides might interact with growth factor receptors (Bremer et al.1986; Nojiri et al. 1991) and certain protein kinases on the cell surface (Tsuji et al. 1985). To clarify the mechanisms, many types of sialoglycoconjugates are required. Easy and efficient sialylation methods will be required in the future.

Utilization of STs from marine bacteria may open the possibility of large-scale preparation of biologically significant gangliosides as we could prove the evidence that there was almost no difference between SPG purified from mammalian sources and in vitro synthesized SPG by these STs in the binding activity to influenza A virus.

In summary, we have described efficient in vitro ganglioside synthesis using two types of STs from marine bacteria. Our assays indicated that these enzymes recognize neolacto- and lacto-series GSLs in which the nonreducing end-galactoside can be modified at the 3 or 6 position with large substituents, such as sialosides, to form gangliosides. Since these recombinant enzymes are stable and have high affinity to the substrate GSLs, large quantities of STs could become available for practical in vitro ganglioside synthesis.

MATERIAL AND METHODS

Materials

Paragloboside (PG) and i antigen GSL were prepared from bovine erythrocytes in our laboratory. I antigen GSL was prepared by treatment of sialyl-I antigen GSL (S-I) with the neuraminidase *Clostoridium perfrigens*(Sigma). GM3, sialyl(α2-3)paragloboside (S2–3PG), and sialyl-i antigen GSL (S-i) were purified from human erythrocytes in our laboratory with the cooperation of the Japan Red Cross Center. Sialyl(α2-6)paragloboside (S2–6PG) was purified from meconium as previously described (Kushi et al.1995). S-I was a generous gift from Dr. T. Taki (Otsuka Pharmaceutical Company, Ltd, Tokushima, Japan). Samples of newborn meconium were collected in the delivery ward (Keiai Hospital, Obihiro, Hokkaido, Japan). Lc_4Cer was kindly provided by Dr. M. Iwamori (Faculty of Science and Technology, Kinki University). All structures of GSLs cited in this study were listed in Table II. The anti-S2–3PG monoclonal antibody (mAb) (SPS-20, IgM) was established previously (Watarai et al. 1995). A mAb specific to S2–6PG (MFH-1, IgM) was recently established in our laboratory (detailed analyses not shown). DU-PAN-2 (mouse mAb, IgM) (Hamanaka et al. 1994) and CMP-NeuAc were purchased from Kyowa medicals (Tokyo, Japan) and Sigma Ltd (St. Louis, MO), respectively. STs isolated from marine bacteria, including #1 (ish-224, 05JTC1), #2 (ish-467, 05JTD2), and #3 (faj-16, 05JTE1), which form α2-3 Sia linkages named α2-3STs, as well as #4 (ISH-224, N1C0), #5 (pda-rec, 05JTB2), and #6 (pda-0160, 05JTA2), which form α2-6 Sia linkages named α2-6STs, were obtained from the Glycotechnology Business Unit, Plant Innovation Center, Japan Tobacco Inc., and enzymatically characterized in detail (Kajihara et al.1996; Yamamoto et al. 1996). These are all recombinant enzymes and substrate specificities for saccharides were already confirmed using pyridylamidated derivatives (Yamamoto et al. 1998; Kajihara et al. 1999).

ST assay using GSLs as acceptor substrates

Assays for the enzyme activity were performed as follows: 3.33 mM CMP-NeuAc, 0.3 mM acceptor substrate (GSLs), 0.3% (w/v) Triton X-100, 10 mM $MnCl_2$, 333 mM sodium cacodyrate (pH 6.5), and 500 mM NaCl were used as a standard method. Each ST was added as 2 μL of enzyme solution (0.1 U) to the incubation mixture. One unit (U) was defined as the amount of enzyme that transferred 1 μmol of NeuAc per minute to lactose, as described previously (Yamamoto et al. 1996). GSLs were dissolved in chloroform/methanol (2:1, by vol.), dried under nitrogen stream, and then suspended in the rest of the reagents used for the assay described above. An aliquot of the total volume (30 μL) was incubated at 37°C for the required incubation times. The alternative ST assay was also performed, by changing temperatures and pHs to investigate the optimal conditions on these STs. The reaction was terminated by adding water (0.5 mL) and applied to Sep-pak C18 cartridge (Waters, Ltd, USA), which was equilibrated with methanol and distilled water. After washing the column with water to remove the nonreacted CMP-NeuAc and the salts, reaction mixtures with 2 mL of distilled water, 1 mL of methanol, and 1 mL of chloroform/methanol (2:1 by vol.) were added. The lipid fractions eluted with the chloroform/methanol mixture was evaporated and an aliquot was used for TLC and TLC/immunostaining (Hashimoto et al. 1984; Kasahara et al. 1994). Ganglioside production on TLC was calculated quantitatively by NIH image software (Becker et al. 1996). For a large-scale gangliosides synthesis, the volume of the ST assay incubation mixtures including the same concentration of substrate and buffers was scaled up to 500 μL in a tube. To the ST assay solution (500 μL), 6–10 μL enzyme solution (0.3–0.6 U) was added and incubated under the appropriate conditions. After ST assay was stopped, the mixtures were further purified, and purified gangliosides were used for GC/MS, NMR, and influenza virus binding assay.

Table II Structures of glycosphingolipids cited in this study

Name	Structure	Nomenclature	Remark
PG or neolacto-tetraosylceramide	Galβ1-4GlcNAcβ1-3Galβ1-4Glcβ1-1′Cer	nLc4Cer	paragloboside
S2-3PG	NeuAcα2-3Galβ1-4GlcNAcβ1-3Galβ1-4Glcβ1-1′Cer	IV3NeuAcα-nLc4Cer	sialy(α2-3)paragloboside, LM1
S2-6PG	NeuAcα2-6Galβ1-4GlcNAcβ1-3Galβ1-4Glcβ1-1′Cer	IV6NeuAcα-nLc4Cer	sialy(α2-6)paragloboside, 6′-LM1
Lactotetraosyl-ceramide	Galβ1-3GlcNAcβ1-3Galβ1-4Glcβ1-1′Cer	Lc4Cer	type 1
i antigen	Galβ1-4GlcNAcβ1-3Galβ1-4GlcNAcβ1-3Galβ1-4Glcβ1-1′Cer	nLc6Cer	
S-i	NeuAcα2-3Galβ1-4GlcNAcβ1-3Galβ1-4GlcNAcβ1-3Galβ1-4Glcβ1-1′Cer	VI3NeuAcα-nLc6Cer	
I antigen	Galβ1-4GlcNAcβ1-3Galβ1-4GlcNAcβ1-3Galβ1-4Glcβ1-1′Cer Galβ1-4GlcNAcβ1-6	IV6kladoLc8Cer	blood group I
S-I	NeuAcα2-3Galβ1-4GlcNAcβ1-3Galβ1-4GlcNAcβ1-3Galβ1-4Glcβ1-1′Cer NeuAcα2-3Galβ1-4GlcNAcβ1-6	VIII3, VI3NeuAcα2-VI6kladoLc8Cer	DG-6 I-type
GM3	NeuAcα2-3Galβ1-4Glcβ1-1′Cer	II3NeuAcα-LacCer	hematoside

Determination of the kinetic parameters K*m* and V_{max}

Lineweaver–Burk plots were obtained by changing the substrate concentrations, and *Km* and V_{max} values were obtained. The apparent kinetics of ST interactions with the acceptor substrate PG was determined using a saturating concentration of CMP-NeuAc. Data points represented the averages of at least duplicate values.

TLC and TLC/immunostaining

TLC and TLC/immunostaining of GSLs were as previously described with slight modifications (Kushi et al. 2001). A high-performance thin-layer chromatography (HPTLC) plate (silica gel 60 HPTLC, Merck, Darmstadt, Germany) was used throughout this study with the following solvent systems: I chloroform/methanol/water (60:35:8, by vol.) and II chloroform/methanol/water (55:45:10, by vol.) containing 0.02% $CaCl_2$. GSLs on TLC plates were visualized by spraying the plate with orcinol-H_2SO_4 (Francois et al. 1962) or resorcinol-HCl (Svennerholm 1957). The developed and dried TLC plates were soaked for 1 min in a 0.02% solution of polyisobutylmethacrylate (Tokyo Kasei Kogyo, Japan) dissolved in hexane and allowed to air dry, and then blocked by incubation in 1% bovine serum albumin/phosphate-buffered saline (PBS) at 37°C for 30 min. They were then rinsed five times with PBS/0.05% Tween 20 (Wako Pure Chemical Industries Ltd, Osaka, Japan) (the washing buffer) and incubated with mAbs, (1:50, in PBS) overnight at 4°C, followed by 90 min horseradish peroxidase-conjugated goat anti-mouse IgM antiserum (Cappel Laboratories, West Chester, PA) at 37°C for 2 h. As a final step, they were washed five times with the washing buffer and incubated with peroxidase substrate solution (Konica immunostain kit, Tokyo).

Mass spectrometry and *1*H-NMR spectroscopy of gangliosides synthesized by STs

Sialylated products were analyzed by in situ PVDF membrane transfer followed by SIMS, as described (Taki et al. 1995; Ishikawa and Taki 2000). GSLs developed by TLC were transferred to a polyvinylidene difluoride (PVDF) membrane (ATTO, Tokyo, Japan) by TLC blotting, after which GSL bands on the membrane were excised and placed on a mass spectrometer probe tip, and a few microliters of triethanolamine was added as the matrix. The sample was analyzed by SIMS. As for the synthesized gangliosides using i antigen GSL as a substrate, matrix-assisted laser desorption ionization/time-of-flight mass spectrometry (MALDI/TOF-MS) analysis was performed using a autoflex TOF/TOF (Bruker Daltonics, Billerica, MA). The samples for this analysis were prepared by scraping from the TLC plate and extracting the synthesized gangliosides with the solvent mixture in chloroform/methanol (2:1, by vol.) after TLC and GSL samples (about 200 ng/μL) dissolved in chloroform/methanol (2:1, by vol.) were mixed with matrix solution (10 mg 2,5-dihydroxybenzoic acid in 1 mL water) (1:1, v/v) and the suspensions were loaded on a sample plate. Positive mass spectra for substrate i antigen GSL and negative mass spectra for the synthesized gangliosides were measured in reflector mode with 100 nsec delayed extraction.

^{1}H-NMR spectra of the isolated newly synthesized gangliosides were obtained in 0.4 mL of dimethylsulfoxide (DMSO)-d_6 containing 2% deuterium oxide (D_2O) with a JEOL JNM-ECA500, 500 MHz ^{1}H-NMR spectrometer at 25°C. Tetramethylsilane was used as an internal standard for chemical shifts. For the analysis of amido proton, DMSO-d_6 was used at 25°C.

Permethylation analysis and neuraminidase treatment

After separating the products from the reaction mixtures containing nonreacted substrates with the small DEAE-Sephadex (acetate form) column, the synthesized gangliosides were confirmed by permethylation analysis. Partially methylated alditol acetates of the synthesized gangliosides were prepared; detailed conditions of GC/MS analysis have been previously described (Ciucanu and Kerek 1984; Månsson et al. 1986; Nohara-Uchida and Ohashi 1987). Neuraminidase digestion of the newly synthesized gangliosides was performed in the acetate buffer (50 mM, pH5.5) containing neuraminidase of *Arthrobacter ureafaciens* (Nacalai Tesque, Kyoto, Japan) at 37°C for 3 h. The reaction mixture was then extracted with 2 mL of chloroform/methanol (2:1 by vol). The extracts also underwent permethylation analysis after confirming the digestion of synthesized ganglioside by TLC.

Binding of influenza A virus to gangliosides

Influenza virus A/panama/2007/99 (H3N2) was used for this assay. The titer of infectious virus was determined by limiting dilution in cultures of Madine-Darby canine kidney (MDCK) cells and was expressed as 50% of the tissue-culture infectious dose (TCID50) (Totani et al. 2003). Five hundred microliters of 1000 TCID50 virus was mixed 1:1 (v/v) with ganglioside-containing liposome suspensions and incubated at 37°C for 1 h. The liposomes were adjusted to the concentration of the standard and in vitro synthesized gangliosides (0, 50, 100, 150, and 250 μg) (Watarai et al. 2001).

After incubation, the mixture was centrifuged at 16,000 × *g* for 20 min (4°C) to remove virus-binding liposomes. After centrifugation, the supernatant (viruses not bound to liposomes) was diluted and transferred onto MDCK cell monolayers in 24-well plates at 37°C for 1 h. After washing with Dulbecco's Modified Eagle's medium (DMEM), 1 mL of DMEM was added to each well and then incubated at 37°C for 3 days. Virus-induced cytopathic effects were then monitored by light microscopy. The virus titer was obtained as the reciprocal of the highest dilution giving TCID50.

FUNDING

The Ministry of Education, Science, Culture and Sports of Japan (17405041).

CONFLICT OF INTEREST STATEMENT

None declared.

ACKNOWLEDGMENTS

The authors are greatly indebted to Dr. Masao Iwamori (Faculty of Science and Technology, Kinki University) for the gift of Lc_4Cer and to Dr. K. Maruyama (Keiai Hospital Obihiro, Hokkaido) for the meconium supply. Furthermore, we thank Dr. Akemi Suzuki (Institute of Glycotechnology, Tokai University) for the critical reading and valuable discussion of this paper.

REFERENCES

1. Angata T, Varki A. Chemical diversity in the sialic acids and related α-keto acids: An evolutionary perspective. *Chem Rev.* 2002;102:439-469.

2. Ariga T, Yu RK. Antiglycolipid antibodies in Guillain–Barré syndrome and related diseases: Review of clinical features and antibody specificities. *J Neurosci Res.*2005;80:1-17.

3. Aspinall GO, McDonald AG, Pang H, Kurjanczyk LA,Penner JL.Lipopo lysaccharides of *Campylobacter jejuni* serotype O:19: Structures of

core oligosaccharide regions from the serostrain and two bacterial isolates from patients with the Guillain–Barré syndrome. *Biochemistry* 1994;33:241-249.

4. Bardor M, Nguyen DH, Diaz S,Varki A. Mechanism of uptake and incorporation of the non-human sialic acid *N*-glycolylneuraminic acid into human cells. *J Biol Chem.*2005;280:4228-4237.

5. Becker A, Reith A, Napiwotzki J, Kadenbach B. A quantitative method of determining initial amounts of DNA by polymerase chain reactions cycle titration using digital imaging and a novel DNA stain. *Anal Biochem.* 1996;237:204-207.

6. Bremer Schlessinger EG J, Hakomori S. Ganglioside-mediated modulation of cell growth. Specific effects of GM3 on tyrosine phosphorylation of the epidermal growth factor receptor. *J Biol Chem.* 1986;261:2434-2440.

7. Chiu CPC, Watts AG, Lairson LL, Gilbert M, Lim D, Wakarchuk WW, Withers SG,Strynadka NCJ. Structural analysis of the sialyltransferase CstII from *Campylobacter jejuni* in complex with a substrate analog. *Nat Struct Mol Biol.* 2004;11:163-170.

8. Ciucanu I, Kerek F. A simple and rapid method for the permethylation of carbohydrates. *Carbohydr Res.* 1984;131:209-217.

9. Connor RJ, Kawaoka Y, Webster RG, Paulson JC. Receptor specificity in human, avian, and equine H2 and H3 influenza virus isolate. *Virology.* 1994;205:17-23.

10. Fox KL, Cox AD, Gilbert M, Wakarchuk WW, Li J, Makepeace K, Rich ards JC, MoxonER, Hood DW. Identification of a bifunctional lipopolysaccharide sialyltransferase in*Haemophilus influenzae. J Biol Chem.* 2006;281:40024-40032.

11. Francois C, Marshall RD, Neuberger A. Carbohydrates in protein: 4. The determination of mannose in hen's-egg albumin by radioisotope dilution. *Biochem J.*1962;83:335-341.

12. Gagneux P, Varki A. Evolutionary considerations in relating oligosaccharide diversity to biological function. *Glycobiology.* 1999;9:747-755.

13. Gilbert M, Brisson J-R, Karwaski M-F,Michniewicz J, Cunningham A-M, Wu Y,Young NM, Wakarchuk WW. Biosynthesis of ganglioside mimics in *Campylobacter jejuni* OH4384. *J Biol Chem.* 2000;275:3896-3906.

14. Gilbert M, Karwaski M-F, Bernatchez S, Young NM, Taboada E, Michniewicz J,Cunningham A-M,Wakarchuk WW. The genetic bases for the variation in the lipo-oligosaccharide of the mucosal pathogen, *Campylobacter jejuni. J Biol Chem.*2002;277:327-337.

15. Godschalk PCR, Kuijf ML, Li J, Michael F St, Ang CW, Jacobs BC, Karwaski M-F,Brochu D, Moterassed A, Endtz HP,et al. Structural characterization of*Campylobacter jejuni* lipooligosaccharide outer cores associated with Guillain–Barré and Miller Fisher syndromes. *Infect Immun.* 2007;75:1245-1254.

16. Guerry P, Szymanski CM, Prendergast MM, Hickey TE, Ewing CP,Pattarini DL,Moran AP. Phase variation of *Campylobacter jejuni* 81–176 lipooligosaccharide affects ganglioside mimicry and invasiveness in vitro. *Infect Immun.* 2002;70:787-793.

17. Hakomori S-I. Tumor-associated carbohydrate antigens. *Annu Rev Immunol.*1984;2:103-126.

18. Hakomori S-I. The glycosynapse. *Proc Natl Acad Sci USA.* 2002;99:225-232.

19. Hakomori S, Kannagi R. Glycosphingolipids as tumor-associated and differentiation markers. *J Natl Cancer Inst.* 1983;71:231-251.

20. Hamanaka Y, Hamanaka S, Shinagawa Y, Suzuki T, Inagaki F, Suzuki M, Suzuki A.Ganglioside antigen of DU-PAN-2 in a human pancreatic cancer. *FEBS Lett*1994;353:48-52.

21. Hashimoto Y, Abe M, Kiuchi Y, Suzuki A, Yamakawa T. Genetically regulated expression of UDP-*N*-acetylgalactosamine:GM3 (NeuGc) *N*-acetylgalactosaminyltransferase [EC 2.4.1.92] activity in mouse liver. *J Biochem.*1984;95:1543-1549.

22. Ishikawa D, Taki T. Thin-layer chromatography blotting using polyvinylidene difluoride membrane (Far-Eastern Blotting) and its applications. *Methods Enzymol.*2000;312:145-157.

23. IUPAC-IUB Commission Biochemical Nomeclature (CBN). The nomenclature of lipids. *Eur J Biochem.* 1976;79:11-21.

24. Kajihara Y, Akai S, Nakagawa T, Sato R, Ebata T, Kodama H, Sato K-I. Enzymatic synthesis of Kdn oligosaccarides by a bacterial α-(2→6)-sialyltransferase. *Carbohydr Res.* 1999;315:137-141.

25. Kajihara Y, Yamamoto T, Nagae H, Nakashizuka M, Sakakibara T, Terada I. A novel α2,6-sialyltransferase: Transfer of sialic acid to fucosyl and sialyl trisaccharides. *J Org Chem.* 1996;61:8632-8635.

26. Kasahara K, Guo L, Nagai Y, Sanai Y. Enzymatic assay of glycosphingolipid sialyltransferase using reverse-phase thin-layer chromatography. *Anal Biochem.*1994;218:224-226.

27. Kushi Y, Shimizu M, Watanabe K,Kasama T, Watarai S, Ariga T, Han da S.Characterization of blood group ABO (H)-active gangliosides in type AB erythrocytes and structural analysis of type A-active ganglioside variants in type A human erythrocytes. *Biochim Biophys Acta.* 2001;1525:58-69.

28. Kushi Y, Tsunoda A, Komatsuzaki K, Watanabe K, Kasama T, Handa S.Characterization of blood-group-ABO(H)-active glycosphingolipids in type-AB human erythrocytes. *Eur J Biochem.* 1995;231:862-867.

29. Månsson J-E, Mo HQ, Egge H, Svennerholm L. Trisialosyllactosylceramide (GT3) is a ganglioside of human lung. *FEBS Lett* 1986;196:259-262.

30. Moran AP, Prendergast MM. Molecular mimicry in *Campylobacter jejuni*lipopolysaccharides and the development of Guillain–Barré syndrome. *J Infect Dis.*1998;178:1549-1551.

31. Nohara-Uchida K, Ohashi M. Isolation and characterization of major gangliosides from frog liver. *J Biochem.* 1987;102:923-932.

32. Nojiri H, Stroud M, Hakomori S. A specific type of ganglioside as a modulator of insulin-dependent cell growth and insulin receptor tyrosine kinase activity. *J Biol Chem.* 1991;266:4531-4537.

33. Nojiri H, Takaku F, Terui Y, Miura Y, Saito M. Gangliosides GM3: An acidic membrane component that increases during macrophage-like cell differentiation can induce monocytic differentiation of human myeloid and monocytoid leukemic cell lines HL-60 and U937. *Proc Natl Acad Sci USA.* 1986;83:782-786.

34. Paulson JC. Glycopoteins: What are the sugar chains for? *Trends Biochem Sci.*1989;14:272-276.

35. Prendergast MM, Lastovica AJ, Moran AP. Lipopolysaccharides from*Campylobacter jejuni* O:41 strains associated with Guillain–Barré syndrome exhibit mimicry of GM1 ganglioside. *Infect Immun* 1998;66:3649-3655.

36. Salloway S, Mermel LA, Seamans M, Aspinall GO, Nam Shin JE,Kurjanczyk LA,Penner JL. Miller–Fisher syndrome associated with *Campylobacter jejuni* bearing lipopolysaccharide molecules that mimic human ganglioside GD3. *Infect Immun.*1996;64:2945-2949.

37. Schauer R. Advances Carbohydrate Chemistry and Biochemistry. *Vol.* 40. New York: Academic Press; 1982. Chemistry, metabolism, and biological function of sialic acids; *p.* 131-234.

38. Schauer R. Victor Ginsburg's influence on my research of the role of sialic acids in biological recognition. *Arch Biochem Biophys.* 2004;426:132-141.

39. Skipski VP. Thin-layer chromatography of neutral glycosphingolipids. *Methods Enzymol.* 1975;35:396-425.

40. Suzuki Y, Ito T, Suzuki T, Holland RE Jr, Chambers TM,Kiso M, Ishid a H, KawaokaY. Sialic acid species as a determinant of the host range of influenza A viruses. *J Virol.* 2000;74:11825-11831.

41. Suzuki Y, Nakao T, Ito T, Watanabe N, Toda Y, Xu G, Suzuki T, Kobay ashi T,Kimura Y, Yamada A, et al. Structural determination of gangliosides that bind to influenza A, B, and C viruses by an improved binding assay: Strain-specific receptor epitopes in sialo-sugar chains. *Virology.* 1992;189:121-131.

42. Svennerholm L. Quantitative estimation of sialic acids: II. A colorimetric resorcinol-hydrochloric acid method. *Biochim Biophys Acta.* 1957;24:604-611.

43. Svennerholm L. Chromatographic separation of human brain gangliosides. *J Neurochem.* 1963;10:613-623.

44. Svennerholm L, Mandel P, Dreyfus H, Urban PF. Structure and Functions of Gangliosides. New York: Plenum; 1980.

45. Takakura Y, Tsukamoto H, Yamamoto T. Molecular cloning, expression and properties of an α/β-galactoside α2,3-sialyltransferase from *Vibrio* sp. JT-FAJ-16. *J Biochem.* 2007;142:403-412.

46. Taki T, Ishikawa D, Handa S, Kasama T. Direct mass spectrometric analysis of glycosphingolipid transferred to a polyvinylidene difluoride membrane by thin-layer chromatography blotting. *Anal Biochem.* 1995;225:24-27.

47. Tangvoranuntakul P, Gagneux P, Diaz S, Bardor M, Varki N, Varki A, Muchmore E.Human uptake and incorporation of an immunogenic nonhuman dietary sialic acid.*Proc Natl Acad Sci USA.* 2003;100:12045-12050.

48. Taniguchi N, Honke K, Fukuda M. Handbook of Glycosyltransferases and Related Genes. Tokyo: Springer-Verlag; 2002.

49. Totani K, Kubota T, Kuroda T, Murata T, Hidari KI-P, Suzuki T,Suzuki Y,Kobayashi K, Ashida H, Yamamoto K, et al

50. . Chemoenzymatic synthesis and application of glycopolymers containing multivalent sialyloligosaccharides with a poly (L-glutamic acid) backbone for inhibition of infection by influenza viruses.*Glycobiology.* 2003;13:315-326.

51. Tsuji S, Arita M, Nagai Y. GQ1b, a bioactive ganglioside that exhibits novel nerve growth factors (NGF)like activities in the two neuroblastoma cell lines. *J Biochem.*1983;94:303-306.

52. Tsuji S, Nakajima J, Sasaki T, Nagai Y. Bioactive gangliosides: IV. Ganglioside GQ1b/Ga^{2+} dependent protein kinase activity exists in the plasma membrane fraction of neuroblastoma cell line, GOTO. *J Biochem.* 1985;97:969-972.

53. Tsukamoto H,Katakura Y, Yamamoto T. Purification, cloning and expression of an α/β-galactoside α-2,3-sialyltransferase from a luminous marine bacteria,*Photobacterium phosphoreum. J Biol Chem* 2007;282:29794-29802.

54. Varki A. Biological roles of oligosaccharides: All of the theories are correct.*Glycobiology.* 1993;3:97-130.

55. Varki A. *N*-Glycolylneuraminic acid deficiency in humans. *Biochimie.*2001;83:615-622.

56. Varki A. Glycan-based interactions involving vertebrate sialic-acid-recognizing proteins. *Nature.* 2007;446:1023-1029.

57. Wakarchuk WW, Gilbert M, Martin A, Wu Y, Brisson J-R, Thibault P, Richard JC.Structures of an α-2,6-sialylated lipooligosaccharide from *Neisseria meningitidis*immunotype L1. *Eur J Biochem.* 1998;254:626-633.

58. Watarai S, Kushi Y, Shigeto R,Misawa N, Eishi Y, Handa S, Yasuda T. Production of monoclonal antibodies directed to Hanganutziu-Deicher active gangliosides, *N*-glycolylneuraminic acid containing gangliosides. *J Biochem.* 1995;117:1062-1069.

59. Watarai S, Tana, Inoue K, Kushi Y, Isogai E, Yokota K, Naka K, Oguma K, KodamaH. Inhibition of Vero cell cytotoxic activity in *Escherichia coli* O157:H7 lysates by globotriaosylceramide, Gb3, from bovine milk. *Biosci Biotechnol Biochem.*2001;65:414-419.

60. Yamamoto T, Hamada Y, Ichikawa M, Kajiwara H, Mine T, Tsukamoto H, TakakuraY. A β-galactoside α2,6-sialyltransferase produced by a marine bacterium,*Photobacterium leiognathi* JT-SHIZ-145, is active at pH 8. *Glycobiology*2007;17:1167-1174.

61. Yamamoto T, Ichikawa M, Kakakura Y. Conserved amino acid sequences in the bacterial sialyltransferases belonging to glycosyltransferase family 80. *Biochem Biophys Res Commun.* 2008;365:340-343.

62. Yamamoto T, Nakashizuka M, Kodama H, Kajihara Y, Terada I. Purif ication and characterization of a marine bacterial β-galactoside α2,6-sialyltransferases from*Photobacterium damsela* JT0160. *J Biochem.* 1996;120:104-110.

63. Yamamoto T, Nakashizuka M, Terada I. Cloning and expression of a marine bacterial β-galactoside α2,6-sialyltransferase gene from *Photobacterium damsela*JT0160. *J Biochem.* 1998;123:94-100.

64. Yamamoto T, Takakura Y, Tsukamoto H. Bacterial sialyltransferases. *Trends Glycosci Glycotechnol.* 2006;18:253-265.

65. Yogeewaran G. Cell surface glycolipids and glycoproteins in malignant transformation. *Adv Cancer Res.* 1983;38:289-350.

66. Yu RK, Usuki S, Ariga T. Ganglioside molecular mimicry and its pathological roles in Guillain–Barré syndrome and related diseases. *Infect Immun.* 2006;74:6517-6527.

67. Yuki N. Carbohydrate mimicry: A new paradigm of autoimmune diseases. *Curr Opin Immunol.* 2005;17:577-582.

68. Yuki N, Handa S, Taki T, Kasama T, Takahashi M, Saito K, Miyatake T. Cross-reactive antigen between nervous tissue and a bacterium elicits Guillain–Barré syndrome:molecular mimicry between ganglioside GM1 and lipopolysaccharide from Penner's serotype 19 of *Campylobacter jejuni. Biomed Res.* 1992;13:451-453.

5

CHAPTER

The Structure-Activity Relationship between Marine Algae Polysaccharides and Anti-Complement Activity

Weihua Jin [1,3], Wenjing Zhang [1,2], Hongze Liang [4] and Quanbin Zhang [1,3,*]

[1]*Key Laboratory of Experimental Marine Biology, Institute of Oceanology, Chinese Academy of Sciences, Qingdao 266071, China*

[2]*College of Earth Science, University of Chinese Academy of Sciences, Beijing 100049, China*

[3]*Laboratory for Marine Biology and Biotechnology, Qingdao National Laboratory for Marine Science and Technology, Qingdao 266071, China*

[4]*The School of Materials Sciences and Chemical Engineering, Ningbo University, Ningbo 315211, China*

ABSTRACT

In this study, 33 different polysaccharides were prepared to investigate the structure-activity relationships between the polysaccharides, mainly from marine algae, and anti-complement activity in the classical pathway. Factors considered included extraction methods, fractionations, molecular weight, molar ratio of galactose to fucose, sulfate, uronic acid (UA) content, linkage, branching, and the type of monosaccharide. It was shown that the larger the molecular weights, the better the activities. The molar ratio of galactose (Gal) to fucose (Fuc) was a positive factor at a concentration lower than 10 μg/mL, while it had no effect at a concentration more than 10 μg/mL. In addition, sulfate was necessary; however, the sulfate content, the sulfate pattern, linkage and branching had no effect at a concentration of more than 10 μg/mL. Moreover, the type of monosaccharide had no effect. Laminaran and UA fractions had no activity; however, they could reduce the activity by decreasing the effective concentration of the active composition when they were mixed with the active compositions. The effect of the extraction methods could not be determined. Finally, it was observed that sulfated galactofucan showed good anti-complement activity after separation.

Keywords: marine algae polysaccharides; anti-complement activity; structure-activity relationship

1. INTRODUCTION

The complement system is an important part of the innate immune system that is designed to eliminate "harmful" substances from the body. This elimination is accomplished in five different ways according to a previous study [1]. Inappropriate activation of the complementary target self-tissues causes pathology in a large number of inflammatory, ischaemic, and other diseases [2].

There is a large body of research on naturally-occurring complement inhibitors isolated from animal, plant, and microbial products, such as polysaccharides [3,4,5,6], phenolic acid [7], proteins [8], flavonoids, and steroides [9,10]. Many synthetic molecules, including dextran sulfate [11,12], nafamastat mesilate (FUT-175) [2] and compstatin [1] have been shown to inhibit activation of the complement system. In addition, many studies have also reported [13,14,15] that a sulfated polysaccharide from brown algae, has been found to inhibit complement activation. Structure and activity studies showed that sulfated polysaccharides are potential candidates for screening as inhibitors of the complement system. Marine algae contain numerous different polysaccharides, which are non-toxic. In addition, they also contain polyanions and are majorly substituted by sulfate, which differs from the polysaccharides from land-plants.

Currently, marine algae mainly contain brown algae, red algae, and green algae. Polysaccharides from brown algae [16,17,18,19,20,21] contain three major types of polysaccharides, namely, fucoidan, laminaran, and alginate. Fucoidan includes sulfated heteropolysaccharides and sulfated galactofucan or fucan. The former contains sulfated glucuronomannan, glucuronan, galactan, etc. Sulfated glucuronomannan has a backbone of alternating 2-linked mannopyranose residue and 4-linked glucuronic acid, sulfated at C6 of the mannnopyranose residue. The linkage of glucuronan is 3-linked. Galactan has a backbone of 6-linked galactopyranose residue branched at C4 with galactose, while the latter one is made up of 3-linked fucopyranose residues or alternating 3-linked and 4-linked fucopyranose residues. The major differences between sulfated galactofucan or fucan are the positions of the sulfate and branching units, including the fucose residues, galactose residues, glucuronic acid residues, and so on. Laminaran is a glucan, consisting mainly of 3-linked glucopyranose residue, branched with 6-linked glucopyranose. Alginate contains polymannuronic acid, polyguluronic acid and a mixture of polymannuronic acid and polyguluronic acid. Polysaccharides from red algae [18,21,22,23,24] contain three types of polysaccharides, namely, carrageenan polysaccharides, agar polysaccharides, and an agar-carrageenan intermediate polysaccharide. They have a backbone of alternating 3-linked β-D-galactopyranose and 4-linked α-galactopyranose residues; however, carrageenan polysaccharides have 4-linked α-D-galactopyranose residues while agar polysaccharides have 4-linked α-L-galactopyranose residues. In addition, all the α-galactopyranose residues might exist in the form of 3,6-substituted by ester sulfate, methyl groups, pyruvic acid acetal, etc. Water-soluble polysaccharides from green algae [18,23,25] can be divided into two types of polysaccharides. Type 1 is classified as xylogalactoarabian, which mainly consists of xylose, galactose, and arabinose. It contains a backbone of 4-linked or 5-linked arabinopyranose residues sulfated at C3, 3-linked or 6-linked galactopyranose residues sulfated at C4 or C6, and 4-linked xylopyranose residues. Type 2 is named as glucuronoxylorhamnan, which mainly contains glucuronic acid, xylose, and rhamnose, according to the type of major monosaccharides. It consists of 4-linked rhamnopyranose sulfated at C3 or C2, a 4-linked glucuronic acid residue and a 4-linked xylopyranose residue.

To clarify the structure-activity relationship between marine algae polysaccharides and their anti-complement activity, several types of polysaccharides from brown algae, red algae, and green algae were prepared. In addition, many factors including the extraction methods, fractionations, molecular weight, the molar ratio of galactose to fucose, sulfate, UA content, linkage, branching, and the type of monosaccharides were taken into account.

2. RESULTS AND DISCUSSION

2.1. The Effect of the Extraction Methods on the Anti-Complement Activity

Table 1 displays the chemical composition of two fucoidans, which were derived from Saccharina japonica by hot water (SJW) and diluted hydrochloric acid (SJS). It was shown that the contents of fucose (Fuc) and sulfate contained in SJS were higher than in SJW, while the content of UA contained in SJS was lower. In addition, the molecular weight of SJS was also lower because of degradation during the process of extraction by dilute hydrochloric acid. Moreover, it also indicated that SJW contained a greater number of other monosaccharides, suggesting that the structure of SJW was more complex than that of SJS. In terms of the chemical compositions of HFW (HFW was obtained from brown algae Hizikia fusiforme by hot water) and the HFS (HFS was obtained from brown algae Hizikia fusiforme by dilute hydrochloric acid), as shown in Table 1, Fuc, UA and the molecular weight of the HFW were higher than in HFS, while the sulfate content of HFW was lower. The molar ratios of monosaccharides indicated that both HFS and HFW contained substantial amounts of laminaran, which was also reported in a previous study [26]. To summarize, different extraction methods might influence the chemical compositions of the polysaccharides.

Table 1. Chemical composition (%, dry weight) of the polysaccharides studied.

Sample	Yields (%)	Fuc (%)	UA (%)	SO4 (%)	Monosaccharides (Molar Ratio)							Mw (kDa)
					Man	Rha	GlcA	Glc	Gal	Xyl	Fuc	
SJW	1.21	22.38	10.32	30.60	0.10	0.03	0.15	0	0.63	0.07	1	152.4
SJW-1	10.67	15.84	33.36	15.05	0.55	0.10	0.96	0	1.22	0.37	1	77.8
SJW-2	19.99	17.01	17.35	28.33	0.12	0.03	0.18	0	1.08	0.10	1	88.4
SJW-3	37.89	28.96	0.45	35.10	0.02	0	0.02	0	0.27	0	1	162.7
SJS	1.08	34.97	3.44	36.88	0.06	0.03	0.06	0.01	0.11	0.02	1	106.3
SJS-1	8.98	21.56	13.71	19.75	0.06	0.02	0.06	0.02	0.17	0.02	1	63.8
SJS-2	15.89	31.07	8.21	34.84	0.08	0.03	0.07	0.01	0.28	0.03	1	84.5
SJS-3	40.78	36.94	0	38.43	0.01	0	0.01	0	0.05	0	1	125.4
HFW	2.43	28.04	5.43	28.46	0.11	0	0.08	0.62	0.27	0.09	1	118.3/3.9
HFW-1	18.51	-	-	-	-	-	-	1	-	-	-	3.9
HFW-2	18.16	22.78	17.64	17.33	0.48	0.30	0.27	0.29	0.23	0.42	1	116.7/5.1
HFW-3	28.01	32.85	0	31.62	0.05	0.01	0.04	0.01	0.29	0.04	1	114.2
HFS	1.53	21.89	1.98	30.59	0.07	0.01	0.09	1.22	0.25	0.02	1	97.3/3.6
HFS-1	16.24	-	-	-	-	-	-	1	-	-	-	3.4
HFS-2	10.66	25.79	20.84	25.66	0.27	0.02	0.20	0.03	0.22	0.04	1	42.0/4.3
HFS-3	9.72	40.84	0.98	39.85	0	0	0	0	0.27	0.01	1	99.2
SJS-OS	98.19	33.31	3.31	46.36	0.06	0.03	0.05	0.05	0.20	0.05	1	119.5
SJS-DS	65.41	41.20	6.11	18.12	0.05	0.03	0.05	0.03	0.18	0.05	1	117.5/33.4
SJS-DS-OS	110.1	24.24	3.79	35.30	0.05	0.02	0.04	0.03	0.20	0.05	1	123.6
ANW		25.08	4.16	27.36	0.05	0.03	0.09	0.29	0.09	0.06	1	122.1
SC	5.78	21.80	6.54	28.15	0.15	0.32	0.13	0.22	0.28	0.08	1	111.9
SJW-2-R	38.99	18.56	9.86	29.63	0.10	0.03	0.06	0.10	1.05	0.07	1	79.6
SJW-2-HJ	10.89	14.35	29.36	14.63	0.46	0.10	0.50	0	2.10	0.20	1	50.4
SJW-2-HW	83.10	25.23	8.36	35.69	0.09	0.01	0.10	0	0.56	0.06	1	95.6

HFW-1-S	109.1	-	-	41.38	-	-	-	1	-	-	-	13.7
ES	23.19	-	-	26.33	-	-	-	-	1	-	-	147.9
GF	13.30	-	-	20.42	-	-	-	-	1	-	-	137.2
GL	30.28	-	-	2.74	-	-	-	-	1	-	-	146.0
PY	10.71	-	-	2.72	-	-	-	-	1	-	-	140.7
EP	19.53	-	26.77	18.09	-	1	0.37	0.13	0.06	0.31	-	189.3
UP	18.75		28.39	20.88	-	1	0.45	0.15	0.04	0.33	-	163.4
CR	8.13	-	-	32.84	-	0.03	-	-	0.14	1	-	174.9
CF	7.89	-	-	23.54	-	0.05	-	0.40	1	0.50	-	173.4

Polysaccharides derived from *Saccharina japonica* by hot water (SJW) and dilute hydrochloric acid (SJS). SJW was fractionated by anion exchange chromatography into three components, namely SJW-1, SJW-2 and SJW-3. SJS was also fractionated by anion exchange chromatography into three components, namely SJS-1, SJS-2 and SJS-3. Polysaccharides derived from *Hizikia fusiforme* by hot water (HFW) and dilute hydrochloric acid (HFS). HFW was fractionated by anion exchange chromatography into three components, namely HFW-1, HFW-2 and HFW-3. HFS was also fractionated by anion exchange chromatography into three components, namely HFS-1, HFS-2 and HFS-3. Overly sulfated SJS (SJS-OS) and laminaran (HFW-1-S), desulfated SJS (SJS-DS) and desulfated SJS with sulfation (SJS-DS-OS). Polysaccharides derived from *Ascophyllum nodosum* (ANW), *Acaudina molpadioides* (SC), *Eucheuma spinosum* (ES), *Grateloupia filicina* (GF), *Gracilaria lamaneiformis* (GL), *Porphyra yezoensis* (PY), *Enteromorpha prolifera* (EP), *Ulva pertusa* (UP), *Cladophera rupestris* (CR) and *Codium fragile* (CF). SJW-2-R was obtained through UA reduction, and SJW-2-HJ and SJW-2-HW were fractionated using an activated carbon column.

Figure 1a displays the anti-complement activities of SJS, SJW, HFS and HFW in the classical pathway. Activities of SJS and HFW reached a plateau at a concentration of 10 μg/mL, while SJW and HFS reached a plateau at a concentration of 50 μg/mL. Specifically, the IC_{50} of SJS, SJW, HFS, and HFW were 4.51, 7.26, 24.65, and 5.51 μg/mL, respectively. Thus, it was difficult to determine the effects of the extraction methods because the IC_{50} of SJS was lower than that of SJW, while the $IC_{50}$of HFW was lower than that of HFS.

2.2. The Effect of the Fractionations on the Anti-Complement Activity

SJW was fractionated by anion exchange chromatography into three components, namely, SJW-1, SJW-2, and SJW-3. Chemical analysis indicated that SJW-1 had the highest UA content and SJW-3 had the highest fucose content and sulfate content. In addition, SJW-3 had the highest molecular weight. Previous studies [27,28,29] reported that SJW-3 was mainly sulfated galactofucan, and SJW-1 and SJW-2 were sulfated heteropolysaccharides. Figure 1b shows the anti-complement activities of SJW and its fractions. Compared to SJW and its fractions, it was found that the activity of SJW-3 was similar to the SHW activity, while SJW-2 had lower activity and SJW-1 had no activity. Specifically, the IC_{50} of SJW-3 (3.11 μg/mL) was lower than that of SJW (7.26 μg/mL). The chemical composition of SJS's fractions (SJS was fractionated by anion exchange

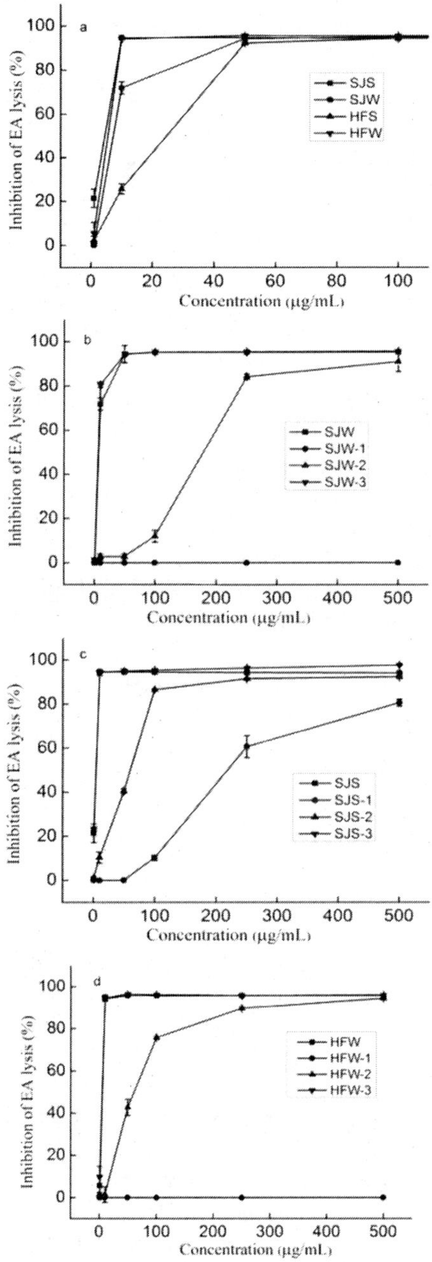
a
Inhibition of EA lysis (%)
Concentration (μg/mL)
SJS
SJW
HFS
HFW
b
Inhibition of EA lysis (%)
Concentration (μg/mL)
SJW
SJW-1
SJW-2
SJW-3
c
Inhibition of EA lysis (%)
Concentration (μg/mL)
SJS
SJS-1
SJS-2
SJS-3
d
Inhibition of EA lysis (%)
Concentration (μg/mL)
HFW
HFW-1
HFW-2
HFW-3

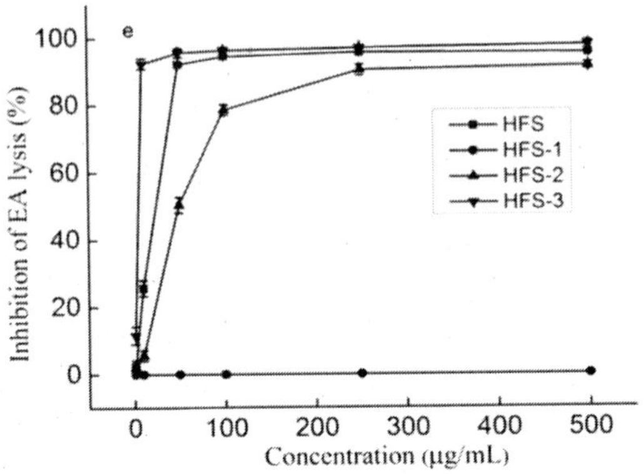

Figure 1. Inhibition of the classical pathway-mediated haemolysis of EA in 1:10-diluted NHS in the presence of increasing amounts of the polysaccharides (**a**) SJS, SJW, HFS, and HFW; (**b**) SJW and its fractions; (**c**) SJS and its fractions; (**d**) HFW and its fractions; (**e**) HFS and its fractions. The results are expressed as percent inhibition of haemolysis. Data are the means from three determinations ± S.E.M. Polysaccharides derived fromSaccharina japonica by hot water (SJW) and dilute hydrochloric acid (SJS). SJW was fractionated by anion exchange chromatography into three components, namely SJW-1, SJW-2 and SJW-3. SJS was also fractionated by anion exchange chromatography into three components, namely SJS-1, SJS-2 and SJS-3. Polysaccharides derived from Hizikia fusiforme by hot water (HFW) and dilute hydrochloric acid (HFS). HFW was fractionated by anion exchange chromatography into three components, namely HFW-1, HFW-2 and HFW-3. HFS was also fractionated by anion exchange chromatography into three components, namely HFS-1, HFS-2 and HFS-3.

chromatography into three components, namely, SJS-1, SJS-2, and SJS-3) were similar to the results of SJW's fractions, suggesting that SJS-3 was a sulfated galactofucan, while SJS-1 and SJS-2 were sulfated heteropolysaccharides. The activities of SJS and its fractions in Figure 1c confirmed the above results. HFW was also fractionated by anion exchange chromatography into three components, namely, HFW-1, HFW-2, and HFW-3. HFW-1 was composed of glucose, HFW-2 had plenty of the other monosaccharides and HFW-3 contained galactose and fucose. According to the chemical compositions inTable 1, it was proposed that HFW-1 was mainly laminaran, HFW-2 was mainly sulfated heteropolysaccharides, and HFW-3 was mainly sulfated galactofucan, which was in accordance with a previous study [26]. Figure 1d also shows that HFW-3 had the best activity. HFS was fractionated by anion exchange chromatography into three components, namely, HFS-1, HFS-2, and HFS-3. Table 1 showed that HFS-1 contained only glucose, HFS-2 had many other monosaccharides, while HFS-3 had galactose and fucose, suggesting that they were similar to the fractions of HFW. The activities of HFS and its fractions inFigure 1e also confirmed the above results. In conclusion, sulfated galactofucan contributed to the anti-complement activities.

2.3. The Effect of Molecular Weight on the Anti-Complement Activity

To study the relationship between the molecular weight and the anti-complement activity of sulfated galactofucan, six sulfated galactofucans (GF90, GF50, GF36, GF30, GF11, and GF8) (the molar ratio of galactose to fucose was approximately 0.08) with different molecular weights were determined. As described in the previous study [30], six sulfated galactofucans, namely GF90, GF50, GF36, GF30, GF11, and GF8, were prepared. The molecular weights of these were 90.1, 50.1, 36.0, 29.5, 11.3, and 8.4 kDa, respectively. It is shown in Figure 2a that the activity increased as the molecular weights increased. To a certain extent, these results were consistent with previous studies [13,30]. Blondin et al. [13] reported that the anti-complement activities of sulfated fucan in the classical pathway increased with increasing molecular weight and reached a plateau at 40 kDa. In addition, the activities also showed dose dependence. GF90 and GF50 had the highest activities at a concentration of 50 μg/mL, while GF8 did not reach the highest activities even at a concentration of 400 μg/mL. Therefore, it was concluded that the molecular weight constituted an important positive factor on the activities.

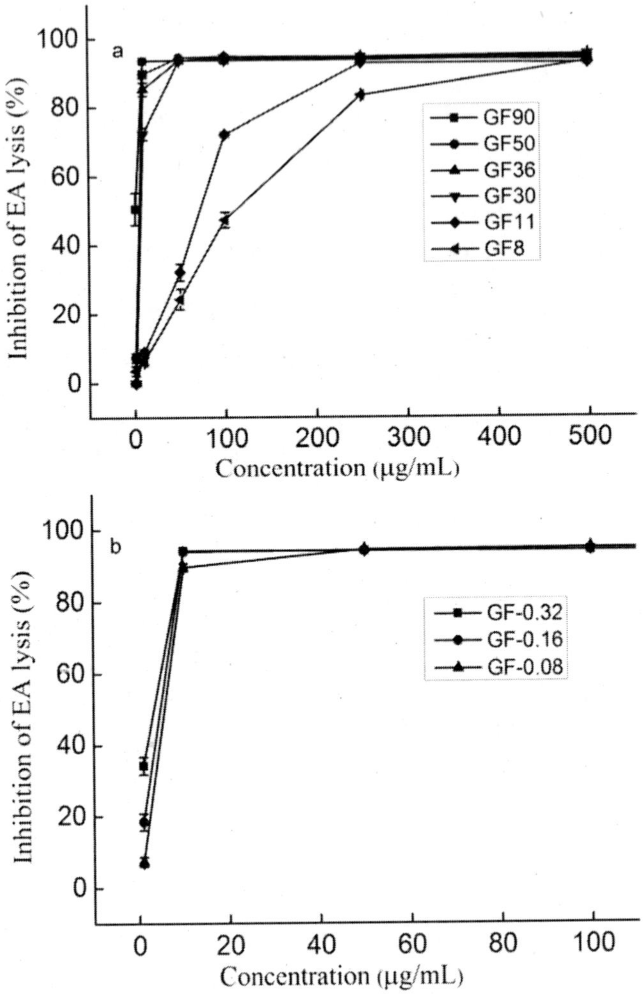

Figure 2. Inhibition of the classical pathway-mediated haemolysis of EA in 1:10-diluted NHS in the presence of increasing amounts of the polysaccharides (**a**) The different molecular weights of sulfated galactofucan; (**b**) Sulfated galactofucan with different molar ratios of galactose to fucose. The results are expressed as percent inhibition of haemolysis. Data are the means from three determinations ± S.E.M. Six sulfated galactofucans weighed 90kDa (GF90), 50kDa (GF50), 36kDa (GF36), 30kDa (GF30), 11kDa (GF11), and 8kDa (GF8). Three sulfated galactofucans had a molar ratio of galactose to fucose of 0.32 (GF-0.32), 0.16 (GF-0.16) and 0.08 (GF-0.08).

2.4. The Effect of the Molar Ratio of Galactose to Fucose on the Anti-Complement Activity

The chemical compositions of GF-0.32, GF-0.16 and GF-0.08 (also sulfated galactofucans) were reported in a previous study [30]. It was shown that GF-0.32, GF-0.16, and GF-0.08 had similar chemical compositions, except for the molar ratio of galactose to fucose. GF-0.32 weighed 49.3 kDa with a molar ratio of galactose to fucose of 0.32; GF-0.16 weighed 49.2 kDa with 0.16, and GF-0.08 weighed 50.1 kDa with 0.08. Figure 2b shows that all samples reached the highest activities at a concentration of 10 μg/mL. There was little difference in the activities of GF-0.32, GF-0.16, and GF-0.08 at a concentration of 1 μg/mL, indicating that the molar ratio of galactose to fucose contributed to the anti-complement activity. In conclusion, the anti-complement activity was attributed to the molar ratio of galactose to fucose.

2.5. The Effect of Sulfate on the Anti-Complement Activity

Previous studies [31,32,33] have demonstrated the importance of sulfate in the biological activities of polysaccharides. Thus, overly sulfated polysaccharide (SJS-OS), desulfated polysaccharide (SJS-DS) and desulfated polysaccharide with sulfation (SJS-DS-OS) were prepared. Chemical compositions, shown in Table 1, indicated that SJS-OS's sulfate increased from 36.88% to 46.36%, SJS-DS's sulfate decreased from 36.88% to 18.12%, and SJS-DS-OS's sulfate increased from 18.12% to 35.30%. Figure 3a indicates that SJS, SJS-OS and SJS-DS-OS reached the highest activities at a concentration of 10 μg/mL, while SJS-DS had no activity. In addition, SJS-OS had the highest activity at a concentration of 1 μg/mL, SJS-DS-OS second, SJS third and SJS-DS had no activity. It was suggested that a certain sulfate content was necessary for the activities. In addition, the activities did not increase along with increasing sulfate content at a concentration of more than 10 μg/mL.

In addition to the effect of the sulfate content on the activities, the effect of the sulfate pattern was also elucidated by comparison of a pair of sulfated galactofucans (SJS-3 and HFS-3), which were fractionated from SJS and HFS by anion exchange chromatography. According to previous studies [26,28,34,35], SJS-3 is mainly sulfated at C-4 of Fuc, while HFS-3 is mainly sulfated at C-2. The molecular weights of SJS-3 and HFS-3 were 125.4 kDa and 99.2 kDa, respectively, as shown inTable 1, suggesting that the effects of the molecular weight on the activity could be ignored. Moreover, the sulfate content of SJS-3 (38.43%) was similar to HFS-3's (39.85%). Figure 3b shows the activities of SJS-3 and HFS-3, which were similar to each other at a concentration of more than 10 μg/mL.

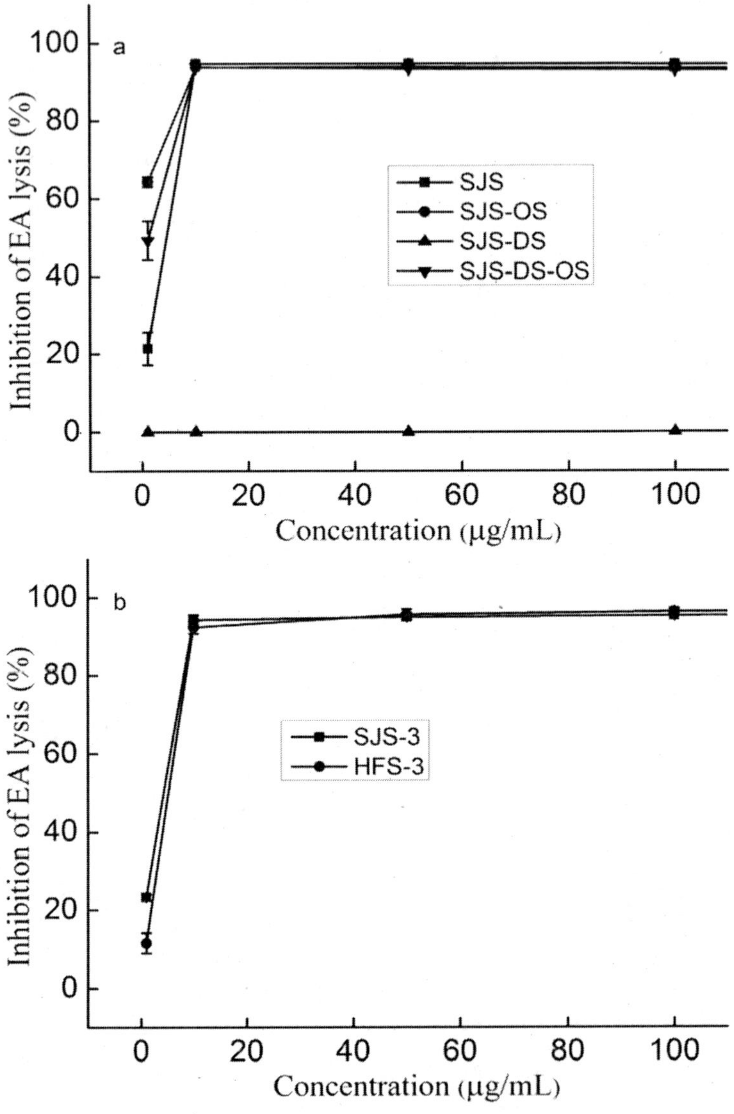

Figure 3. Inhibition of the classical pathway-mediated haemolysis of EA in 1:10-diluted NHS in the presence of increasing amounts of the polysaccharides (**a**) Polysaccharides with different sulfate contents; (**b**) Polysaccharides with different sulfate patterns. The results are expressed as percent inhibition of haemolysis. Data are the means from three determinations ± S.E.M. Overly sulfated SJS (SJS-OS), desulfated SJS (SJS-DS) and desulfated SJS with sulfation (SJS-DS-OS).

Finally, it was concluded that the sulfate content was necessary for the activities; however, the effect of the sulfate on the anti-complement activity was influenced by the concentrations of the samples.

2.6. The Effect of the Content of Uronic acid (UA) on the Anti-Complement Activity

From the above results, it was concluded that the anti-complement activities of sulfated galactofucan were better than the heteropolysaccharides with UA. To study the effect of UA content on the activity, three polysaccharides with different UA contents were obtained. One polysaccharide (SJW-2-R) was obtained through UA reduction, and the other two (SJW-2-HJ and SJW-2-HW) were fractionated using an activated carbon column. After reduction, the most apparent changes were in the UA content, which was decreased from 17.35% to 9.86%, while the molar ratio of glucose (Glc) to Fuc was increased from 0 to 0.10, and the glucuronic acid (GlcA) to Fuc ratio decreased from 0.18 to 0.06. Other chemical compositions, shown in Table 1, did not change as obviously. A comparison of the activities of SJW-2 and SJW-2-R in Figure 4a indicates that the change in UA content did not influence the activity. SJW-2-HJ, which was eluted with ethanol, contained a higher UA content and lower sulfate content than that of SJW-2-HW, which was eluted with water. Moreover, SJW-2-HJ had a smaller molecular weight and a higher content of other monosaccharides. Figure 4a shows that SJW-2-HJ did not have any activity, while SJW-2-HW had better activities than that of SJW-2.

nodosum (ANW) and Acaudina molpadioides (SC).

In summary, the UA fraction had no anti-complement activity; however, it could reduce the activity in other samples by decreasing the effective concentration of the active compositions when it was mixed with the active compositions. This might explain the observed phenomenon that the heteropolysaccharides with high UA content had lower activities.

2.7. The Effect of the Linkage of Sulfated Galactofucan on the Anti-Complement Activity

Crude polysaccharide (ANW) was extracted from brown algae Ascophyllum nodosum. According to previous studies [36,37,38,39], it has a backbone of alternating α (1 → 3) and α (1 → 4)-L-Fucp residues and is sulfated at C-2 and possibly C-3 or C-4. SJW contains mainly α (1 → 3)-L-Fucp residues sulfated at C-4, while HFW is mainly composed of α (1 → 3)-L-Fucp residues sulfated at C-2 [26,27]. Figure 4b shows the anti-complement activities of ANW, SJW, and HFW. They all showed dose dependence and reached a plateau. Specifically, the IC_{50} values of ANW, SJW, and HFW were 0.98, 7.26, and 5.51 μg/mL, respectively. The differences in the activities could be explained by noting that SJW had a higher UA content, while HFW had a higher content of laminaran, which had no anti-complement activity (as discussed in Section 3.9). Thus, it was proposed that there might be little or no effect of the linkage on the anti-complement activity.

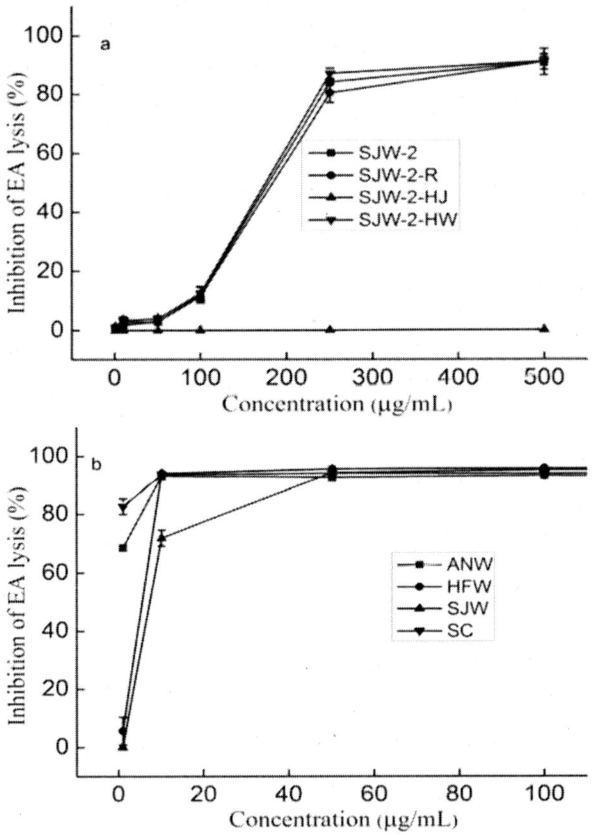

Figure 4. Inhibition of the classical pathway-mediated haemolysis of EA in 1:10-diluted NHS in the presence of increasing amounts of polysaccharides (**a**) Polysaccharides with different linkages; (**b**) Polysaccharides with branching or no branching. The results are expressed as percent inhibition of haemolysis. Data are the means from three determinations ± S.E.M. SJW-2-R was obtained through UA reduction, and SJW-2-HJ and SJW-2-HW were fractionated using an activated carbon column. Polysaccharides derived from Ascophyllum

2.8. The Effect of the Branching of Polysaccharides on the Anti-Complement Activity

Polysaccharide (SC) was extracted from sea cucumber Acaudina molpadioides according to a previous study [40]. It was reported [16] that polysaccharides from marine invertebrates possess a clearly linear backbone, consisting of a regular repeating unit, in contrast with the algal polysaccharides. Thus, the anti-complement activities of the polysaccharides from algae and marine invertebrates were determined to elucidate the effect of the branching. The activity of SC in Figure 4b is similar to that of ANW, indicating that there was little impact of branching on the anti-complement activities. This result contrasts a previous study [41] that suggested that branching of fucoidan oligosaccharides had a major impact on their anti-complement activity. This difference might be explained based on the conformational state of the polysaccharides. The conformational states of the oligosaccharides depend on branching, while the conformational states of the polysaccharides do not. Therefore, it was concluded that there might be little or no effect of the branching of polysaccharides on the anti-complement activity.

2.9. The Effect of the Type of Monosaccharides on the Anti-Complement Activity

To confirm whether Fuc was necessary for the activity, sulfated laminaran (HFW-1-S) was prepared. Figure 5a shows that laminaran (both HFW-1 and HFS-1 were laminaran) had no activity, while sulfated laminaran had good activity, suggesting that Fuc was not necessary for the activity. Thus, it was concluded that laminaran, like the UA fraction, had no activity. However, it could reduce the activity by decreasing the effective concentrations of the active components. In addition, it also confirmed that the fucose residue was not necessary.

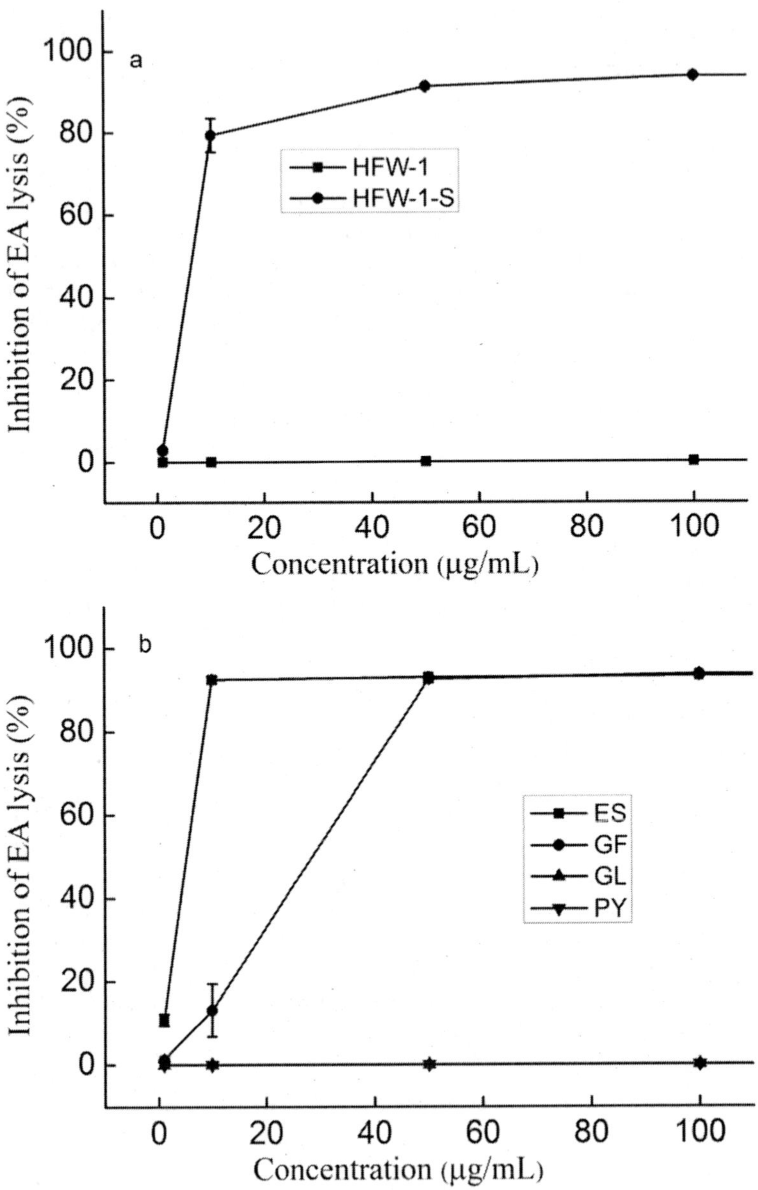
a
Inhibition of EA lysis (%)
100
80
60
40
20
0
HFW-1
HFW-1-S
0
20
40
60
80
100
Concentration (μg/mL)
b
Inhibition of EA lysis (%)
100
80
60
40
20
0
ES
GF
GL
PY
0
20
40
60
80
100
Concentration (μg/mL)

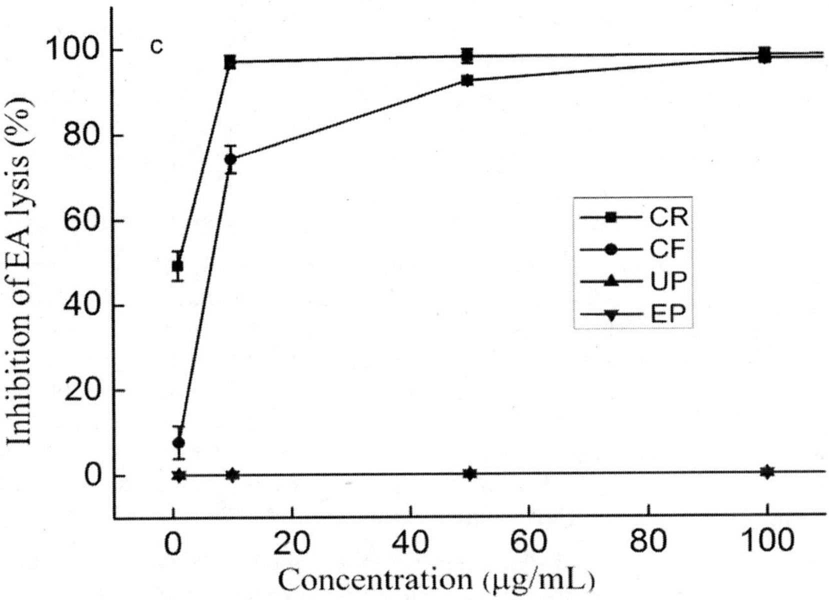

Figure 5. Inhibition of the classical pathway-mediated haemolysis of EA in 1:10-diluted NHS in the presence of increasing amounts of the polysaccharides (**a**) Laminaran and sulfated laminaran; (**b**) Polysaccharides from red algae; (**c**) Polysaccharides from green algae. The results are expressed as percent inhibition of haemolysis. Data are the means from three determinations ± S.E.M. Polysaccharides derived from Eucheuma spinosum (ES), Grateloupia filicina (GF), Gracilaria lamaneiformis (GL), Porphyra yezoensis(PY), Enteromorpha prolifera (EP), Ulva pertusa (UP), Cladophera rupestris (CR) andCodium fragile (CF). SJW-2-R was obtained through UA reduction, and SJW-2-HJ and SJW-2-HW were fractionated using an activated carbon column. Overly sulfated laminaran (HFW-1-S).

Four types of crude polysaccharides from red algae, namely Grateloupia filicina (GF), Eucheuma spinosum (ES), Gracilaria lamaneiformis (GL), and Porphyra yezoensis (PY), were prepared. ES is a carrageenan polysaccharide; GF is an agar-carrageenan intermediate polysaccharide, and GL and PY are agar polysaccharides. The most apparent difference in the chemical compositions, as shown in Table 1, is the sulfate content. Figure 5b shows that ES had the best anti-complement activity, GF second, and GL and PY had no activities, indicating that sulfate was essential for the anti-complement activity. It was also confirmed that Fuc was not necessary.

Four types of crude polysaccharides from green algae, namely Cladophera rupestris (CR), Codium fragile (CF), Enteromorpha prolifera (EP), and Ulva pertusa (UP), were prepared. The analysis of the chemical compositions in Table 1 indicates that the former two belonged to type 1, while the latter two were type 2. Figure 5c shows that CR and CF showed good anti-complement activities, while EP and UP did not. This might be explained by the high content of UA. In addition, the solution of the polysaccharides with high UA content was too thick.

The main polysaccharides from brown algae (Saccharina japonica and Hizikia fusiforme) were laminaran, alginate, and fucoidan. Fucoidan might be classified as a combination of sulfated heteropolysaccharides and sulfated galactofucan. In the above results, it was concluded that sulfated galactofucan was the active component of the anti-complement activity. In addition, the results also showed that laminaran did not have any activity. Moreover, it was indicated that heteropolysaccharides with UA had lower activity. Alginate (data not shown) displayed no activity, which might confirm this.

Thus, it was concluded that the types of monosaccharides (neutral monosaccharides) had no effect on the anti-complement activity.

3. EXPERIMENTAL SECTION

3.1. Materials

The brown algae Saccharina japonica was collected in Rongcheng, Qingdao, China, on 28 May 2014. The brown algaeHizikia fusiforme was purchased from Dong Tou, Zhejiang, China, on 18 July 2014. The green algae Enteromorpha prolifera was collected in Qingdao, China, on 15 June 2012. The green algae Codium fragile and Ulva pertusa were collected in Qingdao, China, on 12 December 2013. The green algae Cladophera rupestris was collected in Rongcheng, Qingdao, China, on 24 July 2014. The red algae Porphyra yezoensis was purchased from Lianyungang, Jiangsu, China. The red algae Grateloupia filicina,Eucheuma spinosum, and Gracilaria lamaneiformis were purchased from Zhanjiang, China. Ascophyllum nodosum and Acaudina molpadioides were purchased from Qingdao, China. These were authenticated by Professor Zuhong Xu.

3.2. Extraction and Preparation of Polysaccharides

Brown algae (Saccharina japonica, Hizikia fusiforme, and Ascophyllum nodosum) (100 g) were cut into pieces and treated with 85% ethanol three times to remove the pigment. Crude polysaccharide was extracted from the residual material with hot water (3 L) for 4 h. The extract solution was filtered with Celite and concentrated. Further elimination of algin was achieved using 20% ethanol with $MgCl_2$ (0.05 M/L). After removing the algin, the supernatant fluid was ultra-filtered. Finally, the dialysate was concentrated, and the crude polysaccharides were obtained by ethanol precipitation and named SJW, HFW, and ANW, respectively.

Green algae and red algae (100 g) were cut into pieces and treated with 85% ethanol three times to remove the pigment. Crude polysaccharides were extracted from the residual material with hot water (3 L) for 4 h. The extract solution was filtered with Celite, concentrated, and dialyzed. Finally, the dialysate was concentrated, and the crude polysaccharides were obtained by ethanol precipitation.

Crude polysaccharides from brown algae (Saccharina japonica and Hizikia fusiforme) were extracted from the residual material with 0.1 M HCl (2 L) at room temperature for 3 h. The extract solution was filtered with Celite, ultra-filtered, and concentrated. Finally, the crude polysaccharides were obtained by ethanol precipitation and named SJS and HFS.

Crude polysaccharide (SC) was extracted from the sea cucumber Acaudina molpadioides as previously described [40]. Briefly, crude polysaccharide was extracted from the fresh body wall (10 g) with hot water (500 mL) for 4 h. The extract solution was filtered with Celite and concentrated. The concentrated solution was hydrolysed with papain. Papain was deactivated by hot water. Then, the solution was centrifuged, concentrated, and dialyzed. Finally, the dialysate was concentrated, and the crude polysaccharide was obtained by ethanol precipitation and named SC.

3.3. Preparation and Purification of Polysaccharides

Crude polysaccharides (SJW and SJS) (8 g) underwent anion exchange chromatography on a DEAE-Bio Gel Agarose FF gel (6 cm × 40 cm) with elution by 0.5 M (5 L) (SJW-1 and SJS-1), 1 M (5 L) (SJW-2 and SJS-2), and 2 M NaCl (5 L) (SJW-3 and SJS-3). The polysaccharides were then ultrafiltrated, concentrated, and precipitated by ethanol.

Crude polysaccharides (HFW and HFS) (8 g) also underwent anion exchange chromatography on a DEAE-Bio Gel Agarose FF gel (6 cm × 40 cm) with elution by water (5 L) (HFW-1 and HFS-1), 0.5 M (5 L) (HFW-2 and HFS-2), and 2 M NaCl (5 L) (HFW-3 and HFS-3). The polysaccharides were then ultrafiltrated, concentrated, and precipitated by ethanol.

Fraction SJW-2 was fractionated using an activated carbon column (2.6 cm × 30 cm) with elution by water (1 L) (SJW-2-HW) and a gradient elution from 50% ethanol to 95% ethanol (SJW-2-HJ). The polysaccharides were concentrated and precipitated by ethanol.

3.4. Preparation of Desulfated Polysaccharides

The desulfated polysaccharides were prepared according to a method reported in the literature [27]. Briefly, SJS (1 g) was dissolved in distilled water (100 mL) and mixed with cationic resin (H^+) for 3 h. After filtration, the solution was neutralized with pyridinium and lyophilized. The solution was dissolved in dimethyl sulfoxide (DMSO): methanol (9:1; v/v, 20 mL). The mixture was heated at 80 °C for 5 h, and the desulfated products (SJS-DS) were dialyzed and lyophilized.

3.5. Preparation of over-Sulfated Polysaccharides

The over-sulfated polysaccharides were prepared according to a method reported in the literature [42]. Briefly, polysaccharides (SJS, SJS-DS and HFW-1) (1 g) in dimethyl formamide (DMF) (20 mL) were treated with a sulphur trioxide pyridine complex (5 g) at 50 °C for 24 h, and then the over-sulfated products (SJS-OS, SJS-DS-OS and HFW-1-S) were neutralized, dialyzed, and lyophilized.

3.6. Preparation of the Polysaccharides with Carboxyl Reduction

Carboxyl reduction was performed as previously described [43]. Briefly, SJW-2 (1 g) was dissolved in distilled water (50 mL), followed by the addition of 1-ethyl-3-(3-dimethyl-aminopropyl) carbodiimide (EDC). The solution was stirred at room temperature for 1 h under constant pH conditions (approximately 4.8) by the addition of 0.1 M HCl. Then, freshly prepared 2 M sodium borohydride (10 mL) was added twice during the next 2 h at 50 °C. Later, the reaction was terminated by the addition of glacial acetic acid. Finally, the polysaccharide with carboxyl reduction (SJW-2-R) was dialyzed, concentrated, and lyophilized.

3.7. Compositional Analysis

The sulfated contents were performed by ion chromatography on a Shodex IC SI-52 4E column (4.0 × 250 mm) and eluted with 3.6 mM Na_2CO_3 at a flow rate of 0.8 mL/min at 45 °C. The molar ratios of monosaccharides and fucose content were determined as described by Zhang et al. [44]. The uronic acid (UA) concentration was determined by a modified carbazole method [45]. The molecular weights of the polysaccharides were evaluated by GPC-HPLC on a TSK G3000 PWxl column (7 μm 7.8 × 300 mm) with elution in 0.05 M Na_2SO_4 at a flow rate of 0.5 mL/min at 40 °C with refractive index detection. Ten different molecular weight dextrans purchased from the National Institute for the Control of Pharmaceutical and Biological Products (Beijing, China) were used as weight standards.

3.8. Anti-Complement Activity

According to previous studies [4,46], the anti-complement activities of the polysaccharides were determined following the classical pathway. In the classical pathway, various dilutions (100 μL) of the polysaccharides were mixed with 1:10 diluted normal human serum (NHS, which was obtained from healthy adult donors) (100 μL), GVB^{2+} (veronal buffer saline [VBS] containing 0.1% gelatin, 0.5 mM Mg^{2+} and 0.15 mM Ca^{2+}) (200 μL) and sensitized erythrocytes (EA) (200 μL). Then, the mixture was incubated at 37 °C for 30 min. The following assay controls were incubated under the same conditions: (1) 100% lysis: EA (200 μL) in water (400 μL); (2) sample control: sample (100 μL) in GVB^{2+} (500 μL); (3) complement: 1:10-diluted NHS (100 μL) and EA (200 μL) in GVB^{2+} (300 μL); and (4) blank: EA (200 μL) in GVB^{2+} (400 μL). After incubation, the mixture was centrifuged (5000 rpm × 10 min) and the erythrocyte lysis was determined at 405 nm. Decreased lysis in the presence of tested polysaccharides indicated anti-complement activity. All of the samples were dissolved in GVB^{2+}. The percent inhibition was calculated using the following equation: inhibition of EA lysis (%) = $(A_{complement} - (A_{sample} - A_{sample\ control}))/A_{complement} \times 100$.

3.9. Statistical Analysis

All data are shown as the mean ± standard deviation (SD). Significant differences between the experimental groups were determined by one-way ANOVA, and differences were considered to be statistically significant if $p < 0.05$. All calculations were performed using SPSS 16.0 statistical software.

4. CONCLUSIONS

In summary, four crude polysaccharides (SJW, SJS, HFS, and HFW) were extracted by water and 0.1 M HCl fromSaccharina japonica and Hizikia fusiforme to determine the effects of the extraction methods. The results could not definitively elucidate the effect of the extraction method. Crude polysaccharides were fractionated using anion exchange chromatography. It was found that sulfated galactofucan maintained its activity, indicating that it was the active composition. To study the effect of molecular weight, several different molecular weight polysaccharides were obtained. It was shown that the larger the molecular weight, the better the activity. However, the activity reached a plateau. In addition, the effect of the molar ratio of galactose to fucose was also considered. It was observed that the molar ratio of galactose to fucose was a positive factor. Moreover, the activities of SJS-OS, SJS-DS, and SJS-DS-OS confirmed that sulfate was an important positive factor. The activities of HFS-3 and SJS-3 suggested that the sulfation pattern had no effect at a concentration of more than 10 μg/mL, like linkage and branching. The type of monosaccharides had no effect. Finally, it was concluded that the UA fraction and laminaran had no effect on the anti-complement activity; however, it could reduce the activity by decreasing the effective concentration of the active composition.

ACKNOWLEDGMENTS

This study was supported by the Shandong Provincial Natural Science Foundation, China (ZR2014DQ024), National Natural Science Foundation of China (41506165), Shandong Province Science & Technology Development Project (2014GHY115017), Qingdao Shinan District S&T Development Fund (2013-12-009-SW), Regional Demonstration Project of Marine Economic Development in 2014, the Ocean Public Welfare Scientific Research Project, the State Oceanic Administration of the People's Republic of China (No. 201405040).

AUTHOR CONTRIBUTIONS

Weihua Jin and Wenjing Zhang conceived and designed the experiments; Weihua Jin and Wenjing Zhang performed the experiments; Weihua Jin, Quanbin Zhang and Hongze Liang contributed reagents, materials and financial support. Weihua Jin wrote the paper. All authors have read and approved the final manuscript.

CONFLICTS OF INTEREST

The authors declare no conflict of interest.

REFERENCES

1. Sahu, A.; Lambris, J.D. Complement inhibitors: A resurgent concept in anti-inflammatory therapertics.Immunopharmacology **2000**, 49, 133–148.
2. Morgan, B.P.; Harris, C.L. Complement therapertics: History and current progress. Mol. Immunol. **2003**, 40, 159–170.
3. Maillet, F.; Maurice, P.; Jean, C.; Kazatchkine, M.D. Structure-function relationships in the inhibitory effect of heparin on complement activation: Independency of the anti-coagulant and anti-complementary sites on the heparin molecule. Mol. Immunol. **1988**, 25, 917–923.
4. Xu, H.; Zhang, Y.; Zhang, J.; Chen, D. Isolation and characterization of an anti-complementary polysaccharide D3-S1 from the roots of Bupleurum smithii. Int. Immunopharmacol. **2007**, 7, 175–182.
5. Fang, X.B.; Yin, X.X.; Yuan, G.F.; Chen, X.O. Chemical and biological characterization of polysaccharides from the bark of Avicennia marina. Eur. Food Res. Technol. **2015**, 241, 17–25.
6. Samuelsena, A.B.; Lunda, I.; Djahromia, J.M.; Paulsena, B.S.; Wolda, J.K.; Knutsen, H.S. Structural features and anti-complementary activity of some heteroxylan polysaccharide fractions from the seeds of Plantago major L. Carbohydr. Polym. **1999**, 38, 133–143.
7. Ni, F.; Liu, L.; Song, Y.; Wang, X.; Zhao, Y.; Huang, W.; Wang, Z.; Xiao, W. Anti-complementary phenolic acids fromLonicera japonica. China J. Chin. Mater. Med. **2015**, 40, 269–274.
8. Sun, Q.; Bao, J. Purification, cloning and characterization of a metalloproteinase from Naja atra venom. Toxicon **2015**,56, 1459–1469.

9. Makrides, S.C. Therapeutic Inhibition of the Complement System. Pharmacological **1998**, 50, 59–88.

10. Xu, X.; Chen, L.; Zhao, X. Review on the research and development of anti-complementary agents from natural products. Nat. Prod. Res. Dev. **2015**, 27, 355.

11. Mauzac, M.; Maillet, F.; Jozefonvicz, J.; Kazatchkine, M.D. Anticomplementary activity of dextran derivatives.Biomaterials **1985**, 6, 61–63.

12. Crepon, B.; Maillet, F.; Kazatchkine, M.D.; Jozefonvicz, J. Molecular weight dependency of the acquired anticomplementary and anticoagulant activities of specifically substituted dextrans. Biomaterials **1987**, 8, 248–253.

13. Blondin, C.; Chaubet, F.; Nardella, A.; Sinquin, C.; Jozefonvicz, J. Relationships between chemical characteristics and anticomplementary activity of fucans. Biomaterials **1996**, 17, 597–603.

14. Zhang, W.; Jin, W.; Sun, D.; Zhao, L.; Wang, J.; Duan, D.; Zhang, Q. Structural analysis and anti-complement activity of polysaccharides from Kjellmaniella crsaaifolia. Mar. Drugs **2015**, 13, 1360–1374.

15. Blondin, C.; Fisher, E.; Boisson-Vadal, C.; Kazatchkine, M.D.; Jozefonvicz, J. Inhibition of complement activation by natural sulfated polysaccharides (fucans) from brown seaweed. Mol. Immunol. **1994**, 31, 247–253.

16. Berteau, O.; Mulloy, B. Sulfated fucans, fresh perspectives: Structures, functions, and biological properties of sulfated fucans and an overview of enzymes active toward this class of polysaccharide. Glycobiology **2003**, 13, 29R–40R.

17. Bilan, M.I.; Grachev, A.A.; Shashkov, A.S.; Kelly, M.; Sanderson, C.J.; Nifantiev, N.E.; Usov, A.I. Further studies on the composition and structure of a fucoidan preparation from the brown alga Saccharina latissima. Carbohydr. Res. **2010**,345, 2038–2047.

18. Jiao, G.; Yu, G.; Zhang, J.; Ewart, H. Chemical structures and bioactivities of sulfated polysaccharides from marine algae. Mar. Drugs **2011**, 9, 196–223.

19. Kusaykin, M.; Bakunina, I.; Sova, V.; Ermakova, S.; Kuznetsova, T.; Besednova, N.; Zaporozhets, T.; Zvyagintseva, T. Structure, biological activity, and enzymatic transformation of fucoidans from the brown seaweeds. Biotechnol. J. **2008**,3, 904–915.

20. Li, B.; Lu, F.; Wei, X.; Zhao, R. Fucoidan: Structure and Bioactivity. Molecules **2008**, 13, 1671–1695.

21. Pomin, V.H.; Mourao, P.A.S. Structure, biology, evolution, and medical importance of sulfated fucans and galactans.Glycobiology **2008**, 18, 1016–1027.

22. Lahaye, M. Developments on gelling algal galactans, their structure and physico-chemistry. J. Appl. Phycol. **2001**, 13, 173–184.

23. Pomin, V.H. Structural and functional insights into sulfated galactans: A systematic review. Glycoconj. J. **2010**, 27, 1–12.

24. USov, A.I. Structural analysis of red seaweed galactans of agar and carrageenan groups. Food Hydrocoll. **1998**, 12, 301–308.

25. Lahaye, M.; Robic, A. Structure and Functional Properties of Ulvan, a Polysaccharide from Green Seaweeds.Biomacromolecules **2007**, 8, 1765–1774.

26. Jin, W.; Zhang, W.; Wang, J.; Ren, S.; Song, N.; Duan, D.; Zhang, Q. Characterization of laminaran and a highly sulfated polysaccharide from Sargassum fusiforme. Carbohydr. Res. **2014**, 385, 58–64.

27. Jin, W.; Zhang, W.; Wang, J.; Ren, S.; Song, N.; Zhang, Q. Structural analysis of heteropolysaccharide from Saccharina japonica and its derived oligosaccharides. Int. J. Biol. Macromol. **2013**, 62, 697–704.

28. Jin, W.; Guo, Z.; Wang, J.; Zhang, W.; Zhang, Q. Structural analysis of sulfated fucan from Saccharina japonica by electrospray ionization tandem mass spectrometry. Carbohydr. Res. **2013**, 369, 63–67.

29. Jin, W.; Wang, J.; Ren, S.; Song, N.; Zhang, Q. Structural analysis of a heteropolysaccharide from Saccharina japonica by electrospray mass spectrometry in tandem with collision-induced dissociation tandem mass spectrometry (ESI-CID-MS/MS). Mar. Drugs **2012**, 10, 2138–2152.

30. Jin, W.; Zhang, Q.; Wang, J.; Zhang, W. A comparative study of the anticoagulant activities of eleven fucoidans.Carbohydr. Polym. **2013**, 91, 1–6.

31. Cumashi, A.; Ushakova, N.A.; Preobrazhenskaya, M.E.; D'Incecco, A.; Piccoli, A.; Totani, L.; Tinari, N.; Morozevich, G.E.; Berman, A.E.; Bilan, M.I.; et al. A comparative study of the anti-inflammatory, anticoagulant, antiangiogenic, and antiadhesive

activities of nine different fucoidans from brown seaweeds. Glycobiology **2007**, 17, 541–552.

32. Karmakar, P.; Pujol, C.A.; Damonte, E.B.; Ghosh, T.; Ray, B. Polysaccharides from Padina tetrastromatica: Structural features, chemical modification and antiviral activity. Carbohydr. Polym. **2010**, 80, 513–520.

33. Pereira, M.S.; Melo, F.R.; Mourão, P.A.S. Is there a correlation between structure and anticoagulant action of sulfated galactans and sulfated fucans? Glycobiology **2002**, 12, 573–580.

34. Li, B.; Xue, X.J.; Sun, J.L.; Xu, S.Y. Structural investigation of a fucoidan containing a fucose-free core from the brown seaweed Hizikia fusiforme. Carbohydr. Res. **2006**, 341, 1135–1146.

35. Wang, P.; Zhao, X.; Lv, Y.; Liu, Y.; Lang, Y.; Wu, J.; Liu, X.; Li, M.; Yu, G. Analysis of structural heterogeneity of fucoidan from Hizikia fusiforme by ES-CID-MS/MS. Carbohydr. Polym. **2012**, 90, 602–607.

36. Chevolot, L.; Foucault, A.; Chaubet, F.; Kervarec, N.; Sinquin, C.; Fisher, A.M.; Boisson-Vidal, C. Further data on the structure of brown seaweed fucans: Relationships with anticoagulant activity. Carbohydr. Res. **1999**, 319, 154–156.

37. Chevolot, L.; Mulloy, B.; Ratiskol, J.; Foucault, A.; Colliec-Jouaultb, S. A disaccharide repeat unit is the major structure in fucoidans from two species of brown algae. Carbohydr. Res. **2001**, 330, 529–535.

38. Daniel, R.; Berteau, O.; Jozefonvicz, J.; Goasdoue, N. Degradation of algal (Ascophyllum nodosum) fucoidan by an enzymatic activity contained in digestive glands of the marine mollusc Pecten maximus. Carbohydr. Res. **1999**, 322, 291–297.

39. Daniel, R.; Berteau, O.; Chevolot, L.; Varenne, A.; Gareil, P.; Goasdoue, N. Regioselective desulfation of sulfated L-fucopyranoside by a new sulfoesterase from the marine mollusk Pecten maximus. Application to the structural study of algal fucoidan (Ascophyllum nodosum). Eur. J. Biochem. **2001**, 268, 5617–5626.

40. Xue, C.; Tang, Q.; Li, D.; Wu, X.; Wang, J. Isolation and characterization of a sea cucumber fucoidan-utilizing marine bacterium. Lett. Appl. Microbiol. **2010**, 50, 301–307.

41. Clement, M.J.; Tissot, B.; Chevolot, L.; Adjadj, E.; Du, Y.; Curmi, P.A.; Daniel, R. NMR characterization and molecular modeling of fucoidan showing the importance of oligosaccharide branching in its anticomplementary activity.Glycobiology **2010**, 20, 883–894.

42. Nifantiev, N.; Ustyuzhanina, N.; Krylov, V.; Grachev, A.; Gerbst, A. Synthesis, NMR and conformational studies of fucoidan fragments, 8: Convergent synthesis of branched and linear oligosaccharides. Synthesis **2006**, 23, 4017–4031.

43. Kariya, Y.; Watabe, S.; Kyogashima, M.; Ishihara, M.; Ishii, T. Structure of fucose branches in the glycosaminoglycan from the body wall of the sea cucumber Stichopus japonicus. Carbohydr. Res. **1997**, 297, 273–279.

44. Zhang, J.; Zhang, Q.; Wang, J.; Shi, X.; Zhang, Z. Analysis of the monosaccharide composition of fucoidan by precolumn derivation HPLC. Chin. J. Oceanol. Limnol. **2009**, 27, 1–5.

45. Bitter, T.; Muir, H.M. A modified uronic acid carbazole reaction. Anal. Biochem. **1962**, 4, 330–334.

46. Klerx, J.P.A.M.; Beukelman, C.J.; Dijk, H.V.; Willers, J.M.N. Microassay for colorimetric estimation of complement activity in guinea pig, human and mouse serum. J. Immunol. Methods **1983**, 63, 215–220.

6

CHAPTER

Perspective on the Use of Sulfated Polysaccharides from Marine Organisms as a Source of New Antithrombotic Drugs

Paulo A. S. Mourão [1,2]

[1]*Connective Tissue Research Laboratory, University Hospital Clementino Fraga Filho, Rio de Janeiro, RJ 21941-590, Brazil; Tel./Fax: +55-21-3938-2090*

[2]*Program of Glycobiology, Institute of Medical Biochemistry, Federal University of Rio de Janeiro, Caixa Postal 68041, Rio de Janeiro, RJ 21941-590, Brazil*

ABSTRACT

Thromboembolic diseases are increasing worldwide and always require anticoagulant therapy. We still need safer and more secure antithrombotic drugs than those presently available. Sulfated polysaccharides from marine organisms may constitute a new source for the development of such drugs. Investigation of these compounds usually attempts to reproduce the therapeutic effects of heparin. However, we may need to follow different routes, focusing particularly in the following aspects: (1) defining precisely the specific structures required for interaction of these sulfated polysaccharides with proteins of the coagulation system; (2) looking for alternative mechanisms of action, distinct from those of heparin; (3) identifying side effects (mostly pro-coagulant action and hypotension rather than bleeding) and preparing derivatives that retain the desired antithrombotic action but are devoid of side effects; (4) considering that sulfated polysaccharides with low anticoagulant action on *in vitro* assays may display potent effects on animal models of experimental thrombosis; and finally (5) investigating the antithrombotic effect of these sulfated polysaccharides after oral administration or preparing derivatives that may achieve this effect. If these aspects are successfully addressed, sulfated polysaccharides from marine organisms may conquer the frontier of antithrombotic therapy and open new avenues for treatment or prevention of thromboembolic diseases.

Keywords: sulfated fucans; sulfated galactans; anticoagulant drugs; glycosaminoglycan-like polysaccharides; fucosylated chondroitin sulfate; fucoidan; carrageenans

1. WE NEED NEW ANTITHROMBOTIC DRUGS

Cardiovascular diseases are the leading cause of mortality worldwide. Among them are the thromboembolic events due to the formation of a thrombus or clot inside the circulatory system. Stasis, changes in blood coagulation, damage of the vascular wall and changes in the concentrations of leukocytes or platelets cause thrombus formation.

The occurrence of thromboembolic processes necessarily requires anticoagulant therapy. Dicumarinic (or warfarin) was the first widely used anticoagulant shortly followed by heparin. Heparin has the highest negative charge density of any known biomolecule described in vertebrate tissues so far. Just behind insulin, heparin is the second-most used naturally occurring drug in medicine. Besides being one of the first biopolymers employed as a medicine, and certainly the principal example of a carbohydrate-based drug, heparin is perhaps one of the oldest natural products still in use [1]. Its clinical use dates back more than 65 years [2]. Although it is primarily destined for treatment and prophylaxis of thromboembolic disorders, extracorporeal circulations during cardiovascular surgeries [3] and hemodialysis [4] also require heparin.

Besides its long-standing and varied clinical applications, heparin long-term therapy causes a number of side effects, such as bleeding, thrombocytopenia, changes in lipid metabolism and osteoporosis [5]. Additionally, heparin originates from animal tissues, and therefore is highly susceptible to contamination by pathogenic particles, such as in the case of spongiform encephalopathy. Furthermore, there is doubt if animal sources of heparin (mostly porcine or bovine intestine) will match the increasing demand for this drug. Nowadays, EUA and European countries obtain heparin exclusively from porcine intestinal mucosa. These countries are now looking for new sources of the drug [6]. However, heparins extracted from different tissues may possess distinct structures and activities, requiring separate monographs in the

Pharmacopeia. This is the case of heparins obtained from bovine and porcine intestinal mucosa [7].

Even more serious are the recent reports of contamination of heparin preparations. In late 2007, the microorganism*Serratia marcescens* contaminated unopened heparin syringes, which leaded to a recall in the USA [8]. In early 2008, there were reports of contamination of heparin preparations with oversulfated chondroitin sulfate. This contaminant has a potent hypotensive effect and induces anaphylactic reactions. These side effects were associated with the death of ~200 patients in the United States [9,10]. Another negative example related to the use of mammalian heparin, which took place in Brazil, was the increase of bleeding when porcine intestinal heparin was replaced by bovine intestinal heparin [7,11].

These observations indicate we still need to search for new antithrombotic drugs, and marine organisms may constitute new sources of heparin analogs.

2. A STEP FORWARD IN THE STUDY OF SULFATED POLYSACCHARIDES FROM MARINE ORGANISMS

Marine organisms are a rich source of new substances with potential applications in medicine, though they are not yet well-explored. In particular, sulfated polysaccharides from marine invertebrates and algae possess unique structures and specific biological effects when tested in mammalian systems. The most abundant sulfated polysaccharides found in algae and marine invertebrates are sulfated fucans (also known as fucoidan when isolated from brown algae) and sulfated galactans (also known as carrageenans when isolated from red algae) [12,13,14]. In general, algal polysaccharides have more complex structures than polymers from marine invertebrates. Figure 1 illustrates this observation by comparing structures of sulfated galactans and sulfated fucans from red or brown algae (highlighted in blue) with those from sea urchins (in red).

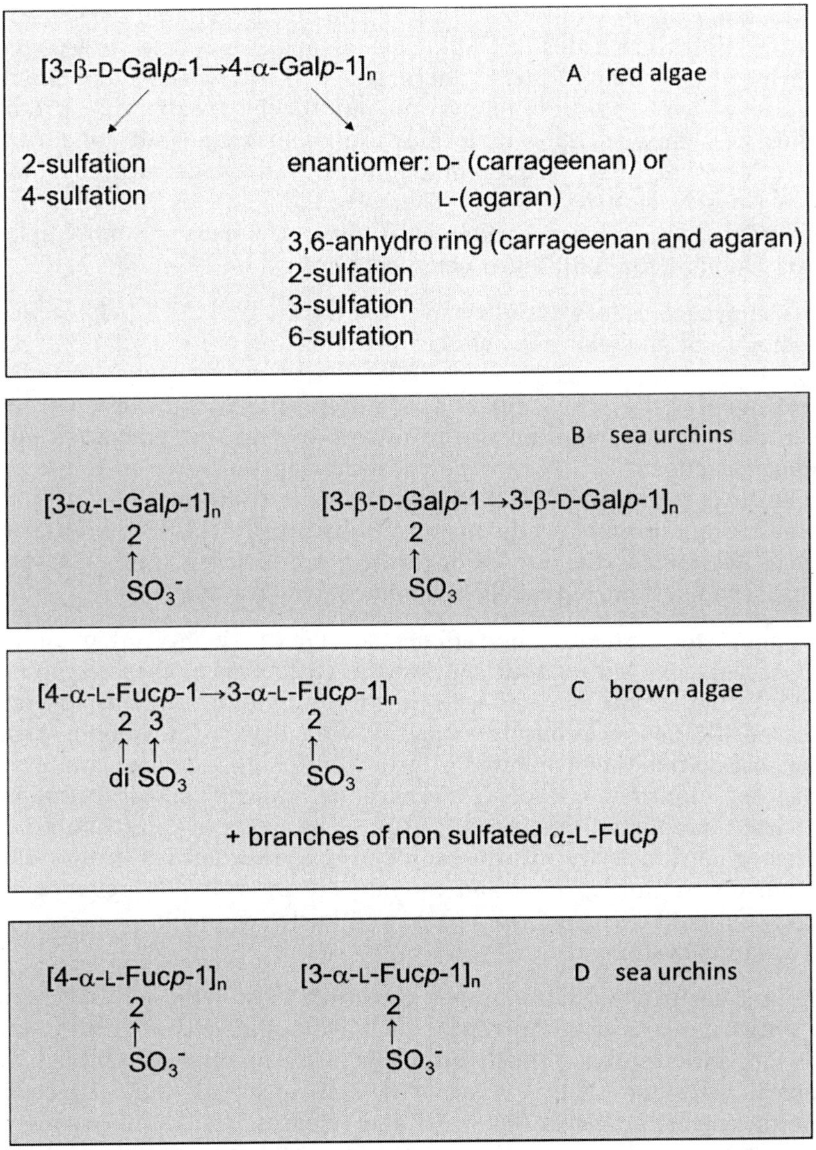

Figure 1. Examples of the structures of sulfated galactans from (**A**) red algae [15,16,17,18,19]; (**B**) sea urchins [20,21]; (**C**) sulfated fucan from brown algae [22,23,24]; and (**D**) from sea urchins [25,26]. Structures of the sulfated polysaccharides from marine algae highlighted in blue and those from sea urchins in red.

Most studies on the effects of anticoagulant sulfated polysaccharides aim at selecting the most active native or chemically sulfated derivatives of natural polysaccharides in biological assays. Thus, the anticoagulant activity of the sulfated fucans increases with increasing sulfate content [27,28]. These studies follow an approach seeking to establish similarities between these molecules and heparin, the gold standard. The objective is to select active compounds based on general coagulation tests (mostly clotting assays). Thereafter, the aim is to compare the effects of these polysaccharides with those of heparin in more specific assays looking for similarities between them.

This approach raises skepticism since it is difficult to believe in the possibility of mimicking the effects of heparin on the coagulation system using sulfated polysaccharides with very distinct structures. The mechanism of the anticoagulant action of heparin constitutes a paradigm of interaction between carbohydrate and protein that triggers a potent biological effect [29]. A rare pentasaccharide sequence in the heparin molecule is responsible for the interaction and conformational activation of antithrombin, which is the major plasma serpin. This is the initial step, which determines the inhibition of serpin-dependent proteases of the coagulation system, especially thrombin and factor Xa.

Attempts to simulate the effect of heparin using other sulfated polysaccharides led to criticism that the structures of these compounds cannot simulate the same effect of heparin due to the obvious absence of the specific pentasaccharide sequence with high antithrombin affinity. The observation that a sulfated galactan, with high anticoagulant activity, does not induce the same conformational change on antithrombin as heparin supports this view [30]. The skeptics attributed the anticoagulant activity of the sulfated polysaccharides from marine organisms to non-specific and low-affinity interactions between their highly sulfated carbohydrate chains and basic regions in proteins of the coagulation system.

Another event which discouraged the study of sulfated polysaccharides as anticoagulants was the report that oversulfated chondroitin sulfate has an anticoagulant effect, although with much less potency than heparin and one that is associated with a strong toxic effect. This polysaccharide activates factor XII and releases bradykinin with potent hypotensive effect. Contamination of heparin preparation with oversulfated chondroitin sulfate has been responsible for ~200 deaths in the USA [9,10]. Sulfated polysaccharides from algae and marine invertebrates may have similar effects, as already reported for a sea cucumber polysaccharide [31]. These observations further increased the disbelief in the possibility of simulating the effects of heparin with other sulfated polysaccharides.

However, it is possible to follow different routes to investigate these sulfated polysaccharides from marine organisms as anticoagulant drugs using a more innovative approach. The essential point is to look for distinct effects of these sulfated polysaccharides in the coagulation system, different from those reported for heparin. Sulfated polysaccharides with new mechanisms of action may open new avenues for therapeutic applications in thromboembolic diseases. This review will focus on some of these possibilities.

3. THE ANTICOAGULANT EFFECT OF SULFATED GALACTANS AND SULFATED FUCANS DEPENDS ON SPECIFIC STRUCTURES

In a study performed in our laboratory, we tested the anticoagulant effects of sulfated galactans and sulfated fucans obtained from marine invertebrates. More than 20 types of sulfated polysaccharides with well-defined and repetitive structures were tested. The substantial number of results obtained allowed us to ensure that the structure of the sulfated polysaccharide correlates with its specific effects on the coagulation system. Three examples of this type of approach are described below.

Initially we compared the anticoagulant effect of a set of sulfated fucans obtained from sea urchins, all composed of linear chains of 3-linked α-L-fucopyranosyl (Fuc*p*) units with different patterns of 2- and 4-sulfation. We observed that the presence of 2,4-disulfated Fuc*p* units is the major requirement for antithrombin-mediated anticoagulant activity [32].

In another approach we compared the antithrombotic effect of two types of polysaccharides rich in 2,4-disulfated α-L-Fuc*p*units. One of them was a fucosylated chondroitin sulfate isoLated from the sea cucumber. This polysaccharide has a backbone like that of vertebrate chondroitin sulfate: [4-β-D-glucuronic acid (GlcA)-1→3-β-D-*N*-acetylgalactosamine (GalNAc)-1]$_n$ but substituted at the 3-position of the β-D-GlcA with branches of 2,4-disulfated α-L-Fuc*p* [33]. Another polysaccharide was a sulfated fucan isolated from the same invertebrate but containing 2,4-disulfated α-Fuc*p* units as part of a linear chain: [α-L-Fuc*p*-2,4(OSO_3^-)-1→3-α-L-Fuc*p*-1→3-α-L-Fuc*p*-2(OSO_3^-)-1→3-α-L-Fuc*p*-2(OSO_3^-)]$_n$ [34] (Figure 2, Panels A and B). The result shown in Figure 2, Panel C, illustrates the effect of these two sulfated polysaccharides in an experimental model of arterial thrombosis. Clearly, occurrence of 2,4-disulfated α-Fuc*p* as branched residues (as in fucosylated chondroitin sulfate) ensures a more potent antithrombotic effect compared to the linear polymer containing the same type of residue [35].

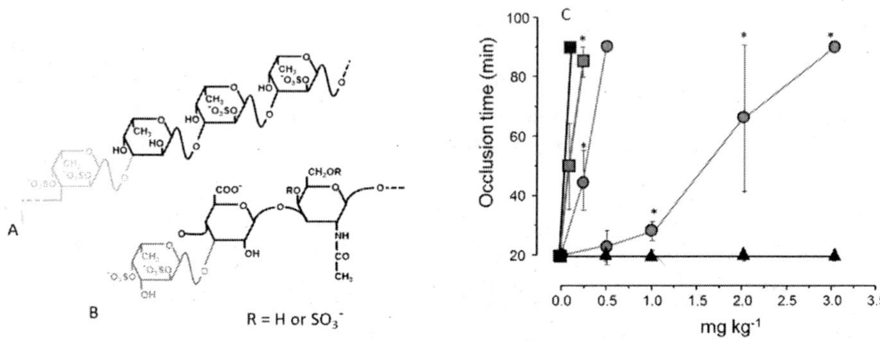

Figure 2. Structures (**A,B**) and antithrombotic effect (**C**) of sulfated polysaccharides rich in 2,4-disulfated α-L-Fuc*p* units. (**A**) The sulfated fucan is composed of [3-α-L-Fuc*p*-2,4(OSO_3^-)-1→3-α-L-Fuc*p*-1→3-α-L-Fuc*p*-2(OSO_3^-)-1→3-α-L-Fuc*p*-2(OSO_3^-)]$_n$repeating units; (**B**) Fucosylated chondroitin sulfate has a chondroitin sulfate-like backbone, but contains branches of 2,4-disulfated α-L-Fuc*p* units linked to the β-D-GlcA residues on the polysaccharide core. The 2,4-disulfated α-L-Fuc*p* units in these two polysaccharides are highlighted in green or blue; (**C**) Antithrombotic effect of sulfated fucan (●), fucosylated chondroitin sulfate (●), unfractionated heparin (■), low-molecular-weight heparin (■) and vertebrate chondroitin 6-sulfate (▲) on an arterial thrombosis model induced in carotid artery of rats by laser irradiation. See Ref. [35] for further details.

Another example of the stereospecificity of the anticoagulant effect of sulfated polysaccharides arises from comparison between a α-L-fucan and a α-L-galactan, both 2-sulfated and 3-linked (Figure 3, Panels A and B). The sulfated α-galactan is significantly more active than the sulfated α-fucan as an antithrombin-mediated thrombin inhibitor (Figure 3, Panel C) [36].

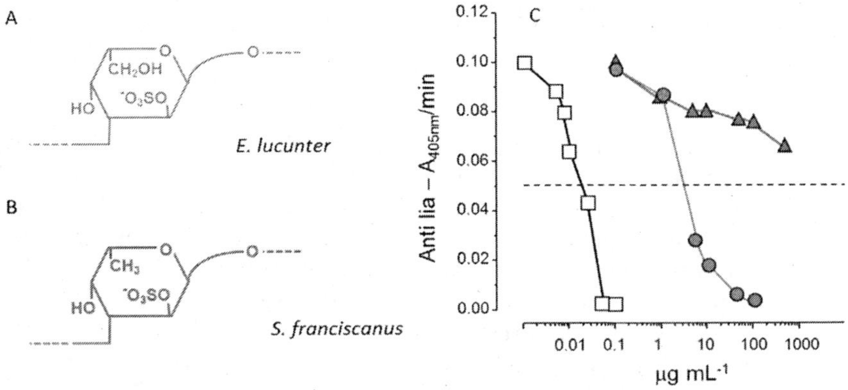

Figure 3. Structures and anticoagulant effect of a sulfated galactan and a sulfated fucan. (**A,B**) Structures of the sulfated α-L-galactan from the sea urchin *E. lucunter* (in blue, Panel A) and sulfated α-L-fucan from *S. franciscanus* (in red, Panel B). Both polysaccharides are 3-linked with a regular sulfation at the 2-position, but differ in their constituent monosaccharides; (**C**) Thrombin inhibition mediated by antithrombin*vs.* concentrations of the sulfated galactan (•), sulfated fucan (▲) and heparin (□). See Ref. [36] for further details.

In conclusion, pattern of sulfation, position of the residue, either internal or non-reducing unit, and type of constituent sugar [α-L-galactopyranosyl (Gal*p*) or α-L-Fuc*p* unit] ensure the anticoagulant and antithrombotic effects. The anticoagulant activity of the polysaccharide is not only a consequence of negative charge density, but of precise structural constituents.

These studies were possible using the sulfated polysaccharides from marine invertebrates that possess well-defined structures. However, these polysaccharides occur in low concentrations, or some of the invertebrate species are scarce. In several cases, we can obtain these polysaccharides only in small amounts, which limits their practical use. A practical alternative to develop antithrombotic drug is to employ sulfated polysaccharides from marine algae, which are abundant. Nevertheless, the invertebrate polymers are model molecules and may reveal a specific structural sequence we need to look for in the algal polysaccharides.

4. SERPIN-INDEPENDENT ANTICOAGULANT EFFECT

Serpins (as antithrombin and heparin cofactor II) mediate the anticoagulant activity of heparin, resulting in the inhibition of the coagulation proteases. Consequently, the effect of that glycosaminoglycan disappears on serpins-depleted plasmas. Sulfated polysaccharides from marine organisms have a similar serpin-mediated activity, demonstrated on assays using purified reagents (proteases and serpins) [16,32,33,36]. It intrigued us when it was observed that the anticoagulant effect of a fucosylated chondroitin sulfate from the sea cucumber [37] and of a sulfated galactan from red algae [38] remains unchanged when tested in serpin-depleted plasma. This observation is even more striking when tests of thrombin or factor Xa generation in serpin-depleted plasma was employed. Clearly, the inhibitory effect of the blood coagulation proteases disappears in tests using heparin, but persists with fucosylated chondroitin sulfate or sulfated galactan [37,38].

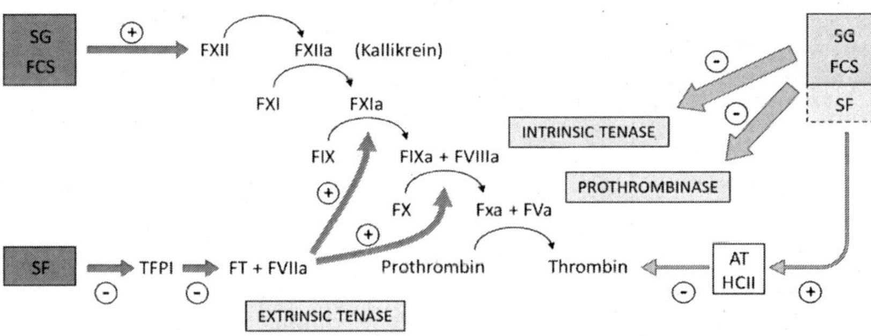

Figure 4. Major target sites for the sulfated polysaccharides from marine organisms on the coagulation system. Blue and red arrows indicate anticoagulant and pro-coagulant effects, respectively. + indicates activation and – indicates inhibitory effects. Anticoagulant effect: sulfated galactans (SG) from marine red algae and fucosylated chondroitin sulfates (FCS) from sea cucumbers inhibit the intrinsic tenase and prothrombinase complexes [37,38]. It is still unclear if sulfated fucans (SF) have similar effects. These polysaccharides also potentiate the inhibitory effect of antithrombin (AT) and/or heparin cofactor II (HCII) on thrombin [16,32,33,36]. Their effects on factor Xa are very modest. The serpin-independent action preponderates on the plasma system. Pro-coagulant effect: SG and FCS activate factor XII [31,39]. This effect may result in severe hypotension (due to bradykinin release) and pro-coagulant (and pro-thrombotic) action. It is unclear if SF activates factor XII. SF inhibits Tissue Factor Protease Inhibitor (TFPI), a specific inhibitor of the extrinsic tenase complex. Consequently, SF has a pro-coagulant effect [40,41].

Further examination revealed that fucosylated chondroitin sulfate and sulfated galactan inhibit the intrinsic tenase and prothrombinase complexes that are critical for the generation of factor Xa and thrombin. Assays using purified proteins of the coagulation system assure this conclusion. The sulfated polysaccharides inhibit the interaction between factor Va and factor X [38]. These results reveal an anticoagulant effect with a distinct mechanism of action compared with heparin or any other known antithrombotic drug. Figure 4 summarizes the preponderant target sites for the sulfated polysaccharides from marine organisms on the coagulation system. Blue arrows at right of the Figure indicate the anticoagulant effects.

Of course, further studies are necessary to investigate whether this distinct mechanism of action may confer favorable effects to the marine polysaccharides for the prevention and treatment of thromboembolic events. In particular, it is necessary to clarify which one of the two mechanisms (serpin-dependent or serpin-independent) is more favorable for an antithrombotic therapy.

5. SULFATED POLYSACCHARIDES WITH LOW ANTICOAGULANT ACTIVITY MAY HAVE POTENT ANTITHROMBOTIC EFFECTS

Screening for new anticoagulant polysaccharides usually requires an initial assessment with general coagulation tests. The compound, which is most active, moves to a more detailed stage of evaluation using animal testing. Following this methodological guideline, our laboratory carried out a study of sulfated polysaccharides from 50 species of marine algae. The work leaded to a sulfated galactan from the red alga *Botryocladia occidentalis* with very high anticoagulant activity [16]. This polysaccharide has a relatively simple structure as compared with that of similar compounds obtained from seaweeds. The presence of 2,3-disulfated α-Gal*p* units was associated with the high anticoagulant activity of the sulfated polysaccharide.

However, we can proceed in a different way, looking for sulfated polysaccharides with low anticoagulant activity. We can propose, conceptually, that a polysaccharide with low activity in coagulation tests using *in vitro* assays may exhibit a potent antithrombotic effect when tested in animal models of thrombosis.

We followed this approach and ended up with a sulfated galactan from the red alga *Acanthophora muscoides* with low anticoagulant activity. Despite this data, we continued to go deeper with the coagulation study. The polysaccharide presented a very complex structure. Its backbone consists of alternating units of α- and β-Gal*p* units, as is the characteristic of seaweed galactans. However, we observed a wide variety of structural modifications, such as sulfation in different positions, the presence of methyl ether, and anhydrous sugars. Another unique feature of this sulfated galactan is its low molecular weight (~20 kDa) [19]. Most seaweed-sulfated galactans have extremely high molecular weights.

We reduced the molecular weight of the sulfated galactan from *B. occidentalis* using mild acid hydrolysis and obtained a derivative with the same molecular weight as that of the native polysaccharide from *A. muscoides*. The clotting assay using these two sulfated galactans, with similar molecular weights, revealed that the compound from *B. occidentalis* was more active (Figure 5, Pane A, closed circles in blue *vs.* closed diamond in red). However, more importantly, it turns out that these two sulfated galactans, with low molecular weights, are practically inactive in coagulation assays using purified coagulation proteases and serpins [19]. That is, its anticoagulant effect, albeit modest as compared to heparin, is independent of serpins. The clotting assays using serpin-depleted plasma confirmed this proposition (Figure 5, Panels B and D, compare closed *vs.* open symbols).

The next step was to test these sulfated galactans, with low anticoagulant activity and serpin-free effects, in animal models of experimental thrombosis. Figure 5, Panel C, shows the test using an experimental model of arterial thrombosis. Surprisingly, the sulfated galactan from *A. muscoides* is very active, whereas that obtained from *B. occidentalis* is completely inactive (closed circles in blue *vs.* closed diamond in red).

These results are very important because they reveal a sulfated polysaccharide, with only modest anticoagulant action on *in vitro* assay but potent arterial antithrombotic effect in the experimental animal model. We attribute this effect to the inhibition of the tenase and prothrombinase complexes that generate the coagulation proteases. This approach consolidates the molecular basis of a new antithrombotic mechanism by a sulfated polysaccharide, involving inhibition of factor Xa and thrombin generations by the tenase and prothrombinase systems.

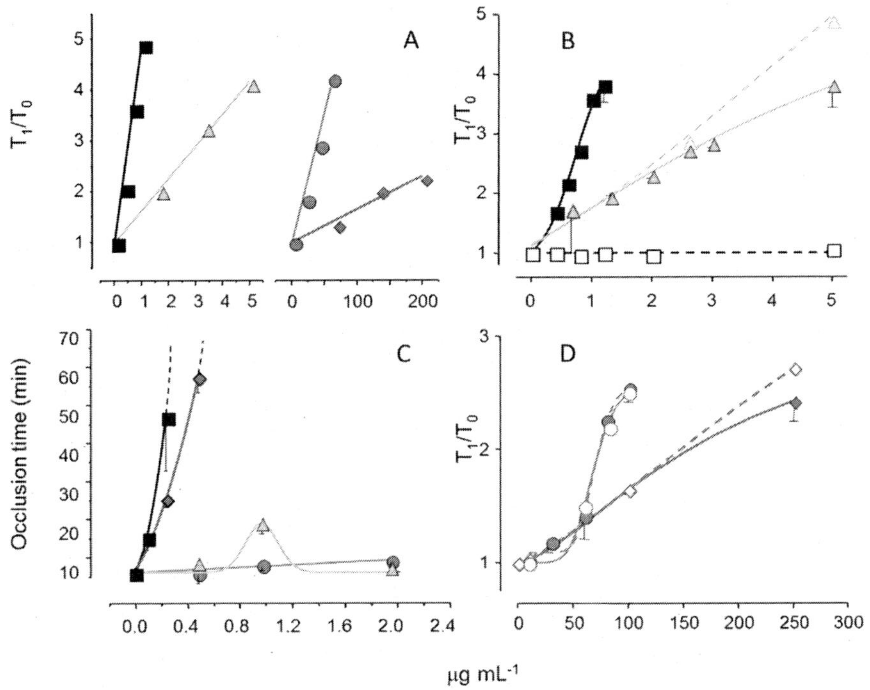

Figure 5. Effect of sulfated galactans on anticoagulant assays (**A,C,D**) and on arterial experimental thrombosis in rats (**B**). Different concentrations or doses of heparin (■,□), native sulfated galactan from *B. occidentalis* (▲,Δ) or from *A. muscoides* (♦,◊) or the sulfated galactan from *B. occidentalis* with reduced molecular weight (●,○) were tested. In Panels C and D closed and open symbols refer to assays performed using normal or serpin-depleted plasma, respectively. Data from Ref. [19].

6. BALANCE BETWEEN PRO- AND ANTICOAGULANT EFFECTS

Sulfated galactans have no bleeding tendency [19,39,42], which is the major side effect of heparin and other antithrombotic drugs. Instead, the major side effect of these polysaccharides is activation of factor XII. This action results in the release of bradykinin, leading to severe hypotension [9,10,31]. Conceptually, we expect that activation of factor XII should trigger coagulation, resulting in pro-thrombotic action. In the case of the sulfated galactan from the red alga *B. occidentalis* we observed a dual effect, as a potent anticoagulant and as an activator of factor XII [39]. This observation is very clear with the use of a recalcification time assay, without the addition of phospholipids (Figure 6, Panel A). Heparin has a potent anticoagulant action (open squares), while the sulfated galactan has a dual effect: up to ~8 μg.mL^{-1}, an anticoagulant effect, and above 8 μg.mL^{-1}, a pro-coagulant effect (closed triangles in Figure 6, Panel A). A 14 kDa fragment shows the same effect as the native polysaccharide but at different concentrations (open circles). Decrease of the molecular weight to ~5 kDa eliminates this dual effect (closed circles).

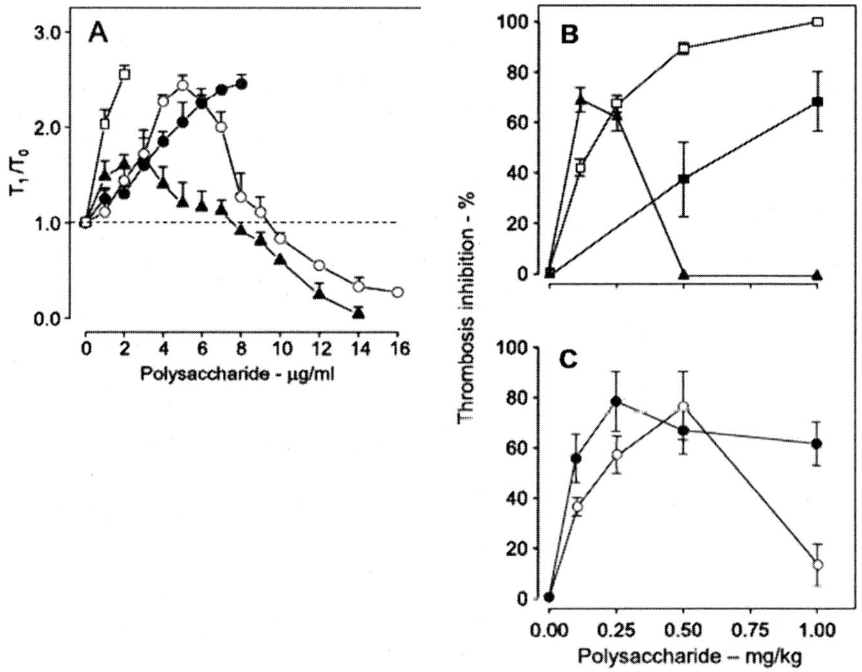

Figure 6. Anticoagulant activity based on recalcification time (**A**) and antithrombotic effect on a stasis thrombosis model in vena cava of rats (**B,C**). Different concentrations (**A**) or doses (**B,C**) of unfractionated heparin (□), low-molecular-weight heparin (■), native sulfated galactan (▲), ~14 kDa (O) and ~5 kDa (•) fragments were tested. In Panel A the anticoagulant activity was expressed as T_1/T_0, which is the ratio between the clotting time in the presence or absence of sulfated polysaccharide. The broken line ($T_1/T_0 = 1$) indicates no effect from the polysaccharide on coagulation. Data from Ref. [39].

Because of the dual effect on coagulation, this sulfated galactan has two distinct effects in the experimental thrombosis model. At lower doses, up to approximately 0.25 mg.kg^{-1} body weight, it is a potent antithrombotic, similar to heparin; at higher doses, above 0.25 mg.kg^{-1} body weight, this effect disappears and is replaced by a pro-coagulant effect (Figure 6, Panel B), due to activation of factor XII.

In order to develop an antithrombotic drug for practical use, it is essential to dissociate the anticoagulant action from the undesirable pro-coagulant effect. We achieved this goal by means of partial acid hydrolysis of the polysaccharide. The sulfated galactan with low molecular weight (~5 kDa) proved to be a promising antithrombotic, devoid of pro-coagulant action (Figure 6, Panel C, closed circles) [39].

Curiously, even fucoidan presents the concomitant anti- and pro-coagulant effects [40,41]. Selected modified derivatives from this polysaccharide, devoid of anticoagulant activity, have been used as a therapeutic option in bleeding disorders, such as hemophilia. The pro-coagulant effect of fucoidan is distinct from the activation of factor XII proposed for the sulfated galactans. It involves a blockade of the extrinsic pathway down-regulator, the tissue factor pathway inhibitor, by the non-anticoagulant fucoidan derivatives. Red arrows at left of Figure 4 indicate the pro-coagulant effects of the sulfated polysaccharides from marine organisms.

Protamine neutralizes the effects of sulfated polysaccharides from marine organisms on the coagulation system, as occurs with heparin. It decreases the pro-coagulant effect of a sulfated galactan from red alga [42] and the anticoagulant action of fucosylated chondroitin sulfate from sea cucumber. These observations are relevant to ensure we can antagonize the effects of the polysaccharides if severe side effects occur. Induction of thrombocytopenia with the sulfated polysaccharides from marine organisms is another important aspect that requires future investigation.

7. SULFATED POLYSACCHARIDE WITH ORAL ANTITHROMBOTIC EFFECTS

The use of sulfated polysaccharides as antithrombotic drug requires intravascular or subcutaneous administration, which limits their use. The new oral antithrombotic agents, which directly inhibit thrombin or factor Xa, have overcome this limitation although they may cause bleeding and have unpredicted responses [43,44].

Sulfated polysaccharides are not promising for oral administration due to their high molecular weight. Even attempts to administer heparin orally were also unsuccessful. Possibly, degradation of heparin by bacteria of the intestinal flora or by enzymes produced by the vertebrate also reduced its absorption after oral administration. However, the different chemical nature of the sulfated galactans and sulfated fucans from marine organisms makes them resistant to degradation by most enzymes that normally act on vertebrate glycosaminoglycans.

A number of studies indicate biological effects after oral ingestion of sulfated fucans from brown algae (fucoidan) [45]. Early work on digestion of these polysaccharides suggested that they were not changed by human bacterial flora and were wholly excreted [46,47]. Following oral administration to rats, we observed that fucosylated chondroitin sulfate has anticoagulant and antithrombotic effects, with low hemorrhagic action [48]. This finding allowed us to overcome the main limitation to the development of new antithrombotics based on sulfated polysaccharides. Of course, we need to extend this kind of study to other types of sulfated polysaccharides from marine organisms, especially sulfated galactans and sulfated fucans.

The observation that a sulfated polysaccharide has antithrombotic effect after oral administration raises a challenging question: how extensive is the gastrointestinal absorption of this polysaccharide? Answering this question requires a careful investigation of the metabolism of sulfated polysaccharides from marine organisms after administration to mammals, including their route of absorption and pharmacokinetics. Studies about the intestinal absorption of fucoidan have been reported recently [49].

8. CONCLUSIONS

In this review, we discussed possible ways to proceed in order to develop new antithrombotic drugs based on sulfated polysaccharides from marine organisms. These compounds occur in high amounts in marine algae. However, their therapeutic use in humans faces several challenges. We need to look for alternative approaches in order to establish a new paradigm, instead of continuing to look at them as "heparin-like molecules." Possibly, we need to consider the following points:

(1) Precisely define the specific structures required for the interaction of sulfated polysaccharides with the proteins and complexes of the coagulation system.

(2) Look for alternative mechanisms of action distinct from those of heparin, including serpin-independent effects. Clarification of the precise target molecules of these sulfated polysaccharides will help to establish new drugs on a more rational basis.

(3) Identify major side effects in order to design the preparation of derivatives that retain the desirable antithrombotic action but are devoid of toxic ones. Pro-coagulant effect and hypotension, both due to activation of factor XII, instead of bleeding, may be the major obstacle for the therapeutic use of sulfated polysaccharides from marine organisms.

(4) Consider that sulfated polysaccharides with low activity on *in vitro* coagulation tests may exhibit potent activity on animal models of experimental thrombosis.

(5) Investigate the effect of these sulfated polysaccharides after oral administration. If unsuccessful, we may attempt to prepare derivatives that achieve this effect. The discovery of new oral anticoagulants derived from the sulfated polysaccharides of marine organisms will put these compounds on the frontier of antithrombotic therapeutics.

ACKNOWLEDGMENTS

The author would like to acknowledge the funding from Conselho Nacional de Desenvolvimento Científico e Tecnológico (CNPq) and Fundação de Amparo à Pesquisa do Estado do Rio de Janeiro (FAPERJ).

CONFLICTS OF INTEREST

The author declares no conflict of interest.

REFERENCES

1. Norma, R. 100 years. Looking back. Skill, drive and luck: The discovery and development of heparin. *CMAJ* **2011**, *13*, 2139–2140
2. Mulloy, B.; Gray, E.; Barrowcliffe, T.W. Characterization of unfractionated heparin: Comparison of materials from the last 50 years. *Thromb. Haemost.* **2000**, *84*, 1052–1056.
3. Tagarakis, G.; Tsilimingas, N.B. Heparin-coated extracorporeal circulation systems in heart surgery. *Recent Pat. Cardiovasc. Drug Discov.* **2009**, *4*, 177–179.

4. Molitor, B.; Klingel, R.; Hafner, G. Monitoring of the heparin therapy during acute haemodialysis. *Hamostaseologie* **2005**, *25*, 272–278.

5. Alban, S. Adverse Effects of Heparin. *Handb. Exp. Pharmacol.* **2012**, *207*, 211–264.

6. Introduction of Proposal for Reintroduction of Bovine Heparin to the U.S Market. Available online: http://www.fda.gov/downloads/UCM399418.pdf (access ed on 4 March 2015).

7. Aquino, R.S.; Pereira, M.S.; Vairo, B.C.; Cinelli, L.P.; Santos, G.R.C.; Fonseca, R.J.C.; Mourão, P.A.S. Heparins from porcine and bovine intestinal mucosa: Are they similar drugs? *Thromb. Haemost.* **2010**, *103*, 1005–1015.

8. Chemaly, R.F.; Rathod, D.B.; Sikka, M.K.; Hayden, M.K.; Hutchins, M.; Tracy Horn, T.; Jeffery Tarrand, J.; Javier Adachi, J.; Kim Nguyen, K.; Trenholme, G.; *et al.* *Serratia marcescens* bacteremia because of contaminated prefilled heparin and saline syringes: A multi-state report. *Am. J. Infect. Control* **2011**, *39*, 521–524.

9. Kishimoto, T.K.; Viswanathan, K.; Ganguly, T.; Elankumaran, S.; Smith, S.; Pelzer, K.; Lansing, J.C.; Sriranganathan, N.; Zhao, G.; Galcheva-Gargova, Z.; *et al.* Contaminated heparin associated with adverse clinical events and activation of the contact system. *N. Engl. J. Med.* **2008**, *358*, 2457–2467.

10. Guerrini, M.; Beccati, D.; Shriver, Z.; Naggi, A.; Viswanathan, K.; Bisio, A.; Capila, I.; Lansing, J.C.; Guglieri, S.; Fraser, B.; *et al.* Oversulfated chondroitin sulfate is a contaminant in heparin associated with adverse clinical events. *Nat. Biotechnol.* **2008**, *26*, 669–675.

11. Melo, E.I.; Pereira, M.S.; Cunha, R.S.; Sá, M.P.; Mourão, P.A.S. Heparin quality control in the Brazilian market: Implications in the cardiovascular surgery. *Rev. Bras. Cir. Cardiovasc.* **2008**, *23*, 169–174.

12. Mourão, P.A.S. Use of sulfated fucans as anticoagulant and antithrombotic agents: Future perspectives. *Curr. Pharm. Des.* **2004**, *10*, 967–981.

13. Pomin, V.H.; Mourão, P.A.S. Structure, biology, evolution, and medical importance of sulfated fucans and galactans. *Glycobiology* **2008**, *18*, 1016–1027.

14. Mourão, P.A.S. A carbohydrate-based mechanism of species recognition in sea urchin fertilization. *Braz. J. Med. Biol. Res.* **2007**, *40*, 5–17.

15. Usov, A.I. Structural analysis of red seaweed galactans of agar and carrageenan groups. *Food Hydrocoll.* **1998**, *12*, 301–308.

16. Farias, W.R.; Valente, A.P.; Pereira, M.S.; Mourão, P.A.S. Structure and anticoagulant activity of sulfated galactans. Isolation of a unique sulfated galactan from the red algae *Botryocladia occidentalis* and comparison of its anticoagulant action with that of sulfated galactans from invertebrates. *J. Biol. Chem.* **2000**, *275*, 29299–29307.

17. Lahaye, M. Developments on gelling algal galactans, their structure and physico-chemistry. *J. Appl. Phycol.* **2001**, *13*, 173–184.

18. Pereira, M.G.; Benevides, N.M.; Melo, M.R.; Valente, A.P.; Melo, F.R.; Mourão, P.A.S. Structure and anticoagulant activity of a sulfated galactan from the red alga, *Gelidium crinale*. Is there a specific structural requirement for the anticoagulant action? *Carbohydr. Res.* **2005**, *340*, 2015–2023.

19. Quinderé, A.L.G.; Santos, G.R.C.; Oliveira, S.N.M.C.G.; Glauser, B.F.; Fontes, B.F.; Queiroz, I.N.L.; Benevides, N.M.B.; Pomin, V.H.; Mourão, P.A.S. Is the antithrombotic effect of sulfated galactans independent of serpin? *J. Thromb. Haemost.* **2014**, *12*, 43–53.

20. Alves, A.P.; Mulloy, B.; Diniz, J.A.; Mourão, P.A.S. Sulfated polysaccharides from the egg jelly layer are species-specific inducers of acrosomal reaction in sperms of sea urchins. *J. Biol. Chem.* **1997**, *272*, 6965–6971.

21. Castro, M.O.; Pomin, V.H.; Santos, L.L.; Vilela-Silva, A.C.E.S.; Hirohashi, N.; Pol-Fachin, L.; Verli, H.; Mourão, P.A.S. A unique 2-sulfated β-galactan from the egg jelly of the sea urchin *Glyptocidaris crenularis*: Conformational flexibily*versus* induction of the sperm acrosome reaction. *J. Biol. Chem.* **2009**, *284*, 18790–18800.

22. Chevolot, L.; Foucault, A.; Kervarec, N.; Sinquin, C.; Fisher, A.M.; Boisson-Vidal, C. Further data on the structure of brown seaweed fucans: Relationships with anticoagulant activity. *Carbohydr. Res.* **1999**, *319*, 154–165.

23. Chevolot, L.; Mulloy, B.; Ratiskol, J.; Foucault, A.; Colliec-Jouault, S. A disaccharide repeat unit is the major structure in fucoidans

from two species of brown algae. *Carbohydr. Res.* **2000**, *330*, 529–535.

24. Bilan, M.I.; Grachev, A.A.; Ustuzhanina, N.E.; Shashkov, A.S.; Nifantiev, N.E.; Usov, A.I. Structure of a fucoidan from the brown seaweed *Fucus evanescens*. *Carbohydr. Res.* **2002**, *337*, 719–730.

25. Vilela-Silva, A.C.E.S.; Alves, A.P.; Valente, A.P.; Vacquier, V.D.; Mourão, P.A.S. Structure of the sulfated α-L-fucan from the egg jelly coat of the sea urchin *Strongylocentrotus franciscanus*: Patterns of preferential 2-*O*- and 4-*O*-sulfation determine sperm cell recognition. *Glycobiology* **1999**, *9*, 225–232.

26. Vilela-Silva, A.C.E.S.; Castro, M.O.; Valente, A.P.; Biermann, C.H.; Mourão, P.A.S. Sulfated fucans from the egg jellies of the closely related sea urchins *Strongylocentrotus droebachiensis* and *Strongylocentrotus pallidus* ensure species specific fertilization. *J. Biol. Chem.* **2002**, *277*, 379–387.

27. Soeda, S.; Ohmagari, Y.; Shimeno, H.; Nagamatsu, A. Preparation of oversulfated fucoidans fragments and evaluation of their antithrombotic activities. *Thromb. Res.* **1993**, *72*, 247–256.

28. Haroun-Bouhedja, F.; Ellouali, M.; Sinquin, C.; Boisson-Vidal, C. Relationship between sulfate groups and biological activities of fucans. *Thromb. Res.* **2000**, *100*, 453–458.

29. Gray, E.; Hogwood, J.; Mulloy, B. The anticoagulant and antithrombotic mechanisms of heparin. *Handb. Exp. Pharmacol.* **2012**, *207*, 347–360.

30. Melo, F.R.; Pereira, M.S.; Foguel, D.; Mourão, P.A.S. Antithrombin-mediated anticoagulant activity of sulfated polysaccharides: Different mechanisms for heparin and sulfated galactans. *J. Biol. Chem.* **2004**, *279*, 20824–20835.

31. Fonseca, R.J.C.; Oliveira, S.N.M.C.G.; Pomin, V.H.; Mecawi, A.S.; Iracema, G.; Araujo, I.G.; Mourão, P.A.S. Effects of oversulfated and fucosylated chondroitin sulfates on coagulation, Challenges for the study of anticoagulant polysaccharides. *Thromb. Haematol.* **2010**, *103*, 994–1004.

32. Pereira, M.S.; Melo, F.R.; Mourão, P.A.S. Is there a correlation between structure and anticoagulant action of sulfated galactans and sulfated fucans? *Glycobiology* **2002**, *12*, 573–580.

33. Mourão, P.A.S.; Pereira, M.S.; Pavão, M.S.G.; Mulloy, B.; Tollefsen, D.M.; Mowinckel, M.C.; Abildgaard, U. Structure and anticoagulant activity of a fucosylated chondroitin sulfate from

echinoderm: Sulfated fucose branches on the polysaccharide account for its high anticoagulant activity. *J. Biol. Chem.* **1996**, *271*, 23973–23984.

34. Ribeiro, A.C.; Vieira, R.P.; Mourão, P.A.S.; Mulloy, B. A sulfated α-L-fucan from sea cucumber. *Carbohydr. Res.* **1994**,*255*, 225–240.

35. Fonseca, R.J.C.; Santos, G.R.C.; Mourão, P.A.S. Effects of polysaccharides enriched in 2,4-disulfated fucose units on coagulation, thrombosis and bleeding. Practical and conceptual implications. *Thromb. Haematol.* **2009**, *102*, 829–836.

36. Pereira, M.S.; Vilela-Silva, A.C.E.S.; Valente, A.P.; Mourão, P.A.S. A 2-sulfated, 3-linked α-L-galactan is an anticoagulante polysaccharide. *Carbohydr. Res.* **2002**, *337*, 2231–2238.

37. Glauser, B.F.; Pereira, M.S.; Monteiro, R.Q.; Mourão, P.A.S. Serpin-independent anticoagulant activity of a fucosylated chondroitin sulfate. *Thromb. Haemost.* **2008**, *100*, 420–428.

38. Glauser, B.F.; Rezende, R.M.; Melo, F.R.; Pereira, M.S.; Francischetti, I.M.B.; Monteiro, R.Q.; Rezaie, A.R.; Mourão, P.A.S. Anticoagulant activity of a sulfated galactan: Serpin-independent effect and specific interaction with factor Xa. *Thromb. Haemost.* **2009**, *102*, 1183–1193.

39. Melo, F.R.; Mourão, P.A.S. Sulfated galactan has an unusual effect on thrombosis due to activation of factor XII and inhibition of the coagulation proteases. *Thromb. Haemost.* **2008**, *99*, 531–538.

40. Liu, T.; Scallan, C.D.; Broze, G.J., Jr.; Patarroyo-White, S.; Pierce, G.F.; Johnson, K.W. Improved coagulation in bleeding disorders by non-anticoagulant sulfated polysaccharides (NASP). *Thromb. Haemostas.* **2006**, *95*, 68–76.

41. Zhang, Z.; Till, S.; Jiang, C.; Knappe, S.; Reutterer, S.; Scheiflinger, F.; Szabo, C.M.; Dockal, M. Structure-activity relationship of the pro- and anticoagulant effects of *Fucus vesiculosus* fucoidan. *Thromb. Haemost.* **2014**, *111*, 429–437.

42. Farias, W.F.; Nazareth, R.A.; Mourão, P.A.S. Dual effects of sulfated d-galactans from the red algae *Botryocladia occidentalis* preventing thrombosis and inducing platelet aggregation. *Thromb. Haemost.* **2001**, *86*, 1540–1546.

43. Holster, I.L.; Valkhoff, V.E.; Kuipers, E.J.; Tjwa, E.T. New oral anticoagulants increase risk for gastrointestinal bleeding: A systematic review and meta-analysis. *Gastroenterology* **2013**, *145*, 105–112.

44. Holster, I.L.; Hunfeld, N.G.; Kuipers, E.J.; Kruip, M.J.; Tjwa, E.T. On the treatment of new oral anticoagulant-associated gastrointestinal hemorrhage. *J. Gastrointestin. Liver Dis.* **2013**, *22*, 229–231.

45. Fitton, J.H. Therapies from fucoidan; multifunctional marine polymers. *Mar. Drugs* **2011**, *9*, 1731–1760.

46. Michel, C.; Lahaye, M.; Bonnet, C.; Mabeau, S.; Barry, J.L. *In vitro* fermentation by human fecal bacteria of total and purified dietary fibers from brown seaweeds. *Br. J. Nutr.* **1996**, *75*, 263–280.

47. Yamada, Y.; Miyoshi, T.; Tanada, S.; Imaki, M. Digestibility and energy availability of Wakame (*Undaria pinnatifida*) seaweed in Japanese. *Nippon Eiseigaku Zasshi* **1991**, *46*, 788–794.

48. Fonseca, R.J.C.; Mourão, P.A.S. Fucosylated chondroitin sulfate as a new oral antithrombotic agent. *Thromb. Haemost.* **2006**, *96*, 822–829.

49. Nagamine, T.; Nakazato, K.; Tomioka, S.; Iha, M.; Nakajima, K. Intestinal absorption of fucoidan extracted from the brown seaweed, *Cladosiphono kamuranus*. *Mar. Drugs* **2015**, *13*, 48–64.

7

CHAPTER

MytiLec, a Mussel R-Type Lectin, Interacts with Surface Glycan Gb3 on Burkitt's Lymphoma Cells to Trigger Apoptosis through Multiple Pathways

Imtiaj Hasan [1,2,†], Shigeki Sugawara [3,†], Yuki Fujii [4], Yasuhiro Koide [1], Daiki Terada [5], Naoya Iimura [1], Toshiyuki Fujiwara [4], Keisuke G. Takahashi [6], Nobuhiko Kojima [1], Sultana Rajia [1,7], Sarkar M. A. Kawsar [1,8], Robert A. Kanaly [1], Hideho Uchiyama [1], Masahiro Hosono [3], Yukiko Ogawa [4], Hideaki Fujita [4], Jiharu Hamako [9], Taei Matsui [9] and Yasuhiro Ozeki [1,*]

[1]*Department of Life and Environmental System Science, Graduate School of NanoBio Sciences, Yokohama City University, 22-2 Seto, Kanazawa-ku, Yokohama 236-0027, Japan*
[2]*Department of Biochemistry and Molecular Biology, Faculty of Science, University of Rajshahi, Rajshahi-6205, Bangladesh*
[3]*Division of Cell Recognition Study, Institute of Molecular Biomembrane and Glycobiology, Tohoku Pharmaceutical University, 4-4-1 Komatsushima, Aoba-ku, Sendai 981-8558, Japan*
[4]*Department of Pharmacy, Faculty of Pharmaceutical Science, Nagasaki International University, 2825-7 Huis Ten Bosch, Sasebo, Nagasaki 859-3298, Japan*
[5]*Graduate School of Medical Life Science, Yokohama City University, 1-7-29 Suehiro-cho, Tsurumi-ku, Yokohama 230-0045, Japan*
[6]*Graduate School of Agricultural Science, Tohoku University, 1-1 Amamiya-machi, Tsutsumidori, Aoba-ku, Sendai 230-0045, Japan*
[7]*Department of Natural Science, Varendra University, Rajshahi-6204, Bangladesh*
[8]*Department of Chemistry, Faculty of Sciences, University of Chittagong, Chittagong-4331, Bangladesh*
[9]*Department of Biology, School of Health Sciences, Fujita Health University, Toyoake, Aichi 470-1192, Japan*

ABSTRACT

MytiLec; a novel lectin isolated from the Mediterranean mussel (Mytilus galloprovincialis); shows strong binding affinity to globotriose (Gb3: Galα1-4Galβ1-4Glc). MytiLec revealed β-trefoil folding as also found in the ricin B-subunit type (R-type) lectin family, although the amino acid sequences were quite different. Classification of R-type lectin family members therefore needs to be based on conformation as well as on primary structure. MytiLec specifically killed Burkitt's lymphoma Ramos cells, which express Gb3. Fluorescein-labeling assay revealed that MytiLec was incorporated inside the cells. MytiLec treatment of Ramos cells resulted in activation of both classical MAPK/ extracellular signal-regulated kinase and extracellular signal-regulated kinase (MEK-ERK) and stress-activated (p38 kinase and JNK) Mitogen-activated protein kinases (MAPK) pathways. In the cells, MytiLec treatment triggered expression of tumor necrosis factor (TNF)-α (a ligand of death receptor-dependent apoptosis) and activation of mitochondria-controlling caspase-9 (initiator caspase) and caspase-3 (activator caspase). Experiments using the specific MEK inhibitor U0126 showed that MytiLec-induced phosphorylation of the MEK-ERK pathway up-regulated expression of the cyclin-dependent kinase inhibitor p21, leading to cell cycle arrest and TNF-α production. Activation of caspase-3 by MytiLec appeared to be regulated by multiple different pathways. Our findings, taken together, indicate that the novel R-type lectin MytiLec initiates programmed cell death of Burkitt's lymphoma cells through multiple pathways (MAPK cascade, death receptor signaling; caspase activation) based on interaction of the lectin with Gb3-containing glycosphingolipid-enriched microdomains on the cell surface.

Keywords: Burkitt's lymphoma cells; caspase-9/3; globotriose (Gb3); JNK; Mytilus galloprovincialis; MEK/ERK; MytiLec; p21: p38 kinase; R-type lectin; TNF-α; β-trefoil

1. INTRODUCTION

During the present decade, genome databases have been established for numerous mollusk species, and are very useful in industrial and ecological studies [1]. "MytiBase" is an impacted EST (expressed sequence tags) database for the Mediterranean mussel, Mytilus galloprovincialis, a worldwide species that plays major roles in marine ecosystems and is commercially important [2,3]. Recent advances in mussel genomics have elucidated the crucial genes in various physiological processes, and shown that certain signal transduction molecules that regulate innate immunity in mussels have analogues in higher animals [4,5,6]. Recent studies have focused on such genes that are capable of being activated by various environmental pollutants. Progress in this area will have important ecological, economic, and human health implications.

Mussels have been found to produce several interesting bioactive compounds. A furan fatty acid isolated from the green-lipped mussel (Perna canaliculus) suppresses inflammatory processes [7] and breast cancer cell growth in vitro [8], and is currently undergoing clinical trials [8,9]. Macromolecules isolated from mussels or other bivalves are useful in food science and as research tools in life sciences [10].

A major focus in glycobiology is the regulation of living systems through molecular interactions between glycans (polysaccharides) and their receptors, which include glycan-binding proteins termed "lectins". Monosaccharides comprising a glycan chain are arranged according to axial or equatorial configuration of hydroxyl groups around each carbon atom. The glycosidic linkages are connections between anomeric carbons of adjacent monosaccharides. Changes in length and branching of glycan chains occur during disease processes, ontogenic development, differentiation, and regeneration [11]. Changes in somatic cells that lead to cancer or reprogramming are often associated with striking changes in glycan chains of glycoproteins and glycosphingolipids to an immature or ancestral form similar to glycans found in Golgi bodies or lower organisms [11,12].

MytiLec is an α-Gal-binding lectin that we isolated in 2012 from the Mediterranean mussel Mytilus galloprovincialis (family Mytilidae). Its primary structure is a 17 kDa polypeptide (149 amino acids, including one Trp and no Cys) containing triple tandem-repeating 50-a.a. subdomains [13,14]. The cDNA sequence coding MytiLec has also been deposited in the MytiBase EST library [15]. Deduced a.a. from cDNA coding a Gal/GalNAc-binding lectin isolated from another mytilid mussel,Crenomytilus grayanus, gives similar primary structure [16,17]. We showed (by frontal affinity chromatography) that MytiLec binds specifically to globotriose (Gb3; Galα1-4Galβ1-4Glc) and is toxic to Burkitt's lymphoma cells (which express globotriaosylceramide-containing Gb3 glycan on the cell surface) [13].

Nevertheless, the primary structure of MytiLec has been quite novel, its 3-D structure obtained by crystallography analysis of MytiLec identified that the lectin had a β-trefoil folding (PDB entry 3wmu at 1.1 Å and PDB entry 3wmv at 1.05 Å [18] that is found in the B-subunit of ricin [19], a representative plant toxin isolated from castor beans (Ricinus communis). Lectins having the β-trefoil folding and similar primary structures with the B-subunit of ricin are included in the "R-type lectin family". In addition to possessing this characteristic folding, MytiLec was a non-covalently bound dimer consisting of two polypeptides having the glycan-binding activity in each sub-domain (Jeremy R.H. Tame, personal communication). Six carbohydrate-binding sites of dimer MytiLec was essential for the cytotoxicity (data not shown) similar to the property of another cytotoxic R-type lectin isolated from mushroom Clitocybe nebularis [20]. Taken together, MytiLec fits in as a new member of the R-type lectin family.

Some R-type lectins have additional domains as toxic subunits. Pierisin, isolated from Pieris rapae (cabbage butterfly), has an ADP-ribosyltransferase domain in the polypeptide and three R-type lectin domains. Pierisin induces apoptosis in HeLa cells by binding to surface Gb3 and Gb4 (GalNAcβ1-3Galα1-4Galβ1-4Glc) glycans [21]. In addition to the original MytiLec, two MytiLec variants (termed MytiLec2 and MytiLec3) containing a pore-forming aerolysin [22]-like domain in the polypeptide that creates pores into infectious organisms and kills them through initiation of innate immunity, according to the recently updated MytiBase [4].

MytiLec does not have additional functional domains or subunits beside glycan-binding domains, in contrast to other R-type lectins, although it is capable of inducing cytotoxicity. It thus occupies a unique category within the R-type lectin family. The mechanisms whereby MytiLec transmits its signals through cells to activate various signal transduction molecules for induction of cancer cell apoptosis are of great interest. We used experimental cell line, Ramos with high levels of Gb3 expression to study apoptosis-inducing molecules (mitogen-activated protein kinases (MAPK) cascade, mitochondria-controlling caspase, and death receptor signal) activated by MytiLec in Burkitt's lymphoma cells.

2. RESULTS AND DISCUSSION

2.1. Glycan-Binding and Cell Agglutination of Recombinant MytiLec

MytiLec agglutinated Burkitt's lymphoma-derived Ramos cells (high Gb3 expression) [23] but did not agglutinate K562 erythroleukemia cells (no Gb3 expression). Strong agglutination was observed for Ramos, as revealed by large cell masses (Figure 1).

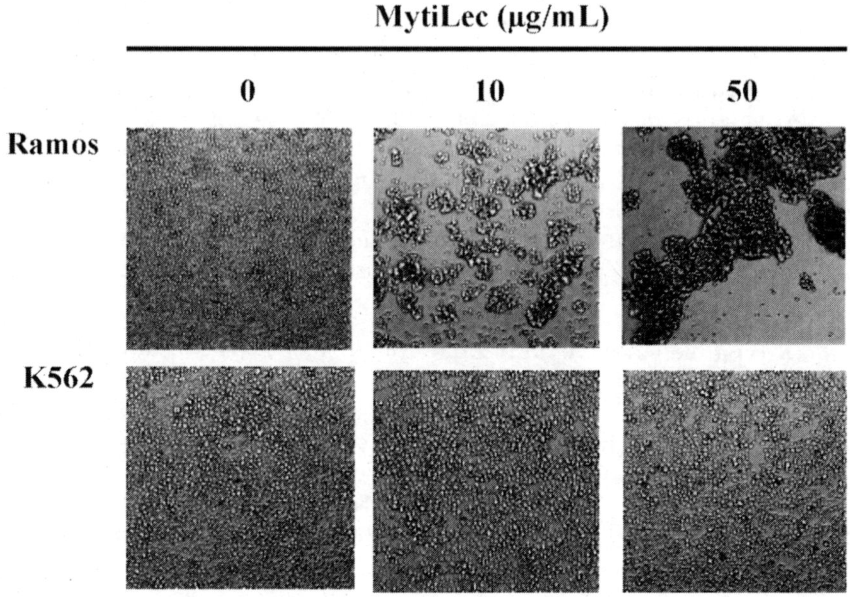

Figure 1. Different cell agglutination activities of MytiLec. MytiLec (0, 10, and 50 μg/mL) was applied to Ramos (5 × 10^5 cells) and K562 (2 × 10^5 cells) cells and observed by phase contrast microscopy.

2.2. Cytotoxic Effects of MytiLec on Burkitt's Lymphoma Cell Lines

Cytotoxic effects of MytiLec administration were evaluated by WST-8 assay rather than trypan blue assay because agglutinated cell masses were not effectively stained by trypan blue reagent. Ramos and K562 cells were cultured for 24 h, treated with MytiLec, and reduction in proportion of living cells was assayed by measuring absorbance at 450 nm. Viability was reduced in comparison with control (nontreated) cells for Ramos treated with 10 μg/mL of MytiLec, indicating a cytotoxic effect. Viability of K562 cells, which do not express Gb3, was unaffected by MytiLec treatment (Figure 2A).

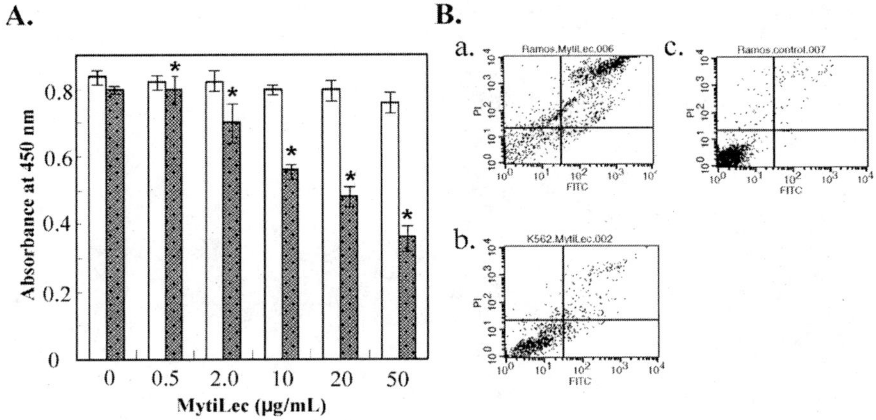

Figure 2. Reduction of cell viability by MytiLec. (**A**) Determination of viability by WST-8 assay. Dotted columns: Ramos. White columns: K562. Cells (2 × 10^5 of Ramos; 5 × $10^5$of K562) were incubated with various MytiLec concentrations as shown. Error bars: SE (n = 3); (**B**) Annexin V-binding and propidium iodide (PI) incorporation in MytiLec-treated cells. Horizontal axis: binding of FITC-labeled annexin V. Phosphatidylserine externalization and PI incorporation were evaluated by FACS analysis using MEBCYTO apoptosis kit. Ramos (**a**,**c**) and K562 (**b**) cells were treated with MytiLec (**a**,**b**: 20 μg/mL; **c**: 0 μg/mL) for 30 min at 4 °C. Data shown are mean values with error bars = SD of triplicate experiments. Asterisks = significant differences ($p < 0.05$) between treated and control groups.

Fluorescence activated cell sorting (FACS) analysis revealed that MytiLec treatment led to deleterious biological processes such as cell membrane inversion and loss of membrane integrity. Horizontal axes in Figure 3B show binding of Fluorescein isothiocyanate (FITC)-labeled annexin V, and vertical axes show incorporation of propidium iodide. Increasing MytiLec concentration was associated with shifting of annexin V-positive and propidium iodate-positive populations into the right and upper portions (respectively) of these histograms. K562 cells were unaffected by MytiLec treatment. The membrane inversion and penetration observed in MytiLec-treated Ramos cells were consistent with results of our previous study on Raji cells, another Burkitt's lymphoma cell line [13]. These effects on the Ramos cell membrane (Figure 2B) appeared to be associated with the cytotoxic effect of MytiLec (Figure 3). MytiLec may increase cell fragility by suppressing biosynthesis of cell surface membrane proteins. The triggering concentration of MytiLec is lower for apoptosis (~10 μg/mL) than for necrosis (>20 μg/mL). These observations may be related to the functions of MytiLec in caspase activation and TNF-α production (Section 2.4).

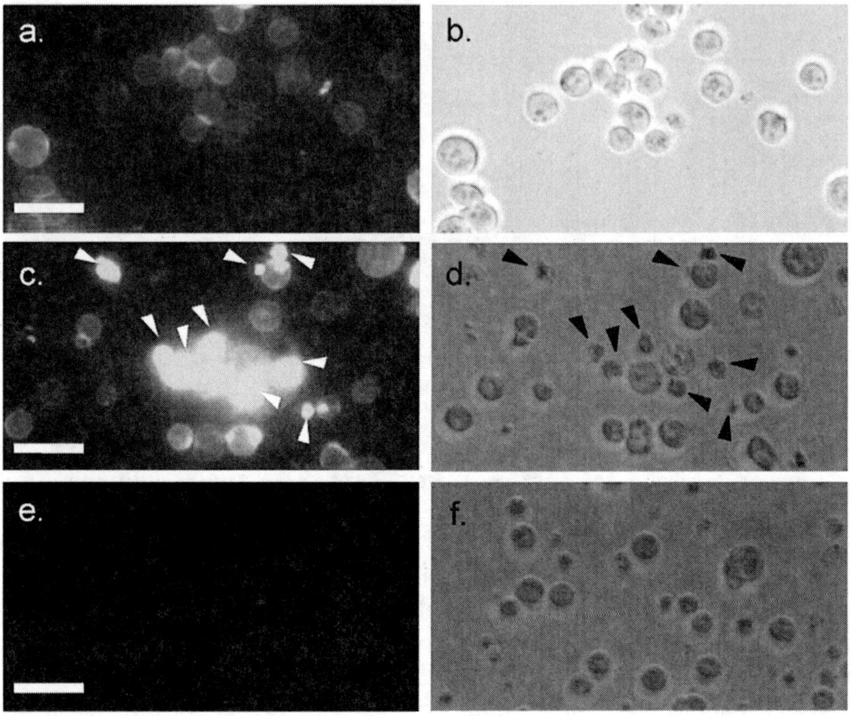

Figure 3. Internalization of FITC-conjugated MytiLec into Burkitt's lymphoma cells. Incubation time: 0 min (**a**,**b**); and 2 h (**c**–**f**). Cells are treated with FITC-MytiLec in the presence of 25 mM D-galactose, as negative control (**e**,**f**). Cells were observed by fluorescence (**a**,**c**,**e**; ex 498 nm and em 522 nm) and phase-contrast (**b**,**d**,**f**), respectively. Arrows in c and d indicated shrunken cells. Bars indicated 50 μm.

2.3. Internalization of MytiLec into Burkitt's Lymphoma Cells

Internalization of fluorescein-conjugated MytiLec (20 μg/mL) by Burkitt's lymphoma cells was demonstrated by confocal microscopy. Cell surface fluorescence was observed at the beginning (Figure 3a) and after 2 h incubation, strong intracellular fluorescence was detected due to the migration of FITC-MytiLec (Figure 3c vs. d). Such internalization was similar to that observed for TF-antigen-binding BEL lectin [24], an R-type lectin purified from mushroom. Cells with internalized MytiLec looked shrunken and irregular shaped with a characteristic rough surface

(Figure 3c,d, arrows). This internalization was totally inhibited by the co-presence of D-galactose, a haptenic sugar of the lectin (Figure 3e). It can be assumed that the internalization of MytiLec activated a number of cell signaling pathways (described in following sections) whether inhibition of this internalization by the sugar gives an idea about the mechanism of Gb3-dependent signaling.

2.4. Activation of MAPK Pathways by MytiLec

MAPKs play essential roles in cell growth and differentiation, cell cycle, and cell death. We found that MytiLec activates several MAPK pathways in Burkitt's lymphoma cells.

2.4.1. Activation of MEK-ERK Pathway

In Ramos, MytiLec activated the classical MAPK pathway of MAPK/extracellular signal-regulated kinase $(MEK)_{1/2}$ and extracellular signal-regulated kinase $(ERK)_{1/2}$ signaling cascade in a dose-dependent manner, as shown by Western-blotting (Figure 4A, P-$MEK_{1/2}$ vs. $MEK_{1/2}$ and P-$ERK_{1/2}$ vs. ERK_1). No such phosphorylation occurred in K562 (data not shown). Phosphorylation of the MEK-ERK pathway by MytiLec resulted in expression of the cyclin-dependent kinase (CDK) inhibitors p21 (Figure 4 column p21). In contrast, up-regulated levels of CDK6 and cyclinD3 were slightly reduced by MytiLec (Figure 4).

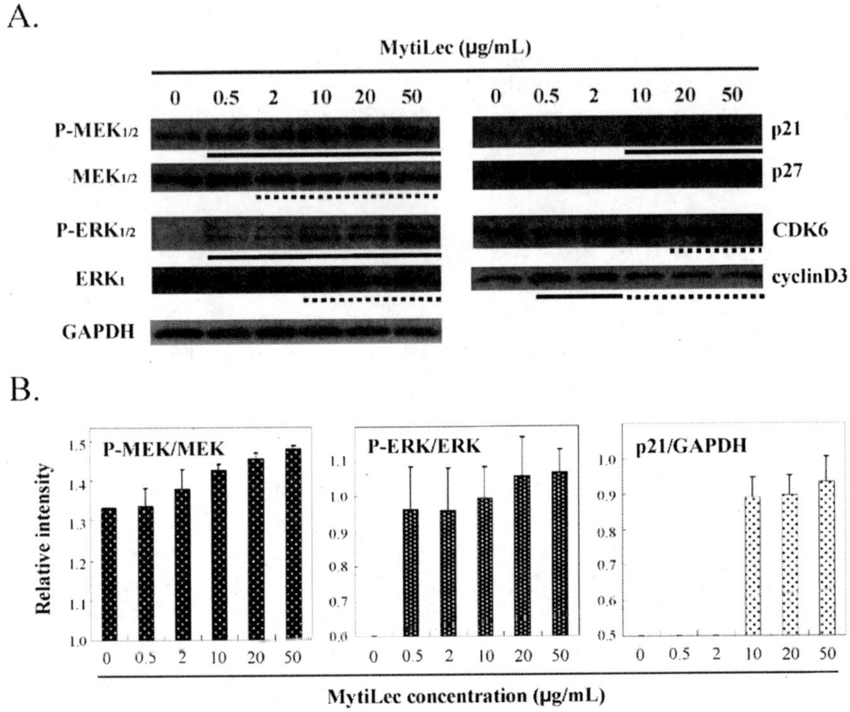

Figure 4. Effects of MytiLec treatment on MEK, ERK, and cell cycle-related molecules in Burkitt's lymphoma Ramos cells. (**A**) Phosphorylation and expression levels of MEK1/2, ERK1/2 and p21, p27, CDK6 and cyclinD3 were shown, respectively. Cells (4×10^5 in each experiment) were treated with various concentrations of MytiLec as shown, and activation levels were evaluated by Western blotting of lysates. Solid and dotted lines indicated increasing and decreasing trends, respectively. GAPDH: Glyceraldehyde 3-phosphate dehydrogenase; (**B**) Relative densitometric quantification of P-MEK/MEK, P-ERK/ERK and p21/GAPDH. Each experiment was repeated three times.

Expression of p21 (which binds to CDK and inhibits its activity) is enhanced by various external stimuli and stress factors, resulting in cell cycle arrest at the $G_{0/1}$ phase [25]. Up-regulation of p21 level by MEK has been documented using specific inhibitors [26]. The above findings indicate that binding of MytiLec and Gb3 on the cells induced MEK-ERK pathway activation and p21 expression, eventually resulting in cell cycle arrest.

2.4.2. Phosphorylation of Stress-Activated MAPK Pathways (JNK, p38 Kinase)

In addition to the classical MAPK pathway (MEK-ERK), MytiLec phosphorylated stress-activated MAPK pathways [c-JunN-terminal kinase (JNK) and p38 kinase] in Ramos (Figure 5, asterisks). Evidently, ERK mediated transduction of the MytiLec/ Gb3 binding signal to JNK and p38 kinase. The MytiLec-generated stimuli were equivalent to signals generated by oxidative stresses [27]. A recent study indicates that the MEK/ERK pathway itself can activate both JNK and p38 kinase [28]. Certain mannose-binding proteins inhibited cell proliferation through activation of these pathways [29,30]. Stress-activated kinases play roles in tumor suppression, apoptosis, termination of cell differentiation, and autophagy [31].

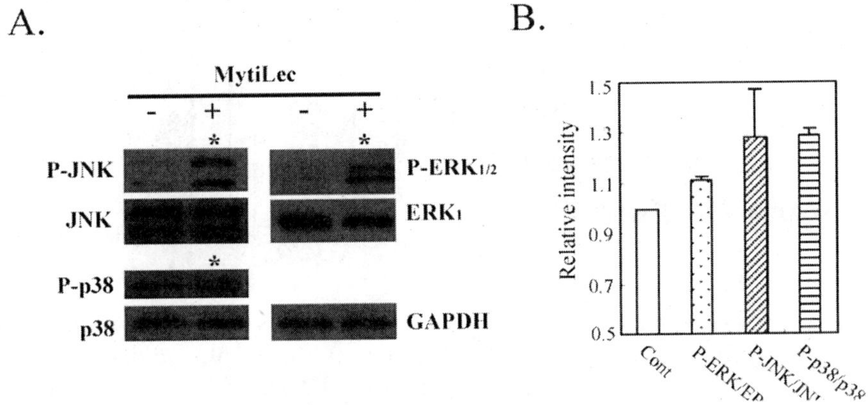

Figure 5. Phosphorylation of JNK and p38 kinase by MytiLec in Burkitt's lymphoma Ramos. Cells (5×10^5) were treated with (+) or without (−) 20 μg/mL MytiLec, and phosphorylation was evaluated by Western-blotting of cell lysates. (**A**) P-ERK, P-JNK, and P-p38: phosphorylated forms of ERK, JNK, and p38 kinase, respectively. Asterisks: increased phosphorylation; (**B**) Relative densitometric quantification of P-ERK/ERK, P-JNK/JNK and P-p38/p38. Each experiment was repeated three times.

2.4.3. MytiLec-Induced Phosphorylation of MEK-ERK Pathway Causes Cell Cycle Arrest through p21 Up-Regulation

Pre-incubation with U0126 (a specific $MEK_{1/2}$ inhibitor) significantly reduced MytiLec-induced phosphorylation of $ERK_{1/2}$in Ramos (Figure 6, P-$ERK_{1/2}$ vs. ERK_1). U0126 also reversed MytiLec-induced enhancement of p21 expression (Figure 6, p21, asterisks). U0126 had no effect on expression of MEK or P-MEK (Figure 6, P-$MEK_{1/2}$ vs. $MEK_{1/2}$), consistent with the finding that U0126 directly inhibits the downstream pathway under P-MEK [32].

In conclusion, MytiLec/Gb3 binding in Ramos promoted both the classical MAPK pathway (MEK-ERK) and stress-activated MAPK pathways (JNK, P38 kinase). MytiLec-induced phosphorylation of the MEK-ERK pathway resulted in up-regulation of p21 expression that might lead to cell cycle arrest.

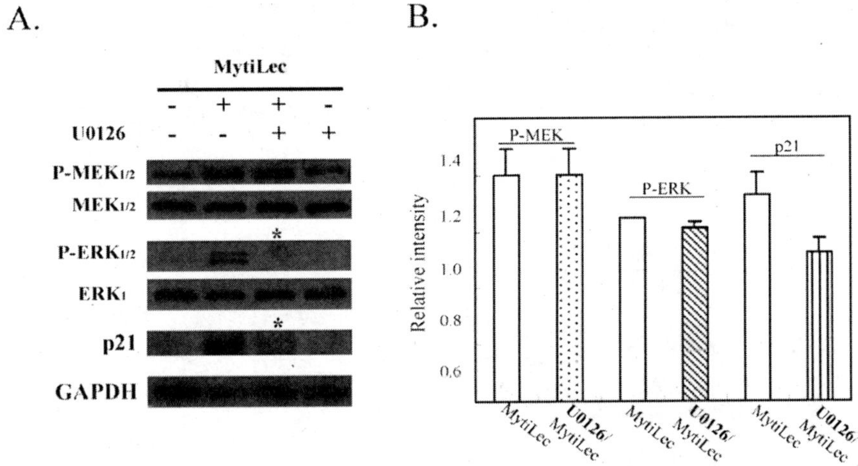

Figure 6. Treatment with MEK inhibitor U0126 reversed up-regulation of p21 expression mediated by MytiLec-induced phosphorylation of MEK-ERK pathway. (**A**) Burkitt's lymphoma Ramos cells (4 × 10^5) were treated with (+) or without (−) 20 μM U0126 for 5 h, and then with 20 μg/mL MytiLec. Expression levels of P-$MEK_{1/2}$, P-$ERK_{1/2}$, and p21 were evaluated by Western-blotting. Asterisks: disappearance of signals; (**B**) Relative densitometric quantification of P-MEK, P-ERK and p21 with (U0126/MytiLec) or without (MytiLec) the MEK inhibitor. Each experiment was repeated three times.

2.5. TNF-α Induction and Caspase Activation in Ramos Cells

MytiLec treatment of Burkitt's lymphoma Ramos cells triggered production of tumor necrosis factor (TNF)-α and activation of caspase-9 and caspase-3 (Figure 7A). The up-regulation of TNF-α by MytiLec was inhibited by P-MEK inhibitor U0126 and by caspase-3 inhibitor Zn-DEVD-FMK (Figure 7B). Association of TNF-α induction with caspase-3 activation was consistent with a previous observation by Burguillos et al. [33], and suggests that TNF-α expression is concurrently regulated by the MEK-ERK pathway and caspase-3 activation.

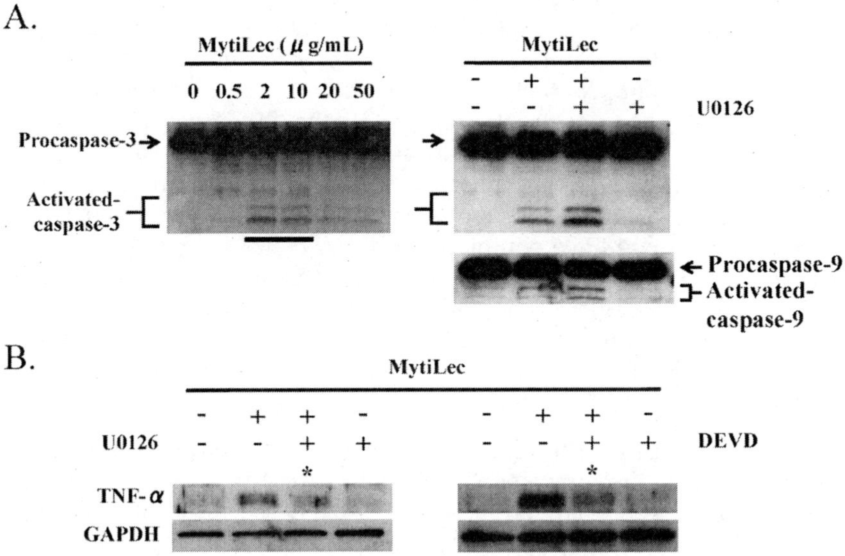

Figure 7. Activation by MytiLec of procaspase-3, procaspase-9, and TNF-α. (**A**) Activation of procaspase-3 in Ramos (4 × 10^5 cells) incubated with various concentrations of MytiLec as shown, evaluated by Western blotting; (**B**) Up-regulation of TNF-α by MytiLec, and inhibition of caspase activation and TNF-α expression by P-MEK inhibitor U0126 and caspase-3 inhibitor Zn-DEVD-FMK (DEVD). Ramos (4 × 10^5cells) were treated with (+) or without (−) 10 μM U0126 or 10 μM DEVD for 2 h, and then with 20 μg/mL MytiLec. TNF-α expression and caspase-3/-9 activation were evaluated by Western blotting. Solid and dotted lines indicated increasing and decreasing trends of phosphorylation, respectively. Asterisks: disappearance of signals. Each experiment was repeated three times.

To elucidate the caspase activation pathway, U0126 was applied to Ramos prior to MytiLec administration. Neither caspase-9 nor caspase-3 was inhibited by U0126, indicating that MytiLec-induced activation of caspase-9 and -3 in these cells is independent of the MEK-ERK pathway (Figure 7A, MytiLec(+)/U0126(+)).

Previous studies have demonstrated stimulation via phosphorylation of various signal transduction molecules by lectins isolated from lower organisms. In mouse macrophage cell lines, a recombinant B-subunit of ricin was shown to stimulate signal transduction pathways through production of inducible nitric oxide synthase, TNF-α, and interleukin-6 [34]. Human TNF-α reduced the phagocytic ability of mussel hemocytes [35]. The normal endogenous role of MytiLec in M. galloprovincialis remains unclear. Genome database analysis suggests that it may function in innate immunity [4]. Certain signal transduction molecules in M. galloprovincialis are also found in vertebrates (including humans) [5,36], suggesting similarities in the fundamental regulatory mechanisms of growth, differentiation, and cell proliferation. A role of native MytiLec in supporting innate immunity could explain its cytotoxic activity against Burkitt's lymphoma cells, since the mussel and vertebrate cells may have common surface glycans such as Gb3, as well as similar cell regulatory mechanisms. Along this line, we are attempting to identify the endogenous ligands of Gb3 and related glycans in the mussel. We previously observed binding of MytiLec to endogenous glycans in mussel tissue [13]. Surface Gb3 expression was observed on cultured cells derived from sea bass (Dicentrarchus labrax) [37].

There are difficulties in studying marine drugs based on proteins. Proteins trigger immune responses, and present logistical research problems because of their large size. However, recent studies based on each of sialic acid-binding and α-galacotside-binding lectins coding genes from fish and sea urchin, respectively successfully recombined the genes into an adenovirus vector and applied the lectins for oncotherapy in vitro [38,39]. MytiLec, with its novel cytotoxic properties, has great potential for similar therapeutic application through Gb3-signaling.

Ponting et al. [40] reviewed studies since 1990 of β-trefoil folding, including the question of why toxins (ricin B-chain [19]), cytokines (fibroblast growth factor [41], interleukin-1 [42]), enzymes (non-catalytic domain of glycosyltransferases [43]) and protease inhibitors (Kunitz-type protease inhibitor [44]) are synchronically assigned to the same 3D structural group even though their primary structures have low similarity (<12%). The polypeptides are highly conserved, with triple-tandem repeating sequences that contain four β-sheets and one α-helix in each subdomain. It appears that β-trefoil folding is one of the most versatile templates for protein conformation, and is involved in a wide range of physiological functions in many organisms. The primary structure of MytiLec is unique in that gene coding for R-type lectins has not been reported for any other Mytilus species. We hypothesize that an ancestral gene was synchronically modified for β-trefoil folding in MytiLec in association with its role in innate immunity.

The cell regulatory properties of Gb3-binding lectins are clearly diverse and merit further study. SAL, a Gb3-specific SUEL/RBL-type lectin isolated from catfish (Silurus asotus) eggs, phosphorylated several signal transduction molecules, resulting in cell cycle delay but had no cytotoxic effect [45], unlike MytiLec. SAL was found to down-regulate the multidrug resistance (MDR) 1 P-glycoprotein (MDR1 P-gp) on Burkitt's lymphoma cells through Gb3 and did not directly influence the viability of the cells. In this study, Gb3 was found to be an effective trigger to regulate cancer cell growth through apoptosis. Therefore, along with Burkitt's lymphoma cells, a number of Gb3-expressing cancer cell lines [46] like HeLa, MCF-7 and T47D, can be good targets to study cell signaling. In particular, detailed investigation of these lectins and their properties will be useful for development of novel anti-cancer drugs and therapeutic strategies.

3. EXPERIMENTAL DESIGN

3.1. Preparation of MytiLec

MytiLec was purified from Mediterranean mussel M. galloprovincialis according to our previous report [13].

3.2. Cell Lines and Culture

Burkitt's lymphoma Ramos cell line was obtained from American Type Culture Collection (CRL-1596). Erythroleukemia K562 cell line were obtained from the Cell Resource Center of Biomedical Research, Institute of Development, Aging and Cancer, Tohoku University (Sendai, Japan). Cells were cultured in RPMI-1640 medium (Nissui Pharmaceutical, Tokyo, Japan) supplemented with 10% FBS and antibiotic-antimycotic solution (Life Technologies, Carlsbad, CA, USA), and maintained at 37 °C in a 95% air/5% CO_2 atmosphere.

3.3. Cytotoxicity and Cell Viability Assays

Cells were maintained in RPMI-1640 supplemented with heat-inactivated FBS (10% v/v), penicillin (100 IU/mL), and streptomycin (100 μg/mL) at 37 °C in a 95% air/5% CO_2 atmosphere. Cytotoxic activity and cell growth following treatment with various concentrations of SAL (0–100 μg/mL) were determined using Cell Counting Kit-8 containing WST-8 (Dojindo Laboratories, Kumamoto, Japan) [47]. To evaluate the inhibitory effects of sugars, sucrose and melibiose (each 200 mM) were co-incubated with SAL (100 μg/mL) for 24 h and applied to the assay system. Cells (5×10^4, in 90 μL solution) were seeded into a 96-well flat-bottom plate and treated with various concentrations of MytiLec (10 μL) for 24 h at 37 °C. The effect of MytiLec on cell growth was assayed by addition of WST-8 solution (10 μL) to each well and incubation for 4 h at 37 °C. The reduction in proportion of living cells was assayed by measurement of absorbance at 450 nm using a GloMax Multi Detection System (Promega, Madison, WI, USA).

3.4. Fluorescein Isothiocyanate (FITC)-Conjugated MytiLec

For confocal microscopy, purified MytiLec (2 mg) was chemically conjugated with NH_2-reactive fluorescein isothioycanate (Dojindo Laboratories, Kumamoto, Japan) according to the manufacturer's protocol. Burkitt's lymphoma cells (100 μL; 2 × 10^5 cells/mL) in RPMI-1640 (Sigma-Aldrich, St. Louis, MO, USA) supplemented with 10% FBS were seeded onto 18-mm round cover slips in Petri dishes and left to attach overnight at 37 °C in a humidified 5% CO_2 atmosphere. The cover slips were washed three times with 2 mL PBS, incubated for 2 h with 50 μg/mL FITC-labeled lectin in PBS, and washed again. Cells were fixed with 4% paraformaldehyde for 15 min. Images at various focal planes were taken with a Leica TCS PS5 confocal microscope. 498 nm lasers were used for excitation of FITC.

3.5. Protein Expression of Signal Transduction Molecules and Their Phosphorylated Forms

Cells (5 × 10^5) were cultured for 12 or 24 h in RPMI-1640 with or without MytiLec (100 μg/mL) at 37 °C in 95% air/ 5% CO_2. Cell lysate was prepared using an AllPrep RNA/Protein Kit (Qiagen, Hilden, Germany), subjected to SDS-PAGE (12.5% separation gel), and electrotransferred onto polyvinylidene difluoride (PVDF) membrane (pore size 0.45 μm) (Hybond-P; GE Healthcare Bio-Sciences AB, Uppsala, Sweden) according to the manufacturer's protocol. The membrane was treated with blocking buffer (Blocking One; Nacalai Tesque, Kyoto, Japan) for 1 h at room temperature and washed with Tris-buffered saline (TBS) containing 0.05% Tween-20. The primary antibodies used were directed to p21 Waf1/Cip1 (1:1000, rabbit mAb, clone 12D1; Cell Signaling Technology, Danvers, MA, USA (CST)), p27 Kip1 (1:2000, rabbit mAb; clone D69C12; CST), CDK6 (1:2000, mouse mAb; clone DCS156; CST), cyclin D3 (1:5000, mouse mAb; clone DCS22; CST), phospho-$MEK_{1/2}$ (1:5000, rabbit mAb; CST), phospho-$ERK_{1/2}$ (1:5000, mouse mAb; BD Biosciences, San Jose, CA, USA), and GAPDH (1:50,000, mouse mAb; clone 6C5; Ambion/ Invitrogen; Carlsbad, CA, USA). These antibodies were applied in immunoreaction enhancer solution (Can Get Signal Solution 1; Toyobo, Osaka, Japan). The membrane was incubated for 16 h at 4 °C. The secondary antibody, horseradish peroxidase (HRP)-conjugated anti-mouse or anti-rabbit IgG (Chemicon International, Temecula, CA, USA), was diluted 1:20,000 in immunoreaction enhancer solution. The membrane was incubated for 1 h at room temperature, exposed to X-ray film (Fuji Film, Tokyo, Japan) using enhanced chemical luminescence, and Western blotted with Enhanced chemiluminescence (ECL) Prime detection reagent (GE Healthcare Bio-Sciences AB, Uppsala, Sweden). 1,4-Diamino-2,3-dicyano-1,4-bis (2-aminophenylthio) butadiene (U0126) (1 μg) (Calbiochem, San Diego, CA, USA), a synthetic inhibitor of $MEK_{1/2}$ [44], was dissolved in 247 μL DMSO to create 10 mM stock solution. Cells were incubated with U0126 (10 μM) in RPMI-1640 with FBS for 2 h, added with MytiLec (100 μg/mL), and incubated for 12 h. Whole cell extracts were separated by SDS-PAGE on 12.5% gel and blotted onto PVDF membranes. Phosphorylated kinases were identified by western blotting with anti-phospho-$MEK_{1/2}$, anti-$MEK_{1/2}$, anti-phospho-$ERK_{1/2}$, anti-ERK_1, and HRP-conjugated anti-mouse IgG antibodies. Signals were detected by X-ray film exposure and western blotting as described above.

3.6. Statistical Analysis

Data are presented as mean ± SE. Differences between means were evaluated by two-tailed Student's t-test, with $p < 0.05$ considered to be statistically significant.

4. CONCLUSIONS

We evaluated the mechanism of the cytotoxic effect of MytiLec purified from Mytilus galloprovincialis on Burkitt's lymphoma Ramos cells. MytiLec triggered activation of various cell death pathways including MAPK cascade (phosphorylation of MEK-ERK, JNK, and p38 kinase pathways), death receptor signal (TNF-α expression), and activation of mitochondria-controlling caspase-9 and caspase-3, through interaction with glycan Gb3 present in glycosphingolipid-enriched microdomains on the cell surface.

ACKNOWLEDGMENTS

This study was supported in part by Grants-In-Aid from the Japan Society for the Promotion of Science (JSPS), the Academic Frontier Project for

Private Universities, the Strategic Project to Support the Formation of Research Bases at Private Universities from the Ministry of Education, Culture, Sports, Science, and Technology (MEXT), The Science Research Promotion Fund from The Promotion and Mutual Aid Corporation for Private Schools, and Yokohama City University, Japan. The authors are grateful to Prof. Shu-Ichi Matsuzawa, Sanford-Burnham Medical Research Institute (La Jolla, CA, USA) for useful suggestions, and to S. Anderson for English editing of the manuscript.

AUTHOR CONTRIBUTIONS

Imtiaj Hasan and Shigeki Sugawara are considered co-first authors and performed the majority of the research. Imtiaj Hasan and Yasuhiro Ozeki wrote the manuscript. Yasuhiro Koide, Daiki Terada, Yuki Fujii, Naoya Imura, Keisuke G.Takahashi and Sultana Rajia were involved in experimental work and data analysis. Toshiyuki Fujiwara, Hideho Uchiyama and Nobuhiko Kojima provided laboratory support. Sarkar M.A. Kawsar, Robert A. Kanaly, Masahiro Hosono, Yukiko Ogawa, HIdeaki Fujita, Jiharu Hamako, and Taci Matsui improved the manuscript through careful review and helpful suggestions.

CONFLICTS OF INTEREST

The authors declare no conflict of interest.

REFERENCES

1. Suárez-Ulloa, V.; Fernández-Tajes, J.; Manfrin, C.; Gerdol, M.; Venier, P.; Eirín-López, J. Bivalve omics: State of the art and potential applications for the biomonitoring of harmful marine compounds. Mar. Drugs **2013**, 11, 4370–4389.

2. Venier, P.; de Pittà, C.; Bernante, F.; Varotto, L.; de Nardi, B.; Bovo, G.; Roch, P.; Novoa, B.; Figueras, A.; Pallavicini, A.;et al. MytiBase: A knowledgebase of mussel (M. galloprovincialis) transcribed sequences. BMC Genomics **2009**, 10.

3. Venier, P.; Varotto, L.; Rosani, U.; Millino, C.; Celegato, B.; Bernante, F.; Lanfranchi, G.; Novoa, B.; Roch, P.; Figueras, A.; Pallavicini, A. Insights into the innate immunity of the

Mediterranean mussel Mytilus galloprovincialis. BMC Genomics **2011**, 12.

4. Gerdol, M.; Venier, P. An updated molecular basis for mussel immunity. Fish Shellfish Immunol. **2015**, 46, 17–38.

5. Buchmann, K. Evolution of Innate Immunity: Clues from Invertebrates via Fish to Mammals. Front. Immunol. **2014**, 5.

6. Toubiana, M.; Gerdol, M.; Rosani, U.; Pallavicini, A.; Venier, P.; Roch, P. Toll-like receptors and MyD88 adaptors in Mytilus: Complete cds and gene expression levels. Dev. Comp. Immunol. **2013**, 40, 158–166.

7. Wakimoto, T.; Kondo, H.; Nii, H.; Kimura, K.; Egami, Y.; Oka, Y.; Yoshida, M.; Kida, E.; Ye, Y.; Akahoshi, S.; et al. Furan fatty acid as an anti-inflammatory component from the green-lipped mussel Perna canaliculus. Proc. Natl. Acad. Sci. USA **2011**, 108, 17533–17537.

8. Sukumaran, S.; Pittman, K.B.; Patterson, W.K.; Dickson, J.; Yeend, S.; Townsend, A.; Broadbridge, V.; Price, T.J. A phase I study to determine the safety, tolerability and maximum tolerated dose of green-lipped mussel (Perna canaliculus) lipid extract, in patients with advanced prostate and breast cancer. Ann. Oncol. **2010**, 21, 1089–1093.

9. Emelyanov, A.; Fedoseev, G.; Krasnoschekova, O.; Abulimity, A.; Trendeleva, T.; Bames, P.J. Treatment of asthma with lipid extract of New Zealand green-lipped mussel: A randomised clinical trial. Eur. Respir. J. **2002**, 20, 596–600.

10. Grienke, U.; Silke, J.; Tasdemir, D. Bioactive compounds from marine mussels and their effects on human health. Food Chem. **2014**, 142, 48–60.

11. Hasehira, K.; Tateno, H.; Onuma, Y.; Ito, Y.; Asashima, M.; Hirabayashi, J. Structural and quantitative evidence for dynamic glycome shift on production of induced pluripotent stem cells. Mol. Cell Proteomics **2012**, 11, 1913–1923.

12. Kovbasnjuk, O.; Mourtazina, R.; Baibakov, B.; Wang, T.; Elowsky, C.; Choti, M.A.; Kane, A.; Donowitz, M. The glycosphingolipid globotriaosylceramide in the metastatic transformation of colon cancer. Proc. Natl. Acad. Sci. USA**2005**, 102, 19087–19092.

13. Fujii, Y.; Dohmae, N.; Takio, K.; Kawsar, S.M.; Matsumoto, R.; Hasan, I.; Koide, Y.; Kanaly, R.A.; Yasumitsu, H.; Ogawa, Y.; et al. A lectin from the mussel Mytilus galloprovincialis has a highly

novel primary structure and induces glycan-mediated cytotoxicity of globotriaosylceramide-expressing lymphoma cells. J. Biol. Chem. **2012**, 287, 44772–44783.

14. UniProtKB-B3EWR1 (LEC_MYTGA). Available online: www.uniprot.org/uniprot/B3EWR1 (accessed on 11 December 2015).

15. Mg_Nor01_51P18 Nor01 Mytilus galloprovincialis cDNA 3-, mRNA sequence. Available online: www.ncbi.nlm.nih.gov/nucest/223022238 (accessed on 11 December 2015).

16. Kovalchuk, S.N.; Chikalovets, I.V.; Chernikov, O.V.; Molchanova, V.I.; Li, W.; Rasskazov, V.A.; Lukyanov, P.A. cDNA cloning and structural characterization of a lectin from the mussel Crenomytilus grayanus with a unique amino acid sequence and antibacterial activity. Fish Shellfish Immunol. **2013**, 35, 1320–1324.

17. Belogortseva, N.I.; Molchanova, V.I.; Kurika, A.V.; Skobun, A.S.; Glazkova, V.E. Isolation and characterization of new GalNAc/Gal-specific lectin from the sea mussel Crenomytilus grayanus. Comp. Biochem. Physiol. C Pharmacol. Toxicol. Endocrinol. **1998**, 119, 45–50.

18. PDBe > 3wmv. Available online: http://www.ebi.ac.uk/pdbe/entry/pdb/3WMV (accessed on 11 December 2015).

19. Rutenber, E.; Robertus, J.D. Structure of ricin B-chain at 2.5 A resolution. Proteins **1991**, 10, 260–269.

20. Pohleven, J.; Renko, M.; Magister, Š.; Smith, D.F.; Künzler, M.; Štrukelj, B.; Turk, D.; Kos, J.; Sabotiè, J. Bivalent carbohydrate binding is required for biological activity of Clitocybe nebularis lectin (CNL), the N,N'-diacetyllactosediamine (GalNAcâ1-4GlcNAc, LacdiNAc)-specific lectin from Basidiomycete C. nebularis. J. Biol. Chem.**2012**, 287, 10602–10612.

21. Matsushima-Hibiya, Y.; Watanabe, M.; Hidari, K.I.; Miyamoto, D.; Suzuki, Y.; Kasama, T.; Kasama, T.; Koyama, K.; Sugimura, T.; Wakabayashi, K. Identification of glycosphingolipid receptors for pierisin-1, a guanine-specific ADP-ribosylating toxin from the cabbage butterfly. J. Biol. Chem. **2003**, 278, 9972–99728.

22. Parker, M.W.; Buckley, J.T.; Postma, J.P.; Tucker, A.D.; Leonard, K.; Pattus, F.; Tsernoglou, D. Structure of the Aeromonas toxin proaerolysin in its water-soluble and membrane-channel states. Nature **1994**, 367, 292–295.

23. Mangeney, M.; Lingwood, C.A.; Taga, S.; Caillou, B.; Tursz, T.; Wiels, J. Apoptosis induced in Burkitt's lymphoma cells via Gb3/CD77, a glycolipid antigen. Cancer Res. **1993**, 53, 5314–5319.

24. Bovi, M.; Cenci, L.; Perduca, M.; Capaldi, S.; Carrizo, M.E.; Civiero, L.; Chiarelli, L.R.; Galliano, M.; Monaco, H.L. BEL β-trefoil: A novel lectin with antineoplastic properties in king bolete (Boletus edulis) mushrooms. Glycobiology **2013**, 23, 578–592.

25. McKenzie, P.P.; Danks, M.K.; Kriwacki, R.W.; Harris, L.C. $p21^{Waf1/Cip1}$ dysfunction in neuroblastoma: A novel mechanism of attenuating G_0-G_1 cell cycle arrest. Cancer Res. **2003**, 63, 3840–3844.

26. Park, J.S.; Carter, S.; Reardon, D.B.; Schmidt-Ullrich, R.; Dent, P.; Fisher, P.B. Roles for basal and stimulated $p21^{Cip-1/WAF1/MDA6}$ expression and mitogen-activated protein kinase signaling in radiation-induced cell cycle checkpoint control in carcinoma cells. Mol. Biol. Cell **1999**, 10, 4231–4246.

27. Klotz, L.O.; Pellieux, C.; Briviba, K.; Pierlot, C.; Aubry, J.M.; Sies, H. Mitogen-activated protein kinase (p38-, JNK-, ERK-) activation pattern induced by extracellular and intracellular singlet oxygen and UVA. Eur. J. Biochem. **1999**, 260, 917–922.

28. Bavaria, M.N.; Jin, S.; Ray, R.M.; Johnson, L.R. The mechanism by which MEK/ERK regulates JNK and p38 activity in polyamine depleted IEC-6 cells during apoptosis. Apoptosis **2014**, 19, 467–479.

29. Wang, Y.; Chen, A.D.; Lei, Y.M.; Shan, G.Q.; Zhang, L.Y.; Lu, X.; Chen, Z.L. Mannose-binding lectin inhibits monocyte proliferation through transforming growth factor-â1 and p38 signaling pathways. PLoS ONE **2013**, 8, e72505.

30. Tamma, S.M.; Kalyanaraman, V.S.; Pahwa, S.; Dominguez, P.; Modesto, R.R. The lectin jacalin induces phosphorylation of ERK and JNK in $CD4^+$ T cells. J. Leukoc. Biol. **2003**, 73, 682–688.

31. Sridharan, S.; Jain, K.; Basu, A. Regulation of autophagy by kinases. Cancers (Basel) **2011**, 3, 2630–2654.

32. Favata, M.F.; Horiuchi, K.Y.; Manos, E.J.; Daulerio, A.J.; Stradley, D.A.; Feeser, W.S.; van Dyk, D.E.; Pitts, W.J.; Earl, R.A.; Hobbs, F.; et al. Identification of a novel inhibitor of mitogen-activated protein kinase kinase. J. Biol. Chem. **1998**,273, 18623–18632.

33. Burguillos, M.A.; Deierborg, T.; Kavanagh, E.; Persson, A.; Hajji, N.; Garcia-Quintanilla, A.; Cano, J.; Brundin, P.; Englund, E.; Venero, J.L.; et al. Caspase signalling controls microglia activation and neurotoxicity. Nature **2011**, 472, 319–324.

34. Xu, N.; Yuan, H.; Liu, W.; Li, S.; Liu, Y.; Wan, J.; Li, X.; Zhang, R.; Chang, Y. Activation of RAW264.7 mouse macrophage cells in vitro through treatment with recombinant ricin toxin-binding subunit B: Involvement of protein tyrosine, NF-êB and JAK-STAT kinase signaling pathways. Int. J. Mol. Med. **2013**, 32, 729–735.

35. Betti, M.; Ciacci, C.; Lorusso, L.C.; Canonico, B.; Falcioni, T.; Gallo, G.; Canesi, L. Effects of tumour necrosis factor á (TNFá) on Mytilus haemocytes: Role of stress-activated mitogen-activated protein kinases (MAPKs). Biol. Cell **2006**,98, 233–244.

36. Toubiana, M.; Rosani, U.; Giambelluca, S.; Cammarata, M.; Gerdol, M.; Pallavicini, A.; Venier, P.; Roch, P. Toll signal transduction pathway in bivalves: Complete cds of intermediate elements and related gene transcription levels in hemocytes of immune stimulated Mytilus galloprovincialis. Dev. Comp. Immunol. **2014**, 45, 300–312.

37. Watanabe, Y.; Tateno, H.; Nakamura-Tsuruta, S.; Kominami, J.; Hirabayashi, J.; Nakamura, O.; Watanabe, T.; Kamiya, H.; Naganuma, T.; Ogawa, T.; et al. The function of rhamnose-binding lectin in innate immunity by restricted binding to Gb3. Dev. Comp. Immunol. **2009**, 33, 187–197.

38. Wu, L.; Yang, X.; Duan, X.; Cui, L.; Li, G. Exogenous expression of marine lectins DIFBL and SpRBL induces cancer cell apoptosis possibly through PRMT5-E2F-1 pathway. Sci. Rep. **2014**, 4.

39. Yang, X.; Wu, L.; Duan, X.; Cui, L.; Luo, J.; Li, G. Adenovirus carrying gene encoding Haliotis disus discus sialic acid binding lectin induces cancer cell apoptosis. Mar. Drugs **2014**, 12, 3992–4004.

40. Ponting, C.P.; Russell, R.B. Identification of distant homologues of fibroblast growth factors suggests a common ancestor for all β-trefoil proteins. J. Mol. Biol. **2000**, 302, 1041–1047.

41. Eriksson, A.E.; Cousens, L.S.; Weaver, L.H.; Matthews, B.W. Three-dimensional structure of human basic fibroblast growth factor. Proc. Natl. Acad. Sci. USA **1991**, 88, 3441–3445.

42. Veerapandian, B. Structure and function of interleukin-1, based on crystallographic and modeling studies. Biophys. J.**1992**, 62, 112–115.

43. Fritz, T.A.; Hurley, J.H.; Trinh, L.B.; Shiloach, J.; Tabak, L.A. The beginnings of mucin biosynthesis: The crystal structure of UDP-GalNAc: Polypeptide á-N-acetylgalactosaminyltransferase-T1. Proc. Natl. Acad. Sci. USA **2004**, 101, 15307–15312.

44. Ravichandran, S.; Sen, U.; Chakrabarti, C.; Dattagupta, J.K. Cryocrystallography of a Kunitz-type serine protease inhibitor: The 90 K structure of winged bean chymotrypsin inhibitor (WCI) at 2.13 Å resolution. Acta Crystallogr. D Biol. Crystallogr. **1999**, 55, 1814–1821.

45. Sugawara, S.; Hosono, M.; Ogawa, Y.; Takayanagi, M.; Nitta, K. Catfish egg lectin causes rapid activation of multidrug resistance 1 P-glycoprotein as a lipid translocase. Biol. Pharm. Bull. **2005**, 28, 434–441.

46. Johansson, D.; Kosovac, E.; Moharer, J.; Ljuslinder, I.; Brännström, T.; Johansson, A.; Behnam-Motlagh, P. Expression of verotoxin-1 receptor Gb3 in breast cancer tissue and verotoxin-1 signal transduction to apoptosis. BMC Cancer **2009**,9.

47. Kawano, T.; Sugawara, S.; Hosono, M.; Tatsuta, T.; Nitta, K. Alteration of gene expression induced by Silurus asotuslectin in Burkitt's lymphoma cells. Biol. Pharm. Bull. **2008**, 31, 998–1002.

8

CHAPTER

Cytotoxicity and Glycan-Binding Properties of an 18 kDa Lectin Isolated from the Marine Sponge Halichondria okadai

Ryo Matsumoto [1,†], Yuki Fujii [1,2], Sarkar M. A. Kawsar [1,3], Robert A. Kanaly [1], Hidetaro Yasumitsu [1], Yasuhiro Koide [1], Imtiaj Hasan [1], Chihiro Iwahara [1], Yukiko Ogawa [2], Chang Hun Im [4], Shigeki Sugawara [4], Masahiro Hosono [4], Kazuo Nitta [4], Jiharu Hamako [5], Taei Matsui [5] and Yasuhiro Ozeki [1]

[1]*Laboratory of Glycobiology and Marine Biochemistry, Department of Genome System Science, Graduate School of NanoBio Sciences, Yokohama City University, 22-2 Seto, Kanazawa-ku, Yokohama 236-0027, Japan*

[2]*Divisions of Functional Morphology and Microbiology, Department of Pharmacy, Faculty of Pharmaceutical Science, Nagasaki International University, 2825-7 Huis Ten Bosch, Sasebo, Nagasaki 859-3298, Japan*

[3]*Laboratory of Carbohydrate and Protein Chemistry, Department of Chemistry, Faculty of Science, University of Chittagong, Chittagong-4221, Bangladesh*

[4]*Division of Cell Recognition Study, Institute of Molecular Biomembrane and Glycobiology, Tohoku Pharmaceutical University, 4-4-1 Komatsusima, Aoba-ku, Sendai 981-8558, Japan*

[5]*Department of Biology, Fujita Health University, Toyoake, Aichi 470-1192, Japan*

ABSTRACT

A divalent cation-independent lectin—HOL-18, with cytotoxic activity against leukemia cells, was purified from a demosponge, Halichondria okadai. HOL-18 is a 72 kDa tetrameric lectin that consists of four non-covalently bonded 18 kDa subunits. Hemagglutination activity of the lectin was strongly inhibited by chitotriose (GlcNAcβ1-4GlcNAcβ1-

4GlcNAc), fetuin and mucins from porcine stomach and bovine submaxillary gland. Lectin activity was stable at pH 4–12 and temperatures lower than 60 °C. Frontal affinity chromatography with 16 types of pyridylaminated oligosaccharides indicated that the lectin had an affinity for N-linked complex-type and sphingolipid-type oligosaccharides with N-acetylated hexosamines and neuramic acid at the non-reducing termini. The lectin killed Jurkat leukemia T cells and K562 erythroleukemia cells in a dose- and carbohydrate-dependent manner.

Keywords: cytotoxicity; frontal affinity chromatography technology; glycoprotein; Japanese black sponge (Halichondria okadai); lectin

1. INTRODUCTION

Lectins are carbohydrate-binding proteins that are present among various living organisms. They have many functions including cell growth regulation, anti-infectious properties and support of natural immunity through their binding to specific oligosaccharides to form glycoconjugates (glycoproteins and glycosphingolipids) [1]. At the same time, administration of some lectins isolated from plants and lower animals have been shown to kill culture cells derived from higher animals [2,3]. Sponges, which are known to be one of the oldest living groups of organisms, produce different types of biological molecules that regulate their growth through symbiotic relationships with microorganisms. In regard to this, unique chemicals that have attractive activities such as okadaic acid and halichondrin B, which are known to act as a phosphatase inhibitor and as an anticancer agent, respectively, are produced by symbiotic microorganisms that are isolated from the Japanese black sponge Halichondria okadai [4,5].

Many types of lectins are produced in different species of sponges whose habitats range from tidal zones to the deep sea.N-acetyl D-galactosamine (D-GalNAc)-binding lectins, known as galectins, are a representative class of animal lectins that have been isolated from the demosponges Geodia cydonium [6] and Suberites domuncula [7]. Additionally, a 14 kDa N-acetyl amino-carbohydrate-specific lectin has been obtained from Axinella corrugate [8]. In another study, cDNAs encoding a polypeptide similar to the lipopolysaccharide-binding lectin of the horseshoe crab were isolated from S. domuncula andEphydatia fluviatilis and interestingly were expressed during sponge development [9,10]. In the Japanese black sponge, H. okadai mentioned above, three lectins (HOL-I, HOL-II and HOL-30) were purified using affinity chromatography on bovine submaxillary mucin-Sepharose, acid-treated

Sepharose 4B and lactosyl-agarose, respectively [11,12]. HOL-I, a 21-kDa polypeptide, recognized N-acetyl amino-carbohydrates such as N-acetyl D-glucosamine (D-GlcNAc) and D-GalNAc while HOL-II, a 42-kDa polypeptide, recognized D-Gal and type-1 N-acetyllactosamine (Galβ1-3GlcNAc). On the other hand, HOL-30, a 30 kDa β-galactoside-binding lectin with a galectin-like primary structure, was indicated to have an affinity to type-1 and type-2 (Galβ1-4GlcNAc) N-acetyllactosamines of branched complex-type oligosaccharides.

Recently, a preliminary investigation suggested the presence of another lectin in H. okadai, which has a smaller molecular mass than HOL-I, HOL-30, and HOL-II and which demonstrates a binding affinity to chitotriose (GlcNAcβ1-4GlcNAcβ1-4GlcNAc). If the isolation of the lectin is successful and the details of its biochemical properties using a glycome approach are accomplished, knowledge of a new sponge lectin may be added to the growing literature in this field. Frontal affinity chromatography technology (FACT) is a method by which the glycan-binding specificity of lectins may be evaluated by using various types of pyridylamino group (PA)-labeled oligosaccharides on an affinity column whereby the delay in the elution of the lectin is indicative of its binding affinity [13,14]. This information in regard to the binding property of the lectin will aid in the understanding of the novel characters of each lectin. In this study, the specific glycan-binding properties and glycan-dependent cytotoxicity of an 18 kDa lectin isolated from a Japanese black sponge were evaluated.

2. MATERIALS AND METHODS

2.1. Animal and Chemicals

The demosponge H. okadai was collected from the tidal zone of the Zushi coast, Kanagawa Prefecture, Japan. The samples were used immediately or stored at −80 °C until studied.

Lactose, melibiose, D-galactose (D-Gal), D-glucose (D-Glc), D-mannose (D-Man) and L-fucose (L-Fuc) were purchased from Wako Pure Chemical Co. Inc., Japan. Chitobiose and chitotriose were purchased from Carbosynth Ltd., UK. Methyl α-N-actyl D-glucosaminide (Me-α-GlcNAc), methyl β-N-actyl D-glucosaminide (Me-β-GlcNAc), methyl α-N-actyl D-galactosaminide (Me-α-GalNAc) and methyl β-N-actyl D-galactosaminide (Me-β-GalNAc) were purchased from Toronto Research Chemicals Inc., Canada. Glycoproteins, bovine submaxillary mucin (BSM), fetuin, porcine stomach mucin (PSM) and standard protein markers for gel permeation chromatography were purchased from Sigma Chemical Co., USA.

Chitotriose-agarose, lactosyl-agarose, horseradish peroxidase-conjugated lectins; wheat germ agglutinin (HRP-WGA), Concanavalin A (HRP-Con A) and Ricinus communis agglutinin (HRP-RCA 120) and a protease inhibitor mixture were purchased from Cosmo Bio Co. Ltd., Japan. Superdex 75, Sephadex G-75, the CM5 sensor chip and a ligand coupling kit were obtained from GE Healthcare, USA. The bicinchoninic acid (BCA) kit was obtained from Pierce Co. Ltd., USA. Pre-stained standard protein marker mixture (EzStandard PrestainBlue), |3,3′,5,5′-tetramethylbenzidine (EzWestBlue) and polyvinylidene difluoride (PVDF) membrane were purchased from ATTO Co. Ltd., Japan. The standard protein marker mixture for sodium dodecyl sulphate-polyacrylamide gel electrophoresis (SDS-PAGE), and pyridylamine (PA) labeled oligosaccharides were obtained from Takara Bio Co. Inc., Japan and Masuda Chemical Co. Ltd., Japan, respectively. Trypan blue (0.5% w/v) was purchased from NACALAI TESQUE Inc., Kyoto, Japan. RPMI 1640 medium was obtained from Nissui Pharmaceutical Co. Ltd., Japan. Fetal calf serum was purchased from Life Technologies Co. Ltd., USA. Penicillin-streptomycin was obtained from Roche Diagnostics KK., Japan.

2.2. Purification of Lectin

Sponges were cut into pieces by stainless steel scalpels and homogenized with 10 volumes (w/v) of tris-buffered saline (TBS) (10 mM tris(hydroxymethyl)aminomethane-HCl plus 150 mM NaCL, pH 7.4) that contained 10 mM of a protease inhibitor mixture. Homogenates were filtered through gauze and filter paper in a glass funnel and centrifuged at 14,720 × gin 500 mL centrifuge bottles for 1 h at 4 °C in a Suprema 21 centrifuge equipped with an NA-18HS rotor (TOMY Co. Ltd., Japan). The supernatant was centrifuged again at 27,500 × g in 50 mL centrifuge tubes to remove debris. The crude supernatant was applied to a chitotriosyl-agarose (5 mL) affinity column connected to Sephadex G-75 and lactosyl-agarose columns to remove HOL-30 (5 mL each), which served as the filter and trap of another lectin, respectively. The chitotriose-agarose column was washed with TBS, the lectin was eluted with 50 mM D-GlcNAc in TBS from the column and 1 mL of each fraction was collected using an automatic fraction collector. The chromatography steps during the wash and elution were monitored using an UV monitor (ATTO Co. Ltd., Japan) that measured absorbance at 280 nm and by SDS-PAGE [15]. The eluted fractions were dialysed against a volume of TBS to remove free sugar. Quantification of the lectin was determined using the BCA protein assay kit with bovine serum albumin as the standard protein. The absorbance at 562 nm [16,17] was measured using an ND-1000 spectrophotometer (Nano Drop Co. Ltd., USA).

2.3. Determination of Molecular Mass Using Gel Permeation Chromatography and SDS-PAGE

The purified lectin was subjected to gel permeation chromatography using a Superdex 75 column (1.0 × 32 cm) connected to an HPLC system that consisted of a PU-2089 intelligent pump and a UV-2027 UV-Vis detector (JASCO Co., Japan). The standard molecular marker proteins and the lectin were separated at a flow rate of 0.5 mL/min. Each protein was detected at an absorbance of 280 nm. To determine polypeptide size, the lectin was mixed with the sample buffer (20 mM Tris-HCl, 0.2% SDS and 10% glycerol in the presence or absence of 2% 2-mercaptoethanol; pH 6.8) and heated at 70 °C for 30 min. Aliquots of 30 μL were applied to the wells of a mini-slab gel (gel size: 80 × 100 × 1 mm; 15% and 5% polyacrylamide were used in the separation and stacking gels, respectively). The purified lectin was treated in the sample buffer at 70 °C for different times (ranging between 5 to 60 min). The molecular mass of the polypeptide was determined using SDS-PAGE at a constant current of 30 mA for 1 h [15]. After electrophoresis, the gel was stained with 0.1% (w/v) CBB R-250 in 40% methanol and 10% acetic acid, followed by de-staining with 40% methanol and 10% acetic acid. One nanomol of purified lectin was subjected to analysis of the N-terminus amino acid sequence using automated Edman degradations by a gas phase protein sequencer, Shimadzu model PPSQ-31A (Shimadzu Corp., Japan), according to a previously described method [11]. The cleavage of the polypeptide into peptide fragments was attempted by using trypsin, lysylendopeptidase and cyanogen bromide according to the manufacturer's instructions.

2.4. Detection of Saccharides in Polypeptide of the Lectin

Sponge lectin that was separated by SDS-PAGE was electrotransferred to a PVDF membrane via the semi-dry method [18]. Lectin blotted onto PVDF membrane was soaked in TBS containing 0.05% Tween-20 at room temperature and incubated with HRP-conjugated lectins (2–5 μg/mL) for 1 h. HRP reaction was used to stain with 3,3′,5,5′-tetramethylbenzidine according to the manufacturer's instructions [19].

2.5. Hemagglutination and Carbohydrate-Binding Specificity

Hemagglutination assays were performed using microtiter plates with V-bottom wells by the 2-fold serial dilution method using erythrocytes of human type A, B, O and rabbit. Carbohydrate-binding specificity of the lectin was evaluated as follows: Twenty microliters of 1% (w/v) trypsin-treated and 0.25% glutaraldehyde-fixed human type A erythrocytes (1% in TBS) were mixed with the serially diluted lectin. After incubation for 1 h at room temperature, the plates were read. One unit of hemagglutination activity was defined as the highest dilution of lectin that yielded a complete hemagglutination (HA unit). For the inhibition tests, a dilution of purified lectin to 16 units of HA was used. The inhibition of hemagglutination was assayed using 2-fold serial dilutions of the following saccharides: D-Gal, Me-α- and Me-β-D-GalNAc, Me-α- and Me-β-D-GlcNAc, D-Glc, D-Man and L-Fuc. BSM, fetuin and PSM were used at concentrations of 50 mM and 1 mg/mL [20].

2.6. Effects of Divalent Cations, Sulfhydryl Preservation Reagent, Temperature and pH

Two hundred and fifty-six (2^8) hemagglutination units (HU) (about 0.5 μg/mL) of diluted lectin were tested for hemagglutination activity by the 2-fold serial dilution method in the presence or absence of divalent cations ($CaCl_2$, $MnCl_2$and $MgCl_2$), chelating agents (EDTA and EGTA) and sulfhydryl preservative (2-mercaptoethanol) reagents (10 mM of each reagent was used). To investigate the tolerance of the lectin to temperature and pH, it was tested over a temperature range from 20–80 °C at 10 °C increments and pH 3–13 for 1 h. Lectin was dissolved in 100 mM sodium acetate-acetic acid (pH 3–5), phosphate buffer (pH 6–7), Tris-HCl (pH 8–8.5) or glycine-NaOH (pH 9–13) at 25 °C and subsequently dialyzed against TBS (pH 7.4). Hemagglutination activity was measured in the titer plates as described above.

2.7. Kinetic Analysis Using Surface Plasmon Resonance

PSM (100 μg/mL) dissolved in 10 mM sodium acetate (pH 4.5) was coupled on a carboxymethyl dextran matrix of a sensor chip CM5 using a ligand coupling kit according to the manufacturer's instructions. Remaining uncoupled activated residues on the chip were masked with 1 M neutralized ethanolamine (pH 8). Different concentrations of purified lectin (0–100 μM) in the HBS buffer were applied to the surface of a PSM-conjugated sensor chip using an auto-sampler and the association and dissociation of the lectin to immobilized PSM on the chip were analyzed for 2 min at a flow rate of 20 μL/min in 25 °C [21]. After application to the sensor chip for the association analysis, the bound lectin was washed with HBS buffer and finally it was removed from the surface of the sensor chip using 200 mM D-GlcNAc containing HBS. The transition of association to dissociation of the lectin to the PSM on the sensor chip was monitored optically as a sensorgram. The dissociation constant estimated by the association rate (k_{ass}) and dissociation rate (k_{diss}) of each concentrated lectin was determined using BIAevaluation software version 3.0.

2.8. Glycan-Binding Property by FACT

The purified lectin (1.7 mg) was dissolved in 0.1 M $NaHCO_3$ (pH 8.3) containing 0.5 M NaCl, then coupled to a NHS-activated Sepharose 4 Fast Flow matrix (250 μL) overnight at 4 °C. After coupling, the uncoupled groups on the gel were masked overnight by 1 M ethanolamine-HCl (pH 8). The lectin-immobilized Sepharose gel was washed with TBS and packed into a miniature column (200 μL). The column was then connected to an HPLC system consisting of a pump (PU-2089, Jasco Co. Ltd., Japan), a fluorescence detector (FP-2020 Plus, Jasco Co. Ltd., Japan) and a data processing integrator to control the systems (ChromNAV, Jasco Co., Ltd., Japan). Sixteen PA-saccharides containing two monosaccharides of PA-GlcNAc and PA-GalNAc were manually applied to the HOL-18 conjugated affinity column at a flow rate of 250 μL/min at 20 °C [14]. PA-rhamnose was used as the inert control. All PA-saccharides were applied into the column at a concentration of 0.6 pmol in 1.6 mL. The front elution volume of the PA-saccharides was detected by measuring the fluorescence detector at 310 nm (excitation) and 380 nm (emission). The retardation volume ($V - V_0$) of each PA-saccharide (V) compared to the inert controls of PA-rhamnose (V_0) was estimated from the conversion of a bar graph. FACT was performed with TBS containing 1 mM EDTA. Analysis was conducted in triplicate in each PA-oligosaccharide and average values were indicated as bars.

2.9. The Cytotoxicity

Jurkat leukemia T cells and K562 erythroleukemia cells were cultured in RPMI 1640 medium (Nissui Pharmaceutical Co. Ltd., Tokyo, Japan) supplemented with 10% fetal bovine serum (FBS), penicillin (100 U/mL), and streptomycin (100 μg/mL) and maintained at 37 °C in a 95% air and 5% CO_2 atmosphere. The cytotoxicity of the lectin was determined using the trypan blue exclusion assay with 0.5% trypan blue solution, according to a previously reported protocol [22]. The cells (2×10^4; 90 μL) were seeded onto 96-well, flat-bottomed plates and treated with 1–25 μg/mL of the lectin at 37 °C for 24 h. The dead cells were counted following staining with trypan blue using a manual cell counter. The termination of cell death by the lectin (25 μg/mL) was tested in the co-presence of D-GlcNAc, D-GalNAc and D-Man (50 mM each). Results are reported as the mean of percentage standard error. Two-tailed Student's t-tests were used as appropriate for statistical analysis. OnlyP values < 0.05 were considered to be statistically significant.

3. RESULTS AND DISCUSSION

3.1. Purification of HOL-18

The supernatant prepared by homogenization of H. okadai with TBS had a strong hemagglutination activity against trypsinized glutaraldehyde fixed-human (type-A) and rabbit erythrocytes. The crude extract was applied to the chitotriose-agarose column and the lectin was specifically eluted with D-GlcNAc-containing TBS (Figure 1A). Using SDS-PAGE under non-reducing and reducing conditions (70 °C for 30 min), the relative molecular mass of the lectin was determined to be 18 kDa (Figure 1B). The 18 kDa chitotriose-binding lectin from H. okadai was consequently defined as HOL-18. Interestingly, this lectin appeared as a molecular mass of 36 kDa, indicating a dimeric form. This dimeric form was retained under heat treatment for only 15 min under either reducing or non-reducing conditions; however, the dimer dissociated within 20–30 min of treatment. It suggested that these two polypeptides of HOL-18 tightly and non-covalently associate to each other. In total, 6 mg of the lectin was successfully purified from 800 g (wet mass) of sponge (Table 1). GPC of HOL-18 indicated a mass of 72 kDa, indicating that the lectin is a tetramer that is non-covalently bonded to four 18 kDa polypeptides (Figure 2). No PTH-amino acid signal was detected by direct sequencing, which suggested that the N-terminus amino acid of the polypeptide was blocked (data not shown). This is different from the case of HOL-30, which was suitably cleaved and sequenced into peptides by endoprotease, the generation of peptides from HOL-18 was not successful by treatment with proteases and cyanogen bromide.

Table 1. Purification of HOL-18.

Steps	Titer (HU)	Volume (mL)	Total activity [a]	Protein conc. (mg mL^{-1})	Specific activity [b]	Purification ratio (fold)	Recovery of activity (%)
Crude extract obtained by TBS	1024	200	204,800	2.3	2.2	1	100
Chitooligo-agarose	4096	10	40,960	0.7	585	292	20

[a] Total activity is shown by Titer × volume; [b] Specific activity is shown by titer/mg of protein.

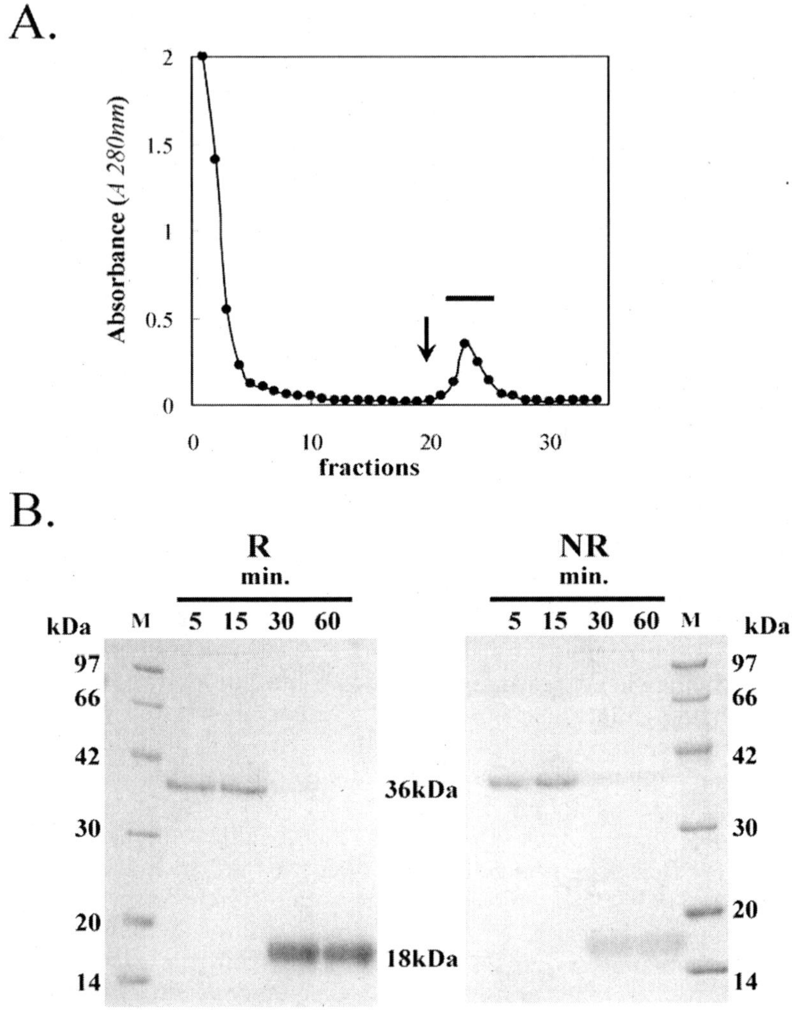

Figure 1. Purification of HOL-18. (**A**) The crude extract of H. okadai (200 g) was applied to the chitotriose-agarose column that had been equilibrated with tris-buffered saline (TBS). The column was washed with TBS and eluted with a TBS solution containing 50 mM D-GlcNAc (arrow). Fractions were collected and dialysed (bar); (**B**) HOL-18 (10 µg) was subjected to SDS-PAGE under reducing (R) and non-reducing (NR) conditions. The numbers under the bar indicate how long (5–60 min) the solution was treated at 70 °C. The following standard marker proteins (M) were used: Phosphorylase b (97 kDa), bovine serum albumin (66 kDa), ovalbumin (42 kDa), carbonic anhydrase (30 kDa), trypsin inhibitor (20 kDa) and lysozyme (14 kDa).

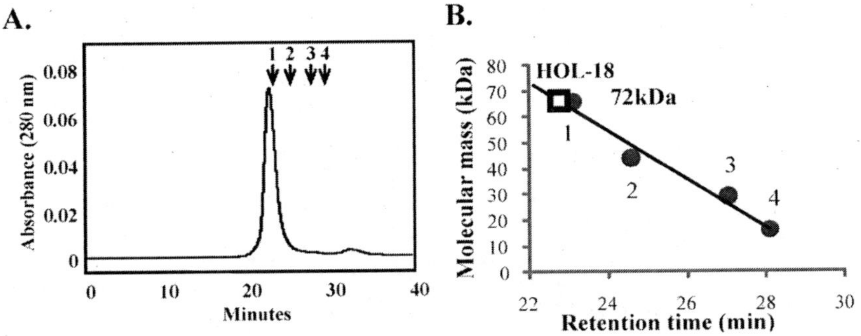

Figure 2. GPC and the calibration curve used to determine the molecular mass of HOL-18. The purified lectin (20 μg) was separated by HPLC on a Superdex 75 column at a flow rate of 0.5 mL/min. (**A**) The numbers indicate the standard molecular mass marker proteins used: (1) bovine serum albumin (66 kDa); (2) ovalbumin (42 kDa); (3) carbonic anhydrase (30 kDa) and (4) myoglobin (17 kDa); (**B**) The calibration curve to determine the molecular mass of the lectin. Square on the curve indicates the evaluated molecular mass of HOL-18 under native conditions.

The hemagglutination activity of HOL-18 was shown to be 1024 and 512 HU to human type-A and rabbit erythrocytes, respectively. However the lectin showed 64 HU to human type-B and type-O erythrocytes (data not shown). HOL-18 maintained maximum activity upon treatment at 60 °C (Figure 3A) and pH 4–12 (Figure 3B). The activity of HOL-18 was not enhanced even in the presence of 10 mM divalent metals or inhibited by the addition of chelating reagents such as EDTA and EGTA (data not shown). The addition of a sulfhydryl preservation reagent (2-mercaptoethanol) did not interfere with the hemagglutination activity of the lectin (data not shown). These results indicate that the activity was independent of the presence of Ca^{2+}, Mn^{2+}, Mg^{2+} or sulfhydryl preservation reagents.

3.3. Glycan Moiety of HOL-18 Polypeptide

HRP-WGA selectively stained PVDF membrane blotted on HOL-18 and the stain was completely blocked by including 50 mM D-GlcNAc-containing TBS (Figure 4). However, other HRP-conjugated lectins (ConA and RCA120) did not stain. It indicated that the polypeptide of HOL-18 contains a covalently extended saccharide chain(s) containing D-GlcNAc residue.

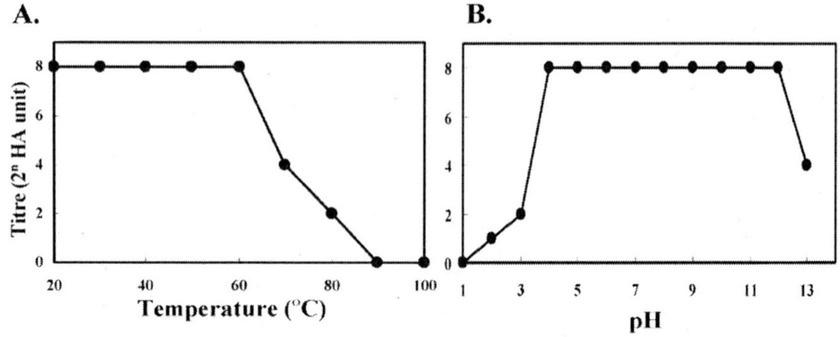

Figure 3. Thermostability and pH stability of the lectin. (**A**) Effects of temperature; (**B**) Effects of pH on hemagglutination activity (vertical axis indicates the titer [2^n HA units]).

3.2. Saccharide-Binding Specificity of HOL-18

The hemagglutination activities of the lectin were effectively inhibited by chitotriose (0.4 mM). These activities were also inhibited by mono-, disaccharide and glycoproteins, including methyl α-D-GlcNAc and methyl β-D-GlcNAc (3.13 mM), methyl α-D-GalNAc (12.5 mM), methyl α-D-GalNAc (6.25 mM), methyl β-D-GalNAc (12.5 mM), chitobiose (0.8 mM), PSM (0.01 mg/mL), BSM (0.125 mg/mL) and fetuin (0.125 mg/mL). The lectin recognized well GlcNAc and GalNAc but it did not selectively recognize their α- and β-anomers. On the other hand, D-Gal, D-Glc, L-Fuc, D-Man, sucrose and lactose did not result in inhibition even at concentrations over 100 mM (Table 2).

3.4. Association and Dissociation Rates of the Lectin

PSM was immobilized on the CM5 sensor chip at 5800 resonance units (RU), and using BIAcore 3000, several concentrations of HOL-18 were applied as an analyte to determine the dissociation constant of HOL-18. SPR analysis indicated that free HOL-18, when used as an analyte, binds to the immobilized PSM on the sensor chip in a dose-dependent manner (Figure 5). The kinetic parameters of the association (k_{ass}) and dissociation rates (k_{diss}) were calculated from the sensorgram response and simulated to mirror a 1:1 affinity model using BIAevaluation software version 3.0. The k_{ass}, k_{diss} and K_D values of HOL-18 were calculated as 1.4×10^3 $M^{-1}s^{-1}$, $6.1 \times 10^{-3} s^{-1}$ and 4.4×10^{-6} M, respectively.

These results indicate that HOL-18 both quickly associates and dissociates from PSM, which is similar to other general lectins.

Table 2. Saccharide and glycoprotein specificity of HOL-18.

Saccharides	Minimum inhibitory concentration (mM)
Chitobiose	0.8
Chitotriose	0.4
D-Galactose	N.I. [a]
Methyl α-*N*-Acetyl D-glucosamine	3.13
Methyl β-*N*-Acetyl D-glucosamine	3.13
Methyl α-*N*-Acetyl D-galactosamine	6.25
Methyl β-*N*-Acetyl D-galactosamine	12.5
D-Glucose	N.I.
D-Mannose	N.I.
L-Fucose	N.I.
Lactose	N.I.
Sucrose	N.I.
Glycoproteins	Minimum inhibitory concentration (mg/mL)
PSM	0.01
BSM	0.125
Fetuin	0.125

[a] Inhibition did not occur even at 100 mM. Titer unit of lectin was previously adjusted to 16.

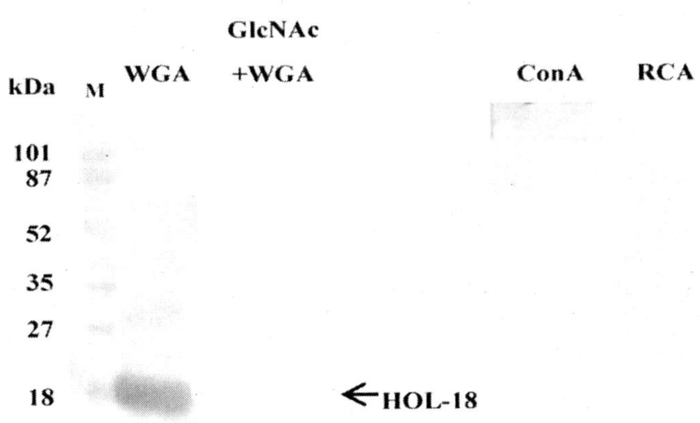

Figure 4. HRP-conjugated lectin blot analysis. HOL-18 (5 μg each) was subjected to SDS-PAGE and blotted onto PVDF membrane. Oligosaccharide of HOL-18 polypeptide was detected by HRP-conjugated WGA, ConA and RCA120 (1 μg each). Column GlcNAc + WGA was co-incubated with GlcNAc (50 mM) in addition to WGA for the cancellation test. M stands for molecular marker. Arrow indicates the position of HOL-18.

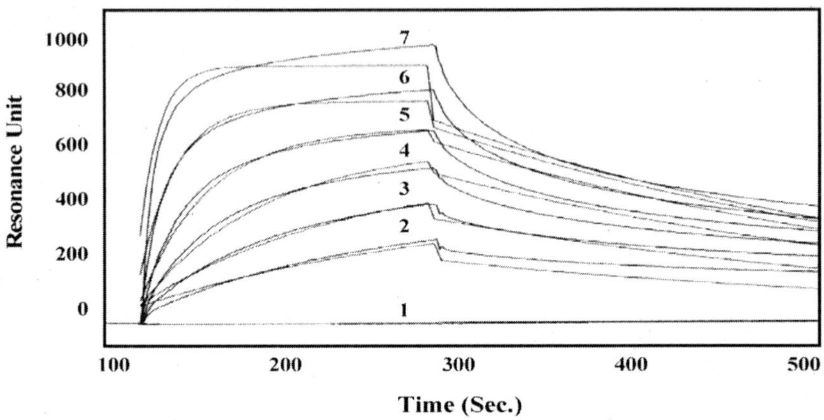

Figure 5. Kinetic analysis of HOL-18 by SPR. HOL-18 was applied to the CM5 sensor chip and coupled with PSM at a concentration of 5.8 ng/mm^2. HOL-18 (0, 2, 4, 6, 8, 10 or 12 nM of each are represented by numbers 1–7 respectively) was applied onto the PSM-conjugated sensor chip at a concentration of 20 μL/min for 2.5 min. The chip was then washed with HBS for 2.5 min. Horizontal axis: Time (s); vertical axis: Resonance unit (indicating the association and dissociation rates of the analyte). The kinetics was analyzed using BIEvaluation software, version 3.0.

3.5. The Glycan-Binding Property

To determine the glycan-binding profile of HOL-18, 14 different PA-oligosaccharides, PA-GlcNAc and PA-GalNAc were used (Table 3). HOL-18 has shown strong affinities to branched N-acetylated glycans of complex-type oligosaccharides in the following hierarchy: Bi-antennary (**012**) > tri-antennary (**013**) > tetra-antennary (**014**). Sialylated complex-type oligosaccharides (two-branched **023** and three-branched (**024**) with N-acetyl group also interacted with HOL-18. With regards to glycosphingolipid-type glycans, Forssman pentasaccharide, which contains two GalNAc residues (**040**), blood type A hexasaccharide that contains an α-GalNAc residue (**047**) and asialo-GM2-tetrasaccharide with β-GalNAc (**027**) were also weakly recognized by the lectin (Figure 6A). Based on the results of FACT, the glycan-binding properties of the lectin indicated that HOL-18 is able to recognize serum glycoprotein-type oligosaccharides that contain D-GlcNAc (agalactosyl) and D-Neu5Ac (sialosyl) clusters and glycosphingolipid oligosaccharides having D-GalNAc at the non-reducing termini (Figure 6B) in addition to O-linked oligosaccharide derived from mucin-type glycoprotein (Table 2).

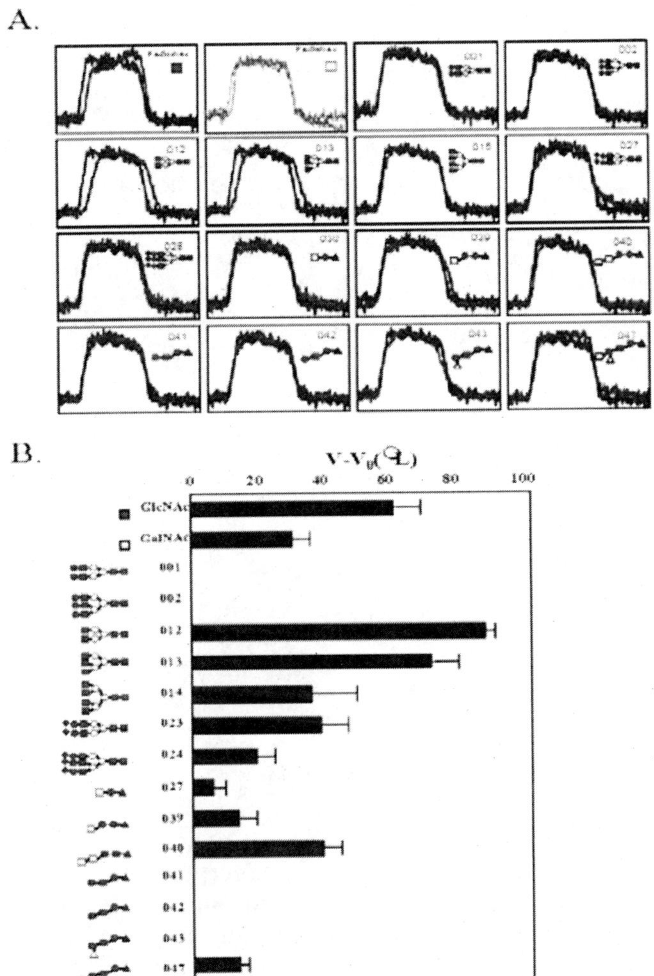

Figure 6. Glycan-binding profile of HOL-18. (**A**) Elution curves of each PA-oligosaccharide (listed in Table 3) obtained from the HOL-18-immobilized column; (**B**) Bar graph representation of the V – V_0 of HOL-18 toward PA-glycans. PA-rhamnose was used as the inert control. Schematic representations of the PA-oligosaccharide for the analysis. Symbols for ▲, ○, ●, ■, □, △ and ◆ represent pyranose rings of D-Glc, D-Man, D-Gal, D-GlcNAc, D-GalNAc, L-Fuc and D-NeuAc, respectively (anomeric carbon, i.e., position 1, is placed at the right side, and 2, 3, 4, 5 and 6 are placed clockwise from positon C-1). Thick and thin bars represent α- and β-linkages, respectively (See Table 3). Results were repeated triplicates and reported as the mean of standard error.

Table 3. List of PA-oligosaccharides for FACT analysis of HOL-18.

PA No.	Glycoprotein type PA-oligosaccharide	PA No.	Glycosphingolipid type PA-oligosaccharide
[001]	Galβ1-4GlcNAcβ1-2Manα1 6 Manβ1-4GlcNAcβ1-4GlcNAc-PA 3 Galβ1-4GlcNAcβ1-2Manα1	[027]	GalNAcβ1-4Galβ1-4Glc-PA
[002]	Galβ1-4GlcNAcβ1-2Manα1 6 Manβ1-4GlcNAcβ1-4GlcNAc-PA Galβ1-4GlcNAcβ1-4 3 Manα1 Galβ1-4GlcNAcβ1-2	[039]	GalNAcβ1-3Galα1-4Galβ1-4Glc-PA
[012]	GlcNAcβ1-2Manα1 6 Manβ1-4GlcNAcβ1-4GlcNAc-PA 3 GlcNAcβ1-2Manα1	[040]	GalNAcα1-3GalNAcβ1-3Galα1-4Galβ1-4Glc-PA
[013]	GlcNAcβ1-2Manα1 6 Manβ1-4GlcNAcβ1-4GlcNAc-PA GlcNAcβ1-4 3 Manα1 GlcNAcβ1-2	[041]	Galβ1-4GlcNAcβ1-3Galβ1-4Glc-PA
[014]	GlcNAcβ1-6 Manα1 GlcNAcβ1-2 6 Manβ1-4GlcNAcβ1-4GlcNAc-PA GlcNAcβ1-4 3 Manα1 GlcNAcβ1-2	[042]	Galβ1-3GlcNAcβ1-3Galβ1-4Glc-PA
[023]	Neu5Acα2-6Galβ1-4GlcNAcβ1-2Manα1 6 Manβ1-4GlcNAcβ1-4GlcNAc-PA 3 Neu5Acα2-6Galβ1-4GlcNAcβ1-2Manα1	[043]	Fucα1-2Galβ1-3GlcNAcβ1-3Galβ1-4Glc-PA
[024]	Neu5Acα2-6Galβ1-4GlcNAcβ1-2Manα1 6 Manβ1-4GlcNAcβ1 4GlcNAc-PA Neu5Acα2-6Galβ1-4GlcNAcβ1-4 3 Manα1 Neu5Acα2-6Galβ1-4GlcNAcβ1-2	[047]	GalNAcα1-3(Fucα1-2)Galβ1-3GlcNAcβ1-3Galβ1-4Glc-PA

3.6. Glycan-Dependent Cytotoxicity

HOL-18 (at concentrations of 1–25 μg/mL) showed 40% and 50% cell death against T-cell leukemia Jurkat and erythroleukemia K562 cells, respectively, as determined by the trypan blue exclusion assay (Figure 7). The cytotoxic activity against both Jurkat and K562 cells of HOL-18 were considerably inhibited by the co-presence of 50 mM D-GlcNAc and D-GalNAc in the medium (Figure 7; column GlcNAc and GalNAc).

HOL-18 had curious biochemical properties in that the 18 kDa polypeptide formed a non-covalent dimer (32 kDa) whose integrity was tightly maintained even if the lectin was heated to 70 °C for 15 min under both reducing and non-reducing conditions (Figure 1B). Finally, it exhibits a tetrameric form (72 kDa) in native conditions (Figure 3). Lectin-glycome analysis showed that HOL-18 was demonstrated to bind to two-branched agalactosyl or sialyl complex-type oligosaccharides of glycoprotein, globo- and lacto-series oligosaccharides such as Forssman and blood type A antigens containing α-D-GalNAc of glycosphingolipid, respectively.

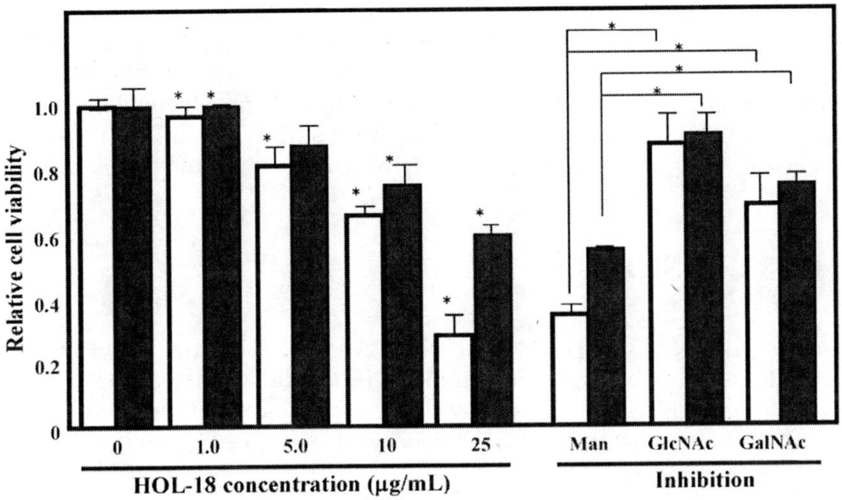

Figure 7. Glycan-dependent cytotoxicity of HOL-18 against Jurkat and K562 cells. Cell viability was determined by using the Trypan blue exclusion assay. Cells were treated with HOL-18 (0–25 μg/mL). White and gray bars indicate cell viability of the Jurkat and K562 cells, respectively. Specific inhibition of the cell viability reduction activity of HOL-18 by the addition of saccharides (Man: Mannose, GlcNAc: N-acetyl-D-glucosamine, GalNAc: N-acetyl-D-galactosamine. 50 mM of saccharides were added to each experiments.). Error bars represents the mean ± SE, n = 3. Only P values < 0.05 were considered to be statistically significant indicated by asterisk (*).

According to the binding profile, HOL-18 showed cytotoxicity to both T-cell leukemia (Jurkat) and erythroleukemia (K562) cells which are known to express D-GlcNAc and D-GalNAc residues, in addition to D-NeuAc, at the non-reducing termini of oligosaccharides derived from cell membrane glycoproteins [23]. From the view of the hemagglutination inhibition test, it appeared that the lectin was inhibited by such glycoproteins as PSM, BSM and fetuin (Table 2). These results indicated that HOL-18 has an affinity for O-linked oligosaccharides in mucins and serum glycoproteins. PSM possesses non-sialylated oligosaccharides, such as GlcNAcβ1-6 (±Fucα1-2Galβ1-3)GalNAc and GlcNAcβ1-6GalNAc. Conversely, BSM possesses Neu5Acα2-6(±GlcNAcβ1-3)GalNAc in addition to sialyl Tn-antigen (Neu5Acα2-6GalNAc) [24]. Fetuin contains O-linked oligosaccharides and also contains sialyl N-acetyllactosamine (Neu5Acα2-3/6Galβ1-4GlcNAc) chains in complex-type N-linked glycans [25].

Recently, a sponge lectin with chemotactic activity toward murine neutrophils was reported and it exhibited N-acetyl sugar-binding activity similarly to HOL-18 [8]. This suggested that the lectin might have cell regulatory functions in addition to toxicity. Chitin and Glycosphingolipids having N-acetyl sugars have been recently isolated from different marine sponges [26,27], which may be binding partner candidates for HOL-18.

4. CONCLUSIONS

A lectin, HOL-18 that consisted of tetrameric 18 kDa subunits was purified from the Japanese demosponge H. okadai and its glycan-binding affinity toward oligosaccharides with N-acetylated sugars was elucidated by a lectin-glycome study. HOL-18 was cytotoxic to cancer cells through a lectin-glycan interaction.

ACKNOWLEDGMENTS

This work was supported by a Grant-in-Aid for Scientific Research (No. 21009100 and 22580226 to YO) from the Japan Society for the Promotion of Science (JSPS), the Japanese Association for Marine Biology (JAMBIO, No. 23-04 and 24-03) from the Ministry of Education, Culture, Sports, Science and Technology of Japan (MEXT) and by a Strategic Research Project Grant (No. G2307) from Yokohama City University.

REFERENCES

1. Ogawa, T.; Watanabe, M.; Naganuma, T.; Muramoto, K. Diversified carbohydrate-binding lectins from marine resources. J. Amino Acids **2011**, 2011.
2. Paunto, P.C.; de Sukva, M.A.; Linardi, A.; Buzin, M.P.; Melo, S.E.S.F.C.; Mello, S.M.; Prado-Franceschi, J.; Hyslop, S. Biological activities of a lectin from Bothrops jararacussu snake venom. Toxicon **2006**, 47, 21–31.
3. Sriwilaijaroen, N.; Kondo, S.; Yagi, H.; Wilairat, P.; Hiramatsu, H.; Ito, M.; Ito, Y.; Kato, K.; Suzuki, Y. Analysis of N-glycans in embrhonated chicken egg chorioallantoic and amniotic cells responsible for binding and adaptation of human and avian influenza viruses. Glycoconj. J. **2009**, 26, 433–443.

4. Bai, R.; Nguyen, T.L.; Burnett, J.C.; Atasoylu, O.; Munro, M.H.; Pettit, G.R.; Smith, A.B.; Gussio, R.; Hamel, E. Interaction of halichondrin B and eribulin with tubulin. J. Chem. Inf. Model **2011**, 51, 1393–1404.

5. Schröder, H.C.; Breter, H.C.; Fattorusso, E.; Ushijima, H.; Wiens, M.; Steffen, R.; Batel, R.; Müller, W.E.G. Okadaic acid, an apoptogenic toxin form symbiotic parasitic annelids in the demosponge Suberites domuncula. Appl. Environ. Microbiol. **2006**, 72, 4907–4916.

6. Pfeifer, K.; Haasemann, M.; Gamulin, V.; Bretting, H.; Fahrenholz, F.; Müller, W.E.G. S-type lectins occur also in invertebrates: High conservation of the carbohydrate recognition domain in the lectin genes from the marine spongeGeodia cydonium. Glycobiology **1993**, 3, 179–184.

7. Schröder, H.C.; Boreiko, A.; Korzhev, M.; Tahir, M.N.; Tremel, W.; Eckert, C.; Ushijima, H.; Müller, I.M.; Müller, W.E.G. Co-expression and functional interaction of silicatein with galectin; matrix-guided formation of siliceous spicules in the marine demosponge Suberites domuncula. J. Biol. Chem. **2006**, 281, 12001–12009.

8. Dresh, R.R.; Zanetti, G.D.; Lerner, C.B.; Trindade, V.M.; Henriques, A.T.; Vozári-Hampe, M.M. ACL-1, a lectin from the marine sponge Axinella corrugate: Isolation, characterization and chemotactic activity. Comp. Biochem. Physiol. C Toxicol. Pharmacol. **2008**, 148, 23–30.

9. Schröder, H.C.; Ushijima, H.; Krasko, A.; Gamulin, V.; Thakur, N.L.; Diehl-Seifert, B.; Müller, I.M.; Müller, W.E.G. Emergence and disappearance of an immune molecule, an antimicrobial lectin, in basal metazoa; a tachylectin-related protein in the sponge Suberites domuncula. J. Biol. Chem. **2003**, 278, 32810–32817.

10. Funayama, N.; Nakatsukasa, M.; Kuraku, S.; Takechi, K.; Dohi, M.; Iwabe, N.; Miyata, T.; Agata, K. Isolation of Efsilicatein and Ef lectin as molecular markers for sclerocytes and cells involved in innate immunity in the freshwater sponge Ephydatia fluviatilis. Zool. Sci. **2005**, 22, 1113–1122.

11. Kawsar, S.M.A.; Fujii, Y.; Matsumoto, R.; Ichikawa, T.; Tateno, H.; Hirabayashi, J.; Yasumitsu, H.; Dogasaki, C.; Hosono, M.; Nitta, K.; Hamako, J.; Matsui, T.; Ozeki, Y. Isolation, purification, characterization and glycan-binding profile of a D-galactoside

specific lectin from the marine sponge, Halichondria okad. Comp. Biochem. Physiol. B Biochem. Mol. Biol. **2008**, 150, 349–357.

12. Kawagishi, H.; Yamawaki, M.; Isobe, S.; Usui, T.; Kimura, A.; Chiba, S. Two lectins from the marine spongeHalichondria okadai. An N-acetyl-sugar-specific lectin (HOL-I) and an N-acetyllactosamine-specific lectin (HOL-II). J. Biol. Chem. **1994**, 269, 1375–1379.

13. Hirabayashi, J.; Arata, Y.; Kasai, K-I. Frontal affinity chromatography as a tool for elucidation of sugar recognition properties of lectins. Methods Enzymol. **2003**, 362, 353–368.

14. Hirabayashi, J.; Hashidate, T.; Arata, Y.; Nishi, N.; Nakamura, T.; Hirashima, M.; Urashima, T.; Oka, T.; Futai, M.; Müller, W.E.G.; Yagi, F.; Kasai, K. Oligosaccharide specificity of galectins: A search by frontal affinity chromatography. Biochim.Biophys. Acta **2002**, 1572, 232–254.

15. Laemmli, U.K. Cleavage of structural proteins during the assembly of the head of bacteriophage T4. Nature **1970**, 227, 680–685.

16. Wiechelman, K.J.; Braun, R.D.; Fitzpatrick, J.D. Investigation of the bicinchoninic acid protein assay: Identification of the groups responsible for color formation. Anal. Biochem. **1988**, 175, 231–237.

17. Smith, P.K.; Krohn, R.I.; Hermanson, G.T.; Mallia, A.K.; Gartner, F.H.; Provenzano, M.D.; Fujimoto, E.K.; Goeke, N.M.; Olson, B.J.; Klenk, D.C. Measurement of protein using bicinchoninic acid. Anal.Biochem. **1985**, 150, 76–85.

18. Kyhse-Andersen, J. Electroblotting of multiple gels a simple apparatus without buffer tank for rapid transfer of proteins from polyacrylamide to nitrocellulose. J. Biochem. Biophys. Methods **1984**, 10, 203–209.

19. Dutta, C.; Henry, H.L. Detection of hemoprotein peroxidase activity on polyvinylidene difluoride membrane. Anal. Biochem. **1990**, 184, 96–99.

20. Gourdine, J.P.; Cioci, G.; Miguet, L.; Unverzagt, C.; Silva, D.V.; Varrot, A.; Gautier, C.; Smith-Ravin, E.J.; Imberty, A. High affinity interaction between a bivalve C-type lectin and a biantennary complex-type N-glycan revealed by crystallography and microcalorimetry. J. Biol. Chem. **2008**, 283, 30112–30120.

21. Shinohara, Y.; Kim, F.; Shimizu, M.; Goto, M.; Tosu, M.; Hasegawa, Y. Kinetic measurement of the interaction between an

oligosaccharide and lectins by a biosensor based on surface plasmon resonance. Eur. J. Biochem. **1994**, 223, 189–194.

22. Kawano, T.; Sugawara, S.; Hosono, M.; Tatsuta, T.; Ogawa, Y.; Fujimura, T.; Taka, H.; Murayama, K.; Nitta, K. Globotriaosylceramide-expressing Burkitt's lymphoma cells are committed to early apoptotic status by rhamnose-binding lectin from catfish eggs. Biol. Pharm. Bull. **2009**, 32, 345–353.

23. Piller, V.; Piller, F.; Fukuda, M. Biosysthesis of truncated O-glycans in the T cell line Jurkat. Localization of O-glycan initiation. J. Biol. Chem. **1990**, 265, 9264–9271.

24. Yamada, K.; Hyodo, S.; Matsuno, Y.K.; Kinoshita, M.; Maruyama, S.Z.; Osaka, Y.S.; Casal, E.; Lee, Y.C.; Kakehi, K. Rapid and sensitive analysis of mucin-type glycans using an in line flow glycan-releasing apparatus. Anal.Biochem.**2007**, 371, 52–61.

25. Baenziger, J.U.; Fiete, D. Structure of the complex oligosaccharides of fetuin. J. Biol. Chem. **1979**, 254, 789–795.

26. Ehrlich, H.; Maldonado, M.; Spindler, K.D.; Eckert, C.; Hanke, T.; Born, R.; Glebel, C.; Simon, P.; Heinemann, S. First evidence of chitin as a component of the skeletal fibers of marine sponges. Part 1 Verongidea (demospongia Porifera).J. Exp. Zool. B Mol. Dev. Evol. **2007**, 308, 347–356.

27. Fujita, Y.; Ohsima, N.; Hasegawa, A.; Schweizer, F.; Takeda, T.; Kiuchi, F.; Hada, N. Synthesis, inhibitory effects on nitric oxide and structure activity relationships of a glycosphingolipid from the marine sponge Aplysinella rhax and its analogues. Molecules **2011**, 16, 637–651.

9

CHAPTER

Structure of an Amino Acid-Decorated Exopolysaccharide Secreted by a Vibrio Alginolyticus Strain

Sophie Drouillard [1], Isabelle Jeacomine [1], Laurine Buon [1], Claire Boisset [1], Anthony Courtois [2], Bertrand Thollas [2], Pierre-Yves Morvan [3], Romuald Vallée [3] and William Helbert [1,*]

[1]*Université Grenoble Alpes, CERMAV, CNRS, 38000 Grenoble, France*

[2]*Polymaris Biotechnology, Aéropôle Centre, 29600 Morlaix, France*

[3]*Codif International, 35610 Roz-sur-Couesnon, France*

ABSTRACT

Vibrio alginolyticus (CNCM I-4994) secretes an exopolysaccharide that can be used as an ingredient in cosmetic applications. The structure was resolved using chromatography and one- and two-dimensional NMR spectroscopy experiments. The results show that the carbohydrate backbone is made of two residues: D-galacturonic acid and N-acetyl-D-glucosamine (GlcNac), which together constitute a tetrasaccharide repetition unit: [→3)-α-D-GalA-(1→4)-α-D-GalA-(1→3)-α-D-GalA-(1→3)-β-GlcNAc(1→]. Two amino acids, alanine and serine, are linked to GalA residues via amido linkages. The position and the distribution of the amino acids were characterized by two-dimensional NMR spectroscopy. To our knowledge, this is the first description of a structure for a marine exopolysaccharide decorated with an amino acid.

Keywords: exopolysaccharide; Vibrio alginolyticus; amino acids; ingredient

1. INTRODUCTION

Many marine bacteria secrete extracellular polysaccharides, called exopolysaccharides (EPSs), which constitute an immense reservoir of novel molecules whose structure and functional properties are largely unknown. The structural diversity of EPSs—as for all other polysaccharides—lies in the stereochemistry of the carbohydrate residues, the glycosidic linkage between the residues, and, in turn, the identity of the (oligo)saccharide repetition units. Homopolysaccharides have the simplest structure, being composed of one type of carbohydrate residue; heteropolysaccharides are generally composed of 2 to 8 different residues, with some exceptionally highly complex structures containing up to 10 to 12 different residues. A second level of EPS complexity is provided by the decoration of the carbohydrate backbone with organic (e.g., acetate, pyruvate) or inorganic (e.g., sulfate ester) derivatives [1].

The structural diversity of marine EPSs is associated with a wide range of biological and physico-chemical properties. EPSs create a micro-environment that facilitates bacterial reproduction and protects bacterial cells from strong environmental variation. Other functions attributed to EPSs include attachment of bacterial cells to surfaces, the formation of biofilms, as well as the concentration or sequestration of specific molecules (e.g., signaling, nutriments) [2]. In vitro, several marine EPSs have been shown to have biological activity, including anti-tumor activity, immunostimulatory activity, anticomplementary activity, and involvement in bone and tissue regeneration [3,4,5]. Although marine EPSs have aroused considerable interest, their use in industry is still very limited in contrast to, for example, terrestrial bacterial EPSs, such as xanthan (Xanthomonas campestris) or gellan (Sphingomonas paucimobilis) which are employed in the food industry for their viscosifying and gelling properties.

Production costs and compliance with regulations hamper the industrial-scale use of marine EPSs as hydrocolloids for food applications, and competition with the already marketed plant and algal polysaccharides is currently insurmountable [1,6]. However, the production of marine EPSs has several technical advantages: their production can be easily controlled and is independent of seasonal variation and they can be very highly purified, making them appropriate for niche applications in the biomedical and cosmetics sectors. One example is the EPS secreted by the Gram-negative marine bacteriaVibrio alginolyticus (CNCM I-4994), which is produced on an industrial scale as an ingredient for cosmetic applications for its matifying and anti-inflammatory properties [7]. This strain was isolated from deep water in the Aber Wrac'h estuary

(Brittany, France). It lacks the tdh and trh genes coding for thermostable direct hemolysin (TDH) and/or TDH-related hemolysin (TRH), respectively, involved in pathogenicity of numerous Vibrio strains [8]. Here, we report the chemical structure of V. alginolyticus EPS (VA-EPS), which has a very original glucosaminogalactan backbone decorated with alanine and serine amino acids.

2. RESULTS AND DISCUSSION

2.1. Purification and Composition of the EPSs

The EPS was purified in several steps of filtration and precipitation with ethanol. The Bradford assay indicated the absence of any detectable proteins, thereby attesting to the high purity of the polysaccharides. However, alanine and serine were detected in high-performance liquid chromatographs after complete hydrolysis of the sample, suggesting that they were not free amino acids. Size-exclusion chromatography (SEC) coupled to multi-angle laser light scattering confirmed that the purified EPS was likely composed of one species of molecule having a molecular weight of 1.16×10^6 Da with a narrow polydispersity index of 1.34.

Composition analysis using gas-chromatography based on the trimethysilyl method on the residues obtained after complete hydrolysis of the polysaccharides revealed only two residues: galacturonic acid (GalA) and N-acetyl-glucosamine (GlcNAc) in a ratio of about 3.0:1.0 GalA:GlcNAc. Elementary analysis did not uncover any inorganic derivatives such as sulfate or phosphate ester groups. Methylation analyses, shown in Table 1, indicated that the GalA residues were involved either in 1,3-, 1,4- or 1,3,6-linkages and the GlcNAc residue in 1,3-linkages.

The ^{13}C NMR spectrum of the polysaccharide (Figure 1) showed signals at 95–105 ppm and were attributed to four anomeric carbons. Three signals were observed in the region of the methyl groups (15–25 ppm): the first corresponded to the methyl group of an acetyl group (22.68 ppm) and the two other signals (18.44 and 18.91 ppm) were attributed to methyl groups of amino acids. Eight different signals were observed in the region of the carbonyl groups between 175 and 180 ppm and likely corresponded to C6 of free uronic acids, carboxyl groups of amino acids and amide groups.

Table 1. Methylation analysis of the Vibrio alginolyticus exopolysaccharide (VA-EPS) after carboxyl reduction.

Type of Sugar	Partially Methylated Alditol Acetate	Deduced Linkage
Galactose	1,3,5-Tri-*O*-acetyl-1-deuterio-2,4,6-tri-*O*-methyl-D-Galactitol	→3)-Gal*p*-(1→
	1,4,5-Tri-*O*-acetyl-1-deuterio-2,3,6-tri-*O*-methyl-D-Galactitol	→4)-Gal*p*-(1→
	1,3,5,6-Tetra-*O*-acetyl-1-deuterio-3,4-di-*O*-methyl-D-Galactitol	→3,6)-Gal*p*-(1→
N-acetyl Glucosamine	1,3,5-Tri-*O*-acetyl-2-(acetylmethylamino)-2-deoxy-1-deuterio-4,6-di-*O*-methyl-D-Galactitol	→3)-GlcNAc-(1→

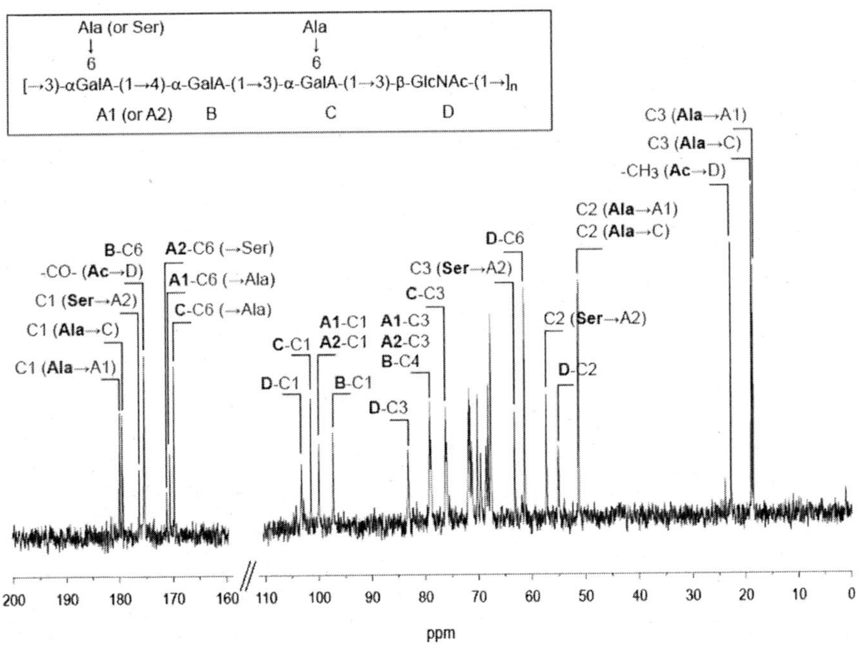

Figure 1. ^{13}C NMR of Vibrio alginolyticus exopolysaccharide recorded at 353 K. Inset: chemical structure of the polysaccharide.

Inspection of ^1H NMR spectra (Figure 2) and $^1H/^{13}C$ heteronuclear single-quantum correlations (HSQC) of the EPS (not shown) confirmed a repetition unit made of four distinct residues of which three (A–C) adopt an α-anomeric configuration and the fourth (D) has a β-anomeric configuration. ^1H NMR spectra confirmed that the repetition moiety is a tetrasaccharide decorated with N-acetyl groups and amino acids.

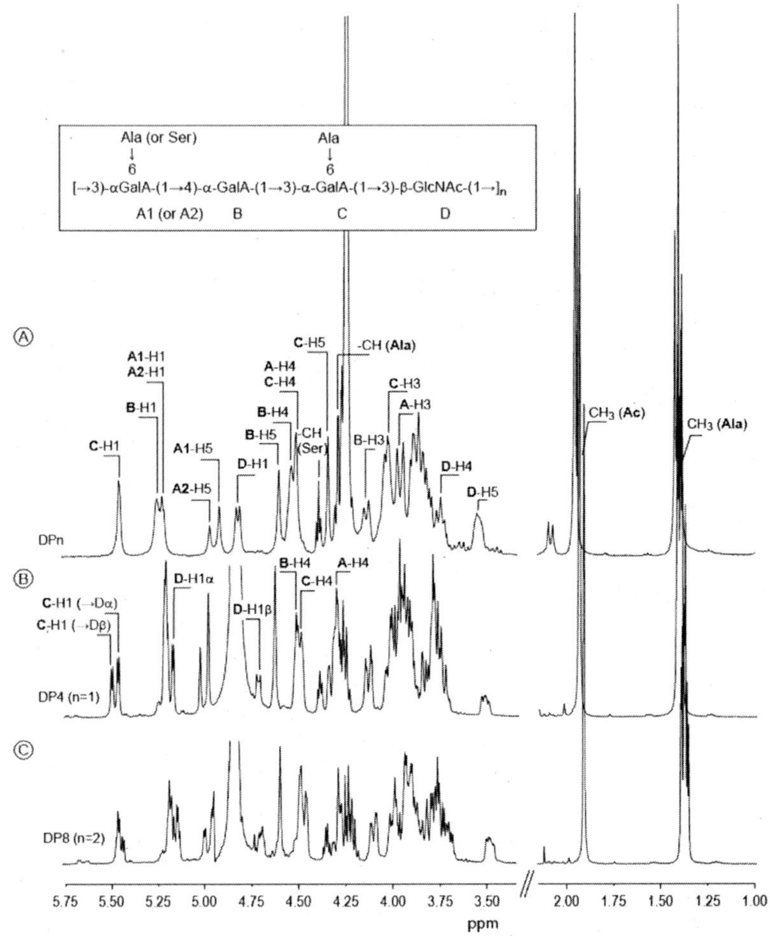

Figure 2. ^{1}H NMR of the Vibrio alginolyticus exopolysaccharide and its oligosaccharides (DP4, DP8) purified after acid hydrolysis. (**A**) The ^{1}H NMR of the polysaccharide was recorded at 353 K; (**B**,**C**) ^{1}H NMR of DP4 and DP8, respectively, recorded at 293 K. Inset: chemical structure of the polysaccharide.

2.2. Detailed Structure Analysis of the EPS Using NMR

We assigned the ring protons of the three α-anomeric residues (A–C) starting from the anomeric protons A-H1, B-H1 and C-H1 by successfully combining correlation spectroscopy (COSY) and total COSY (TOCSY) analyses (Table 2). The chemical shifts of the corresponding carbons were determined using $^1H/^{13}C$ heteronuclear single-quantum correlation (HSQC) and are reported in Table 3.

Table 2. Proton chemical shifts (ppm) of the Vibrio alginolyticus exopolysaccharide (VA-EPS) and the purified oligosaccharides (DP4, DP8). A–D refer to the residues composing the repetition moieties of the polysaccharide. When the residues are located at the reducing end or at the non-reducing end, they are labelled r and nr, respectively.

	VA-EPS: $[ABCD]_n$	**DP4: $A^{nr}BC^{r'}D^{r}$**	**DP8: $A^{nr}BCDABC^{r'}D^{r}$**
A1: 6Ala-GalA/A2:6Ser-GalA			
H1	5.18/5.19		5.20/5.20
H2	3.84/3.84		3.80/3.80
H3	4.00/4.00		4.04/4.04
H4	4.49/4.51		4.54/4.54
H5	4.91/4.96		4.99/5.04
$A1^{nr}$: 6Ala-GalA/$A2^{nr}$:6Ser-GalA			
H1		5.22/5.23	5.20/5.22
H2		3.78/3.78	3.80/3.80
H3		3.97/3.99	3.99/3.99
H4		4.33/4.36	4.34/4.34
H5		4.99/5.04	4.99/5.04
B: GalA			
H1	5.22	5.22	5.23
H2	3.93	3.93	3.95
H3	4.10	4.14	4.15
H4	4.51	4.52	4.53
H5	4.58	4.64	4.65
C: 6Ala-GalA			
H1	5.43		5.50
H2	3.97		3.98
H3	4.03		4.02
H4	4.48		4.50
H5	4.31		4.32

C$^{r'}$: 6Ala-GalA			
H1-D^r α/β		5.48/5.51	5.48/5.51
H2-D^r α/β		3.98/3.97	3.98/3.97
H3-D^r α/β		4.02/4.00	4.03/4.00
H4		4.50	4.50
H5		4.31	4.32
D: GlcNAc			
H1	4.77		4.75
H2	3.85		3.87
H3	3.81		3.87
H4	3.70		3.75
H5	3.51		3.53
H6′	3.93		3.95
H6′	3.77		3.82
NHAc:CH_3	1.95		1.95
D^r: GlcNAc			
H1 α/β		5.18/4.73	5.18/4.73
H2 α/β		4.03/3.80	4.03/3.81
H3 α/β		3.99/3.81	4.00/3.81
H4 α/β		3.75/3.73	3.75/3.73
H5 α/β		3.92/3.52	3.93/3.53
H6 α/β		3.84/3.91	3.95/3.82
H6′		3.78	3.82
NHAc: CH_3		1.95	1.95

Table 3. Carbon chemical shifts (ppm) of the Vibrio alginolyticus exopolysaccharide (VA-EPS) and the purified oligosaccharides (DP4, DP8).

	VA-EPS: $[ABCD]_n$	DP4: $A^{nr}BC^{r'}D^r$	DP8: $A^{nr}BCDABC^{r'}D^r$
A1: 6Ala-GalA/A2: 6Ser-GalA			
C1	99.82/99.94		100.07/100.22
C2	67.99/67.99		68.81/68.81
C3	78.97/78.97		79.14/79.23
C4	70.12/70.12		70.34/70.34
C5	71.43/71.60		71.92/72.02
C6	170.50/171.05		170.94/171.37
$A1^{nr}$: 6Ala-GalA/$A2^{nr}$: 6Ser-GalA			
C1		100.07/100.22	100.07/100.22
C2		68.84/68.84	68.81/68.81
C3		69.78/69.78	69.76/69.76
C4		70.47/70.43	70.47/70.43
C5		71.93/72.04	71.92/72.04
C6		171.10/171.54	171.10/171.54
B: GalA			
C1	97.20	97.63	97.66
C2	68.33	68.34	68.34
C3	69.34	69.25	69.25
C4-A1/A2	78.68/78.85	78.93/79.15	78.80/78.93
C5	71.82	71.73	71.73
C6	175.34	175.80	175.83

C: 6Ala-GalA			
C1	101.32		101.06
C2	67.40		67.27
C3	76.06		76.37
C4	67.61		67.92
C5	71.72		71.66
C6	169.68		170.05
C′′: 6Ala-GalA			
C1-D′ α/β		101.39/101.04	101.37/101.06
C2-D′ α/β		67.28/67.44	67.27/67.44
C3-D′ α/β		76.41/76.36	76.37/76.26
C4-D′ α/β		67.90/67.84	67.92/67.86
C5-D′ α/β		71.67/71.73	71.51/71.66
C6-D′ α/β		170.01/170.14	170.05/170.15
D: GlcNAc			
C1	103.17		103.55
C2	54.84		54.85
C3	82.89		81.74
C4	71.03		70.93
C5	75.87		75.93
C6	61.13		60.91
NHAc: CH_3/CO	22.68/175.25		22.76/175.64
D′: GlcNAc			
C1 α/β		91.76/95.73	91.76/95.73
C2 α/β		53.11/55.80	53.11/55.80
C3 α/β		80.37/81.89	80.37/81.90
C4 α/β		71.09	71.08
C5 α/β		72.15/76.28	72.15/76.26
C6 α/β		61.02/61.17	61.01/61.14
NHAc: CO		174.90/175.20	174.88/175.19
NHAc: CH_3		22.44/22.70	22.46/22.72

The chemical shifts as well as the coupling constants were in agreement with three α-D-galacturonic residues. For example, the A-H4, B-H4 and C-H4 protons observed at about 4.50 ppm were characteristic of the chemical shifts observed for uronic residues. Detailed inspection of the coupling of the H5 protons with C6 carboxyl groups or amide groups by $^1H/^{13}C$ heteronuclear multiple-bond correlation (HMBC) revealed that there were four instead of three α-D-galacturonic residues. The A-H5 protons (A1-H5: 4.91, A2-H5: 4.96 ppm) correlated with two carbon signals corresponding to two different amide groups (A1-C6: 170.50 ppm, A2-C6: 171.05 ppm) suggesting that these residues were likely bound to two different amino acids (see detailed analyses below). The B-

H5 proton (4.58 ppm) was coupled to the carbon of a free carbonyl group (B-C6: 175.34 ppm), indicating that the B residue is an underived α-D-GalA. The C-H5 proton (4.31 ppm) coupled to a single carbon (C-C6: 169.68 ppm) belonged to an amide group.

Assignment of the protons and carbons of the D residue was conducted similarly using COSY, TOCSY and HSQC analyses. The chemical shifts reported in Table 2 and Table 3 were in agreement with a β-GlcNH (or β-GlcNAc) residue with in particular the characteristic D-H5 proton observed at 3.51 ppm. Coupling of the nitrogen (D-NAc) on the acetyl group with D-H1 (4.77 ppm) and D-H2 (3.85 ppm) revealed by rotating frame nuclear Overhauser effect spectroscopy (ROESY) experiments confirmed that the acetyl group is bound to the β-D-GlcNH. Proton NMR and ROESY experiments suggested that all the β-D-GlcNH were acetylated, giving only β-D-GlcNAc residues.

HMBC experiments to analyze the $^{1}H/^{13}C$ heterenuclear coupling relationships highlighted linkage between most of the residues (Figure 3A). The A-C3 carbon (78.97 ppm) was coupled to the D-H1 proton (4.77 ppm), indicating that the β-D-GlcNAc (D) residue is linked to the α-D-GalA (A) residue via a β(1,3) linkage. Similarly, the correlation between B-C4 (78.68/78.85 ppm) and A-H1 (5.18/5.19 ppm) suggested that the linkage between the two α-D-GalA residues A and B occurs through an α(1,4) linkage. The linkage between the two α-D-GalA residues B and C was also demonstrated (correlations C-C3: 76.06/B-H1: 5.22 ppm and B-C1: 97.20/C-H3: 4.03 ppm) and B is linked to C via an α(1,3) linkage. The linkage between the α-D-GalA residue C with the β-GlcNAc (D) could not be determined here.

2.3. Preparation and Analysis of Standard Oligosaccharides

After mild acid hydrolysis, the degradation of the EPS was monitored by SEC (Figure 4). After 20 min in 1 M TFA at 100 °C, degradation led to three main peaks centered at 130 min (shoulder), 140 min and 160 min elution times. ^{1}H NMR of the fractions collected at elution times of 150–165 min is given in Figure 2. The spectra showed all the characteristic signals of the polysaccharide, along with additional signals attributed to the reducing and non-reducing ends of the oligosaccharide. The signal of D-H1 (polymer) at 4.77 ppm shifted to 4.73 ppm for D-H1β (DP4). A new D-H1α signal appeared at 5.18 ppm indicating that cleavage occurred at the reducing end of the D residues. The C-H1 (polymer) at 5.43 ppm was split into two doublet signals at 5.48 ppm and 5.51 ppm (DP4). This observation suggested the α-galacturonic residue C was linked to the β-GlcNAc (D) and that the splitting distinguished C(α) from C(β) residues linked to the β- or α-GlcNAc residues, respectively. $^{1}H/^{13}C$ HMBC showed the correlation between D-C3 (α/β) (80.37 ppm/81.89 ppm) and C-H1 (Dα/β) (5.48 ppm/5.51 ppm), confirming that residue C is linked to the reducing end of β-GlcNAc (D) via an α(1,3) linkage.

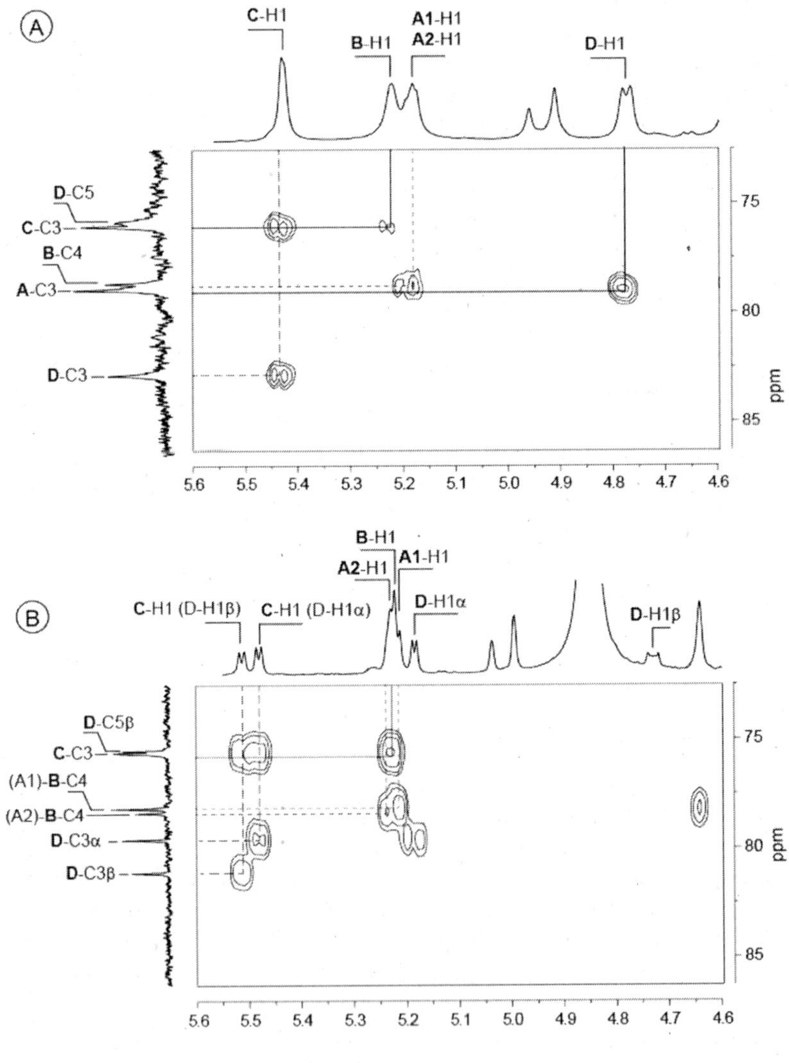

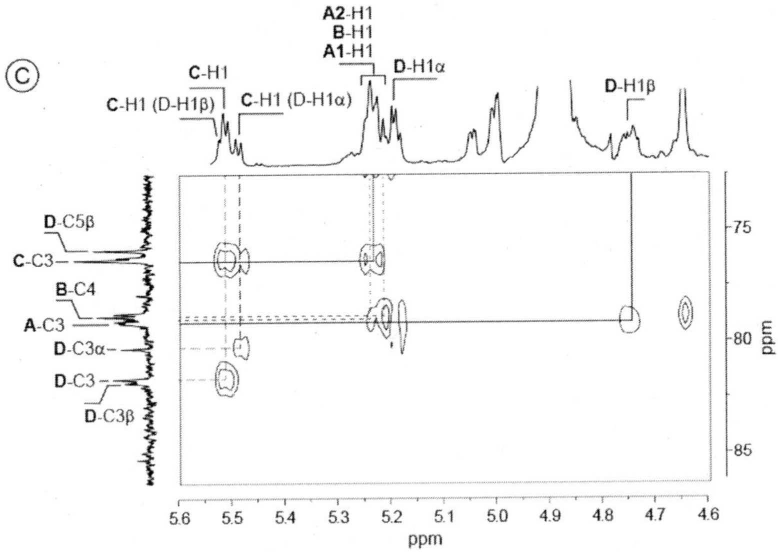

Figure 3. ^{1}H-^{13}C Heteronuclear chemical shift correlated spectrum (HMBC) of theVibrio alginolyticus exopolysaccharide (**A**) and its oligosaccharides: DP4 (**B**) and DP8 (**C**). Correlations demonstrating the linkages between residues in the repetition moieties are indicated.

The fractions collected at elution times of 135–145 min contained a single oligosaccharide whose spectra showed similar characteristics to the polymer and the DP4. The α- and β-anomeric protons of the D-GlcNAc were present, as observed for the DP4, at 5.18 ppm and 4.73 ppm. The anomeric protons of the α-D-GalA (C) residues linked to the D-GlcNAc were also present, but signal intensity was different. Integrations of C-H1 (Dα) and C-H1 (Dβ) revealed a 1:2.3 ratio (Figure 2B) instead of a 1:0.6 ratio (Figure 2C) observed for the DP4, suggesting that one C residue was linked to D-GlcNAc (α- and β-anomer) located at the reducing end and one C residue was linked to D-GlcNAc (β-anomer) located within the oligosaccharide. The COSY spectra distinguished α-GalA (A) residues located at the non-reducing end from those linked toD-GlcNAc (the corresponding signals are indicated in Figure 2C). Altogether, integration of the anomeric protons, as well as complete assignment of the proton and carbon chemical shifts (Table 2 and Table 3) confirmed that this oligosaccharide was an octasaccharide made of two DP4 repetition moieties.

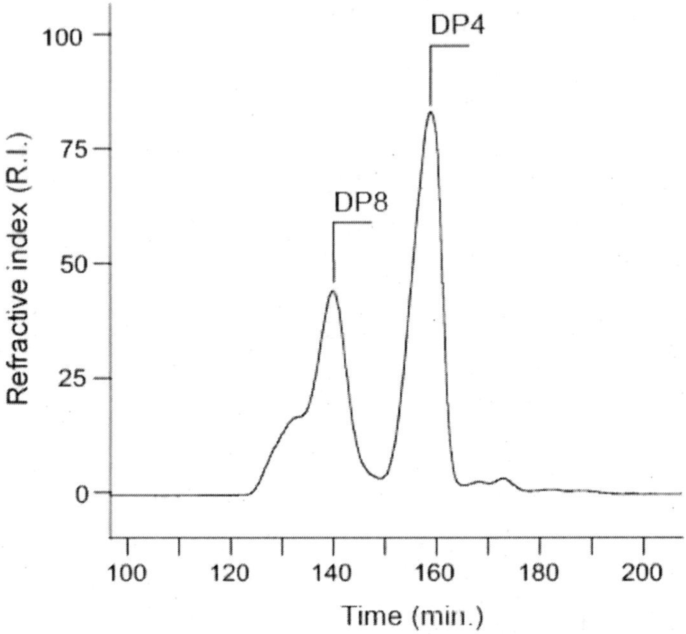

Figure 4. Size-exclusion chromatogram of the acid-hydrolyzed Vibrio alginolyticusexopolysaccharide. The two main products, DP4 and DP8, are indicated.

2.4. Position and Distribution of D-Alanine and D-Serine

The two amino acids, D-alanine and D-serine, detected in the chromatography analyses were also observed by NMR. The proton and carbon chemical shifts were attributed unambiguously using COSY, HMBC and HSQC experiments. The chemical shifts reported in Table 4 are in agreement with known D-alanine and D-serine chemical shifts. The acid treatment and purification steps required to prepare the oligosaccharides DP4 and DP8 did not remove the amino acid, which remained bound to the oligosaccharides; their chemical shifts are reported in Table 4.

Table 4. Proton and carbon chemical shifts of the alanine and serine amino acids linked to the carbohydrate backbone. Ala-A1: alanine linked to residue A1; Ala-C: alanine linked to residue C; Ser-A2: serine linked to residue A2.

	VA-EPS: |ABCD|n	**DP4: $A^{nr}BC^{r'}D^{r}$**	**DP8: $A^{nr}BCDABC^{r'}D^{r}$**
Ala-A1			
H2	4.25	4.27	4.26
H3	1.37	1.41	1.41
C1	179.92	180.44	180.46
C2	51.09	51.29	51.31
C3	18.44	18.43	18.40/18.35

The position of the amino acids on the polysaccharide and on the purified oligosaccharides was deduced from the 2D HMBC spectra. Figure 5 shows a close-up of the HMBC recorded on the polysaccharide highlighting the coupling of the H2 protons of alanine and serine to the C6 carbons involved in the amide linkages (A1-C6, A2-C6 and C-C6). Two different D-alanines were observed, the first was linked to the A1 GalA residue and the second to the C galacturonic acid residue. The D-serine was coupled only to the A2 GalA residue. Integration of the A1-H5 (4.91 ppm) and A2-H5 (4.96 ppm) protons indicated that about 65% of A residues were decorated with D-alanine and other 35% with D-serine.

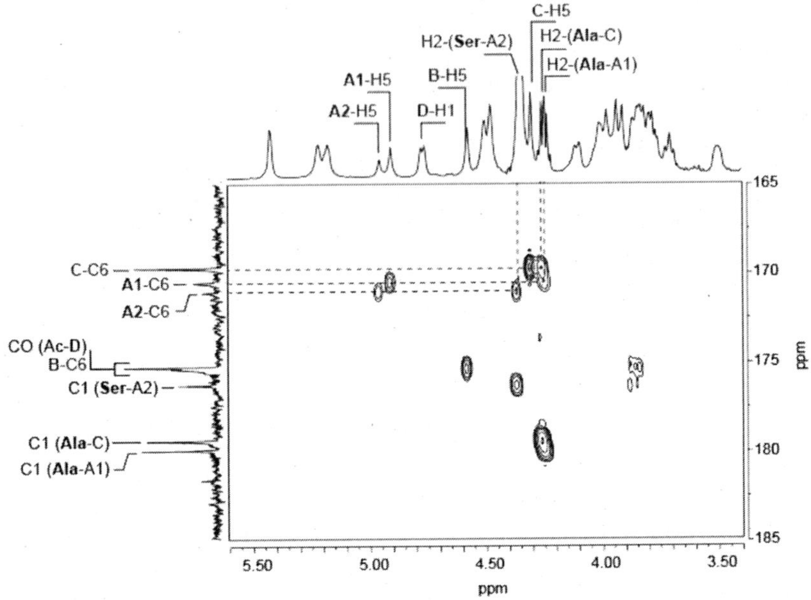

Figure 5. ^{1}H-^{13}C Heteronuclear chemical shift correlated spectrum (HMBC) of theVibrio alginolyticus exopolysaccharide. Correlations between the amino acids (alanine and serine) with the carbohydrate backbone of the polysaccharide are highlighted.

The position of the amino acids were also confirmed by long-distance ROESY coupling experiments, which showed coupling of the A1-NH proton (7.72 ppm) of the alanine amine to the Ala-A1-H2 and Ala-A1-H3 protons of the D-alanine, and more interestingly to the A1-H5 proton (4.91 ppm). Similar observations on the D-serine amine, with the A2-NH proton (7.83 ppm) coupled to the A2-H5 proton (4.96 ppm), confirmed that the GalA A residue is linked either to D-alanine (A1) orD-serine (A2). A second coupling system for D-alanine was also observed, whereby the C-H5 proton (4.31 ppm) of the GalA residue was coupled to the amine proton of alanine (7.80 ppm), in agreement with an amide linkage of D-alanine on this residue.

3. EXPERIMENTAL SECTION

3.1. Production, Isolation and Purification of VA-EPS Exopolysaccharide

VA-EPS was produced by Vibrio alginolyticus (CNCM I-4994) in a 30 L fermenter containing marine broth medium (30 g/L sea salts, 1 g/L yeast extracts, 4 g/L peptone) supplemented with glucose (30 g/L) at 25 °C. The culture medium was inoculated at 10% (v/v) with a bacterial suspension in the exponential growth phase. The pH was adjusted and maintained at 7.2 by automatic addition of 1 M NaOH. The medium was oxygenated at 15 L/min with an agitation rate of 350 rpm. After 72 h of fermentation, bacterial cells were removed from the culture medium by centrifugation (16,000× g, 30 min). The supernatant, containing the excreted VA-EPS, was then purified by filtration through a cellulose membrane (0.7 μm) and then by ultrafiltration (100 kDa). The sample was freeze-dried and stored at room temperature away from light and moisture.

3.2. Monosaccharide Analysis

The molar ratio of monosaccharides was determined according to [9] modified by [10]. 400 μg of EPS were hydrolyzed with 500 μL of 3 M MeOH/HCl at 110 °C for 4 h. After neutralization with silver carbonate a re-N-acetylation step was made with 50 μL Ac_2O overnight at room temperature. After centrifugation, and evaporation of the supernatant, the methyl glycosides were converted to their corresponding trimethylsilyl derivatives by adding 80 μL pyridine and 80 μL of BSTFA:TMCS (99:1; Supelco, Bellefonte, PA, USA). Separation and quantification of the per-O-trimethylsilyl methyl glycosides were performed by gas-liquid chromatography (GLC) using an Agilent system equipped with a HP-5ms capillary column (Agilent 0.25 mm × 30 m). The trimethylsilyl derivatives were analyzed using the following temperature program: 120 °C for 1 min, 120 °C → 180 °C at 3 °C/min, 180 °C → 200 °C at 3°C/min, 200 °C for 5 min.

3.3. Methylation Analysis

Glycosyl-linkage positions were determined as described in [11]. 20 mg of the native EPS in 10 mL H_2O were carboxyl-reduced by treatment with 650 mg of N-cyclohexyl 1-N'[β (N-methyl-morpholino)-ethyl] carbodiimide P-toluene sulfonate and with 3 mg of $NaBD_4$ for 4 h at room temperature [12]. After dialysis against distilled water and lyophilization, the carboxyl reduced EPS was dissolved in 500 μL DMSO. Hydroxyl groups were then methylated using 500 μL of 2.5 N butyl lithium in hexanes and 500 μL of methyl iodide [13]. Methylation reaction was stopped with 1 mL H_2O and the methylated compounds were extracted with CH_2Cl_2. The methylated products were then hydrolyzed in 500 μL of 2 M TFA for 2 h at 120 °C. After evaporation of TFA the corresponding monomers were reduced with 20 mg of $NaBD_4$ in a NH_4OH solution for 30 min at 80 °C, and finally acetylated with 200 μL of 1-methyl imidazole and 2 mL of pyridine for 10 min at room temperature. GLC-mass spectrometry (MS) was performed on an Agilent instrument fitted with a HP-5ms capillary column (Agilent, Santa Clara, CA, USA, 0.25 mm × 30 m). The temperature program was 90 °C for 1 min, 90 °C → 300 °C at 5 °C/min, 300 °C for 1 min. Ionization was carried out in electron impact mode (EI, 70 eV).

3.4. Amino Acid Composition

VA-EPS (1% w/v) in water was hydrolyzed by addition of an equal volume of 11 N HCl for 5 h at 105 °C. After a cooling step, neutralization was carried out with 10 N NaOH, then with 1 N NaOH, using pH-indicator paper to reach pH 6–7. The final volume was adjusted to 25 mL with ultrapure water and filtered on a 0.2 μm sterilizing grade membrane.

The neutralized samples and an amino acid hydrolysate standard (Waters, Milford, MA, USA) were derivatized by adding 6-aminoquinolyl-N-hydroxysuccinimidyl carbamate using AccQ•Tag™ Ultra Derivatization kit (Waters, Milford, MA, USA). The derivatized samples (10 μL) were injected on a AccQ Tag Ultra C18 column (1.7 μm 2.1 × 100 mm, Waters, Milford, MA, USA) mounted on a Waters Acquity H class ultra-high-performance liquid chromatography apparatus equipped with UV detector working at 260 nm.

A gradient elution was conducted for 10 min at 55 °C starting with 100% eluent A (acetonitrile/formic acid) to 100% of eluent B (acetonitrile). Signals were compared to derivatized amino acid standards. According to experimental conditions, serine and alanine eluted at 2.3 min and 4.4 min, respectively.

3.5. Sulfate Content

Elemental analysis was performed by the CNRS Institute of Analytical Sciences (Lyon, France). Potential sulfate content (sodium salt) was evaluated.

3.6. Molecular Weight Determination

The molecular weight of VA-EPS was determined by high-performance size-exclusion chromatography (HPSEC) using an eighteen-angle light scattering detector, coupled with refractive index and specific refractive index increment dn/dc (DAWN™ HELEOS, Wyatt, Santa Barbara, CA USA). Elution was performed on Shodex OHpak SB-805 HQ and OHpak SB-806 HQ placed in series (Phenomenex, fractionation range <2.10^7 g/mol) with 0.1 M $NaNO_3$ as the eluent. To calculate the molecular mass, the dn/dc value used was 0.145 mL/g. The polydispersity index was calculated from the Mw/Mn ratio.

3.7. Acid Hydrolysis and Oligosaccharides Purification

VA-EPS underwent mild acid hydrolysis, 20 min in 1 M TFA at 100 °C. The resulting oligosaccharides were fractionated by SEC on a Toyopearl HW-40 (Tosoh, fractionation range <10^4 Da) with 0.1 M $NaCO_3$ as the eluent.

3.8. NMR

Carbon-13 and proton NMR spectra were recorded with a Bruker Avance 400 spectrometer operating at a frequency of 100.618 MHz for ^{13}C and 400.13 MHz for ^1H. Samples were solubilized in D_2O at a temperature of 293 K for the oligosaccharides and 353 K for the polysaccharide. Residual signal of the solvent was used as internal standard: HOD at 4.85 ppm at 293 K and 4.25 at 353 K. ^{13}C spectra were recorded using 90° pulses, 20,000 Hz spectral width, 65,536 data points, 1.638 s acquisition time, 1 s relaxation delay and between 8192 and 16,834 scans. Proton spectra were recorded with a 4006 Hz spectral width, 32,768 data points, 4.089 s acquisition times, 0.1 s relaxation delays and 16 scans. The ^1H and ^{13}C NMR assignments were based on ^1H-^1H homonuclear and ^1H-^{13}C heteronuclear correlation experiments (correlation spectroscopy, COSY; heteronuclear multiple-bond correlation, HMBC; heteronuclear multiple quantum correlation, HMQC). They were performed with a 4006 Hz spectral width, 2048 data points, 0.255 s acquisition time, 1 s relaxation delay; from 32 up to 512 scans were accumulated.

4. CONCLUSIONS

The structure of the exopolysaccharide secreted by Vibrio alginolyticus (CNCM I-4994) was completely resolved. It is a linear polysaccharide made up of two residues: N-acetyl-glucosamine and galacturonic acid, whose repetition unit—according to linkage and decoration—is a tetrasaccharide. During structural analyses of the polysaccharide, we observed that acid hydrolysis produced an oligosaccharide series having a size equal to (DP4) or a simple multiple (DP8) of the repetition unit, suggesting strong reactivity differences between the glycosidic linkages of the polysaccharide. A second, very interesting feature was the discovery of amino acids naturally bound to uronic residues via amide linkages. Several organic derivatives (pyruvate, succinic acid, acetate) or inorganic acid (sulfate, phosphate) have already been found in many EPSs. However, decoration of the carbohydrate backbone by one amino acid bond via an amide linkage appeared until now, limited to O-antigen lipopolysaccharides [14,15,16] or capsular polysaccharides [17,18]. Therefore, to our knowledge, this is the first report of a polysaccharide secreted by a Gram-negative bacteria carrying two different amino acids (alanine and serine).

ACKNOWLEDGMENTS

The research leading to these results received funding from Brittany Regional Council and OSEO as part of the FUI-AAP14 "Poly-mer" project.

AUTHOR CONTRIBUTIONS

B.T., A.C., P.-Y. M. and R. V. produced and purified the EPS. Preliminary sugar content determination and amino acid composition were also done by these authors. C.B. and L.B. determined the sugar content and the linkages by chemical methods. They also measured the molecular weight of the EPS. S. D. and I. J. performed NMR experiments and analysis. W.H. lead the project and wrote the article.

CONFLICTS OF INTEREST

The authors declare no conflict of interest.

REFERENCES

1. Sutherland, I.W. Microbial polysaccharides from gram-negative bacteria. Int. Dairy J. **2001**, 11, 663–674.
2. Decho, A.W. Microbial exoplomymer secretion in ocean environments: Their role(s) in food webs and marine processes. Oceanogr. Mar. Biol. Annu. Rev. **1990**, 28, 73–153.
3. Arena, A.; Maugeri, T.L.; Pavone, B.; Iannello, D.; Gugliandolo, C.; Bisignanoc, G. Antiviral and immunoregulatory effect of a novel exopolysaccharide from a marine thermotolerant Bacillus licheniformis. Int. Immunopharmacol. **2006**, 6, 8–13.
4. Zanchetta, P.; Lagarde, N.; Guezennec, J. Systemic effects on bone healing of a new hyaluronic acid-like bacterial exopolysaccharide. Calcif. Tissue Int. **2003**, 73, 232–236.
5. Courtois, A.; Berthou, C.; Guézennec, J.; Boisset, C.; Bordron, A. Exopolysaccharides isolated from hydrothermal vent bacteria can modulate the complement system. PLoS ONE **2014**, 9, e94965.
6. Freitas, F.; Alves, V.D.; Reis, M.A.M. Advances in bacterial exopolysaccharides: From production to biotechnological applications. Trends Biotechnol. **2011**, 29, 389–398.
7. Gedouin, A.; Vallee, R. Compositions à Visée Cosmétique d'exopolysaccharides Issus de Bactéries Marines. Patent FR_2975906_A1, 2012.
8. Nishibuchi, M.; Kaper, J.B. Thermostable direct hemolysin gene of Vibrio parahaemolyticus: A virulence gene acquired by a marine bacterium. Infect. Immun. **1995**, 63, 2093–2099.
9. Kamerling, J.P.; Gerwig, G.J.; Vliegenthart, J.F.; Clamp, J.R. Characterization by gas-liquid chromatography-mass spectrometry and proton-magnetic-resonance spectroscopy of pertrimethylsilyl methyl glycosides obtained in the methanolysis of glycoproteins and glycopeptides. Biochem. J. **1975**, 151, 491–495.
10. Montreuil, J.; Bouquelet, S.; Debray, H.; Fournet, B.; Spik, G.; Strecker, G. Glycoproteines. In Carbohydrates Analysis: A Practical Approach; Chaplin, M.F., Kennedy, J.K., Eds.; IRL Press: Oxford, UK, 1986; pp. 143–204.

11. Hakomori, S.I. A rapid permethylation of glycolipid, and polysaccharide catalyzed by methylsulfinyl carbanion in dimethyl sulfoxide. J. Biochem. **1964**, 55, 205–208.

12. Taylor, R.L.; Conrad, H.E. Stoichiometric depolymerization of polyuronides and glycosaminoglycans to monosaccharides following reduction of their carbodiimide activated carboxyl groups. Biochemistry **1972**, 11, 1383–1388.

13. Kvernheim, A.L. Methylation analysis of polysaccharides with butyllithium in dimethyl sulfoxide. Acta Chem. Scand.**1987**, B41, 150–152.

14. Linnerborg, M.; Wollin, R.; Widmalm, G. Structural studies of the O-antigen polysaccharide from Escherichia coli O167.Eur. J. Biochem. **1997**, 246, 565–573.

15. Liu, B.; Knirel, Y.A.; Feng, L.; Perepelov, A.V.; Senchenkova, S.N.; Wang, Q.; Reeves, P.R.; Wang, L. Structure and genetics of Shigella O antigens. FEMS Microbiol. Rev. **2008**, 32, 627–653.

16. Knirel, Y.A.; Perepelov, A.V.; Kondakova, A.N.; Senchenkova, S.N.; Sidorczyk, Z.; Rozalski, A.; Kaca, W. Structure and serology of O-antigens as the basis for classification of Proteus strains. Innate Immun. **2011**, 17, 70–96.

17. Jann, B.; Kochanowski, H.; Jann, K. Structure of the capsular K96 polysaccharide (K96 antigen) from Escherichia coliO77:K96:H and comparison with the capsular K54 polysaccharide (K54 antigen) from Escherichia coli O6:K54:H10.Carbohydr. Res. **1994**, 253, 323–327.

18. Gorshkova, R.P.; Isakov, V.V.; Nedashkovskaya, O.I.; Nazarenko, E.L. Structure of carbohydrate antigens fromMicrobulbifer sp. KMM 6242. Chem. Nat. Compd. **2011**, 46, 837–840.

10

CHAPTER

Fucanomics and Galactanomics: Marine Distribution, Medicinal Impact, Conceptions, and Challenges

Vitor H. Pomin

Program of Glycobiology, Institute of Medical Biochemistry, and University Hospital Clementino Fraga Filho, Federal University of Rio de Janeiro, Rio de Janeiro, RJ, 21941-590, Brazil

ABSTRACT

Glycomics turned out to be a very extensive project where its subdivision is consequently emerging. This is seen by the growing number of terminologies used to define subprojects concerning particular classes of bioactive carbohydrates. Sulfated fucans (SFs) and sulfated galactans (SGs) are relatively new classes of sulfated polysaccharides (SPs) that occur mostly in marine organisms, and exhibit a broad range of medicinal effects. Their structures are taxonomically dependent, and their therapeutic actions include benefits in inflammation, coagulation, thrombosis, angiogenesis, cancer, oxidation, and infections. Some red algae, marine angiosperm and invertebrates express SPs of unique structures composed of regular repeating oligomeric units of well-defined sulfation patterns. This fine pattern of structural regularity is quite rare among any naturally occurring long SPs, and enables accurate structure-biofunction correlations. Seeing that, fucanomics and galactanomics may comprise distinguished glycomics subprojects. We hereby discuss the relevance that justifies the international recognition of these subprojects in the current glycomics age associated with the beneficial outcomes that these glycans may offer in drug development.

Keywords: fucanome; fucanomics; galactanome; galactanomics; glycome; glycomics; sulfated fucan; sulfated galactan

1. FUCANOME AND GALACTANOME: GETTING INTO GLYCOMICS

After many "ome" projects such as genome, metagenome, transcriptome, proteome, lipidome and metabolome, glycome has ultimately launched itself into the world of biology. Glycome is the specific project that deals with carbohydrates or glycosylated molecules [1]. It is generally defined as a set of glycans or glycoconjugates expressed by a living form (cell, tissue, organ, or organism) under certain circumstances [1]. Likewise, glycomics is defined as the comprehensive structural and functional studies of glycome(s) of individual or group of organisms [1,2]. Based on clear evidences, glycomics are so far bringing more challenges than the first "omics" [3]. This is particularly due to: (i) higher structural complexity and the flexibility of glycans, which consequently make data interpretation more difficult [3,4]; (ii) lack of a clear automatic sequencing method for complex carbohydrates as opposed to those routinely employed for nucleic acids (in genomics) and proteins (in proteomics) [3]; (iii) existence of innumerable, possible determinants in the biosynthesis of glycans which usually allows multiple questionings [1,2,3]; and (iv) existence of many sub-classes of glycans or glycoconjugates [3,5].

Because of these generally overwhelming challenges, a tendency to undertake or provide sectorized emphasis into sub-classes of glycans has consequently and naturally emerged in glycomics [3,5,6,7,8,9,10,11,12,13]. Empirical subprojects defined by novel terminologies are spreading across the community, and they aim to deal essentially with particular sub-classes of bioactive carbohydrates. Recent nomenclatures that are coming into common use are Proteoglycanome [6], Glycosaminoglycanome [7], Heparanome [8,9,10], Glycolipidome [13], Glycoproteome [12], and Sialome [11]. Glycomics seems perhaps to be ultimately evolving more as a conjunction of isolated or individualized works [3,5,11,14] rather than just a single compact ongoing project. The propensity to adopt such subdivisions and terminologies seems to be less evident than in the other "omics".

Sulfated fucans (SFs) and sulfated galactans (SGs) are classes of sulfated polysaccharides (SPs) composed essentially of α-L-fucopyranosyl (Fucp), or α-L-, α-D-, β-D-galactopyranosyl (Galp) units, respectively. SFs and SGs are distinct from the majority of other glycans in the following aspects:

i. relatively young and short literature timeline;

ii. basically found in marine organisms in which the structures are taxonomically related;

iii. fine and very rare patterns of structural regularity in certain cases;

iv. broad range of therapeutic actions with high levels of effectiveness in certain assays;

v. well-defined SFs and SGs achieve advanced structure-biofunction correlations more easily.

In accordance with the actual tendency for assuming subprojects in the current glycomics age, and due to the five above-cited features, we herein propose the dissemination of the terms Fucanome and Galactanome, and of the respective projects Fucanomics and Galactanomics, each concerning SFs and SGs, respectively. These SPs have gained some international recognition and are now under investigation in laboratories of many countries. From now on, we hereby discuss in detail the main characteristics of SFs and SGs that justify the investments into their research in the current glycomics age. We also give appropriate arguments for international assimilation of such projects into a closer collaborative network, and suggested topics for discussion in future glycoscientific events.

1.1. SFs and SGs: Young Molecules of Glycobiology

The bulk interest in structural and biological studies of SFs and SGs effectively began in the early 90's [15]. This particularly results from the worldwide advancement of glycobiology, particularly due to the dissemination and progress of powerful analytical tools such nuclear magnetic resonance (NMR) spectroscopy and mass spectrometry (MS), associated with the promising medicinal properties of SFs and SGs [15,16,17,18,19,20,21]. Research from some two decades has not been enough to make these SPs as internationally prominent as other bioactive carbohydrate classes, such as sialylated glycans or glycosaminoglycans (GAGs). Even though the SFs and SGs may show distinguished structural and biomedical properties, as discussed here, the latter classes are more widely recognized because of their natural physiological roles in humans. Specific terms therefore arose to define particular subprojects of these latter glycans: Sialomics [11], and Glycosaminoglycanomics [7], respectively. An even hierarchically lower subproject of Glycosaminoglycanomics, and in turn of Proteoglycanomics, have appeared: Heparanomics, which aims to deal with the bioactive domain of heparan sulfates [3,6,8,9,10]. The similar use of Fucanomics and Galactanomics still remain, but shall be disseminated across the international glyco-society, because of the uniqueness in both structures and functions of SFs and SGs.

1.2. Marine Taxonomic Distribution of SF and SG Structures

The richest sources of SFs and SGs are marine organisms, and their lack of occurrence in terrestrial mammals, especially humans, is two of the reasons for the narrow international fame of these glycans. Considering only the marine environment, SFs and SGs are expressed by macroalgae, marine angiosperms and invertebrates [16]. However, few outside works have reported the occurrence of SGs in bacteria and fungi as well [22]. Virtually ubiquitous in cell walls of all macroalgae, SFs have been synthesized in brown seaweeds (Phaeophyceae), in which they are also commonly known as fucoidans [16,18,19,20]. Whereas in green (Chlorophyceae) and red seaweeds (Rhodophyceae) only SGs have been found so far [16,19,21]. However, SFs and SGs can both be found in invertebrate animals, such as ascidians (also known as sea squirts or tunicates) (Urochordata, Ascidiaceae), sea cucumber (Echinodermata, Holothuroidea) and sea urchins (Echinodermata, Echinoidea) [16,20,21]. In ascidians and sea cucumbers, these SPs are known to participate in the structural assemblies of their body walls [15], most probably assembling the cell walls in macroalgae. In sea urchins, these SPs have been found building the jelly coat that surrounds the female gametes, and are known to participate in species-recognition during the initial steps of the fertilization process of these animals [15]. Marine angiosperms (Angiospermae, Spermatophyta) are another potential source for SGs, as evidenced by the work of Aquino et al. [23]. However, the lack of additional reports compromises the definitive assertion of marine superior plants as a confident source for this material. Again, as in algae, the angiosperm SGs contribute to build-up of cell walls in different tissues [24]. With no exceptions so far, SFs and SGs have unequivocally been shown to exist as essential components of the extracellular matrices in these marine organisms [16].

The structures of SFs and SGs are directly dependent on the species in which they occur, but some general aspects are still restricted to the phyla [15,16]. SFs from brown algae are usually the most complex molecules, even though mostly consisting of L-Fucp units. The presence of other monosaccharide types associated with occasional sparse branches enhances structural complexity. The occurrence of repetitive units in brown algal SFs is somewhat still uncertain, but evidence supporting such a concept has appeared along the past few years, at least in certain species [16]. This is most likely a consequence of the advances in instrumentation and methods capable for the structural analysis of complex carbohydrates. Nonetheless, the currently proposed oligomeric repeating motifs of certain fucoidans still show high degrees of heterogeneity (Table 1). Regardless of structural patterns, brown algal SFs are the most abundant SPs in the sea, and perhaps across the entire globe, since brown seaweeds by far dominate the sea environment in both number of species (1.5 to 2 thousand) and biomass [16], as the sea environment totals more than two-thirds of the planet.

Table 1. Illustrative examples of repetitive units currently assumed for famous phaeophyceae species.

Brown Seaweed Species	Proposed Repetitive Structure	Ref.
Ascophylum nodosum	[→4-α-L-Fuc*p*-2,3di(SO_3^-)-1→3-α-L-Fuc*p*-2(SO_3^-)-1→]$_n$ + branches of non-sulfated α-L-Fuc*p*	[24]
Fucus evanescences		[25]
Fucus vesiculosus		[24]
Ecklomia kurome	[→3-α-L-Fuc*p*-2R,4(SO_3^-)-1→]$_n$, where R = H or SO_3^-	[26]
Chorda filum	[→3-α-L-Fuc*p*-2R^1,4R^2-1→3-α-L-Fuc*p*-2R^1,4R^2-1→3-α-L-Fuc*p*-2R^1,4R^2-1→ 3-α-L-Fuc*p*-2(→1-α-L-Fuc*p*-2,4diR^2), 4R^2-1→3-α-L-Fuc*p*-2R^1,4R^2-1→]$_n$, where R^1 = H or SO_3^- or $COCH_3$, and R^2 = H or SO_3^-	[27]

The discovery of SGs in green algae is quite recent. Codium has been the genus most studied so far [15,28,29,30]. Green algal SGs usually show structures that are much less complex than those of brown algal SFs, but still more complex than those of red algal SGs, as discussed below [16,21]. Although no clear evidence that regular sequences really exist in green algal SG backbones, some clues have favored the concept of chains predominantly composed of 4-sulfated 3-linked β-D-Galp units [15,29,30]. However, these chains may still bear high degrees of heterogeneity, like pyruvylated non-reducing terminal residues, and occasional branches. Sulfation positions other than those at C4 may also occur and increase complexity as well [29,30].

Among the three macroalgal classes, red seaweeds are most likely the only class able to really express SPs in regular backbones [21]. Like GAGs, red algal SG backbones are normally composed in disaccharide repeating units, but of alternating 3-linked β-D-Galp and 4-linked α-D- or α-L-Galp units [16,21]. The possible presence of an extra-bond between C3 and C6 of the same ring leads to the 3,6-anhydro-Galp (3,6-AnGalp) unit that can occur only at the 4-linked Galp unit. The enantiomeric variation, D- or L-, in this 4-linked unit respectively results in the nomenclature "carrageenan" or "agaran". The names carrageenose or agarose are respectively related to these molecules when 3,6-AnGalp units occur along them [21]. Sulfated esters and/or occasionally methyl esters may occur at the 2- and/or 4-position(s) of the 3-linked Galpunits. These same substituents may be placed at 2-, 3-, and/or 6-position(s) of the 4-linked Galp units as well. All these structural variations comprise the main heterogeneities in red seaweed SGs. But since the sugar chains of these polymers are regularly composed of repeating disaccharides, the difficulties in structural characterization are significantly diminished, compared to those from the other algal classes. In works concerning structural characterization of red algal SGs, these glycans have usually been extensively characterized, generally through a combination of NMR spectroscopy, particularly ^{13}C-based spectra, with data analysis generated from chemical reactions [31,32,33].

1.3. Rare Structural Regularity among Polysaccharides of High-Molecular Weights

According to what has been stated before, the structural simplicity in macroalgal SPs rises in the following order: brown algal SFs, green algal SGs and red algal SGs. But even though red algae express SGs in disaccharide repeating units, there are still certain degrees of heterogeneity that impairs the arrival of a totally regular structural design. However, most likely through the steps of evolution, this complete structural pattern of regularity became noticeable in SPs from superior plants and from some marine invertebrates [21,23], probably due to a more organized or even simpler biosynthetic machinery in these latter organisms [15,16].

Table 2 depicts some illustrative examples of SF and SG of well-defined structures. It is clearly seen that the structures occur through a species-specific manner, varying in sulfation patterns (but always restricted to 3-O-, 2-O- and/or 4-O-positions), in glycosidic linkages [α(1→3), α(1→4), and β(1→3)], in repetitive oligomeric lengths (tetrasaccharides, trisaccharides, disaccharides, and monosaccharides), and sometimes in the presence of short branching segments such as the structure of Styela plicata, but still conserving the fine pattern of regularity. The molecular weight (MW) of these polymers, although quite polydisperse, are commonly very high, frequently ranging above 100 kDa. In polymers composed of a repetitive tetrameric unit as observed for echinoderms Ludwigothuria grisea and Lytechinus variegatus (Table 2), the chain extension of such glycans would range approximately over 100 tetrameric units. Certain red algae express SGs with quite regular structures, and thus must be gathered at the list of Table 2 as well. Overall, these fine patterns of regularity are very rare among naturally-occurring long polysaccharides.

Table 2. Oligomeric repetitive units of SFs and SGs from the echinoderms: sea urchins (Echinoidea), and sea cucumber (Holothuroidea); from red algae (Rhodophyta): marine superior plant (Angiospermae), and ascidians, also known as tunicates (Ascidiacea).

Species (Class)	Structure	Occurrence	Ref.
Ludwigothuria grisea (Holothurioidea)	[→3)-α-L-Fucp-2,4(OSO_3^-)-(1→3)-α-L-Fuc*p*-(1→3)-α-L-Fuc*p*-2(OSO_3^-)-(1→3)-α-L-Fuc*p*-2(OSO_3^-)-(1→]$_n$	Brazil	[34]
Strongylocentrotus purpuratus II (Echinoidea)	[→3)-α-L-Fuc*p*-2,4di(OSO_3^-)-(1→3)-α-L-Fuc*p*-4(OSO_3^-)-(1→3)-α-L-Fuc*p*-4(OSO_3^-)-(1→]$_n$	USA	[35]
Strongylocentrotus purpuratus I (Echinoidea)	80% [→3)-α-L-Fuc*p*-2,4di(OSO_3^-)-(1→]$_n$ and 20% [→3)-α-L-Fuc*p*-2(OSO_3^-)-(1→]$_n$	USA	[35]
Strongylocentrotus franciscanus (Echinoidea)	[3)-α-L-Fuc*p*-2(OSO_3^-)-(1→]$_n$	USA	[36]
Strongylocentrotus droebachiensis (Echinoidea)	[→4)-α-L-Fuc*p*-2(OSO_3^-)-(1→]$_n$	USA, Norway	[37]
Strongylocentrotus pallidus (Echinoidea)	[→3)-α-L-Fuc*p*-2(OSO_3^-)-(1→3)-α-L-Fuc*p*-2(OSO_3^-)-(1→3)-α-L-Fuc*p*-(1→3)-α-L-Fuc*p*-(1→]$_n$	USA, Norway	[37]
Lytechinus variegatus (Echinoidea)	[→3)-α-L-Fuc*p*-2(OSO_3^-)-(1→3)-α-L-Fuc*p*-2(OSO_3^-)-(1→3)-α-L-Fuc*p*-4(OSO_3^-)-(1→3)-α-L-Fuc*p*-2,4di(OSO_3^-)-(1→]$_n$	Brazil	[34]
Arbacia lixula (Echinoidea)	[→4)-α-L-Fuc*p*-2(OSO_3^-)-(1→4)-α-L-Fuc*p*-2(OSO_3^-)-(1→4)-α-L-Fuc*p*-(1→4)-α-L-Fuc*p*-(1→]$_n$	Brazil	[38]
Echinometra lucunter (Echinoidea)	[→3)-α-L-Gal*p*-2(OSO_3^-)-(1→]$_n$	Brazil	[38]
Glyptosidaris crenularis (Echinoidea)	[→3)-β-D-Gal*p*-2(OSO_3^-)-(1→3)-β-D-Gal*p*-(1→]$_n$	Japan	[39]
Botryocladia occidentalis (Rodophyta)	[→3)-β-D-Gal*p*-2R_1-3R_2-(1→4)-α-L-Gal*p*-2R_3-3R_4-(1→]$_n$, where R_{1-4} = OSO_3^- or OH, R_1 and R_2 = OSO_3^- in ~66%, and ~33%, respectively.	Brazil	[40]
Gelidium crinale (Rodophyta)	[→3)-β-D-Gal*p*-2R_1-4R_2-(1→4)-α-L-Gal*p*-2R_3-3R_4-(1→]$_n$, where R_{1-4} = OSO_3^- or OH, R_1 and R_2 = OSO_3^- in ~60%, and ~15%, respectively.	Brazil	[41]
Gigartina skottsbergii and *G. chamissoi* (Rodophyta)	[→3)-β-D-Gal*p*-2R_1-4R_2-(1→4)-α-D-Gal*p*-2R_3-6R_4-(1→]$_n$, where λ-carrageenan R_1, R_3 = OSO_3^- and R_2, R_4 = OH; μ-carrageenan R_1, R_3 = OH, and R_2, R_4 = SO_3^-; ν-carrageenan R_1 = OH and R_2, R_3, R_4 = SO_3^-; κ-carrageenan R_1, R_3 = OH and R_2 = OSO_3^-; ι-carrageenan R_1 = OH and R_2, R_3 = OSO_3^-. κ- and ι-carrageenans have 4-linked α-D-Gal*p* units cyclized as 3,6-anhydro: whereas μ- and ν- are just partially cyclized.	Argentina, others	[42,43]
Rupia maritma (Angiospermae)	[→3)-β-D-Gal*p*-2(OSO3$^-$)-(1→4)-α-D-Gal*p*-(1→4)-α-D-Gal*p*-(1→3)-β-D-Gal*p*-4(OSO3−)-1→]$_n$	Brazil	[23]
Styela plicata (Ascidiacea)	{→4)-α-L-Gal*p*-2[→1)-α-L-Gal*p*]-3(OSO_3^-)-(1→}$_n$	Brazil	[44]
Hedmania monus (Ascidiacea)	[→4)-α-L-Gal*p*-3(OSO_3^-)-(1→]$_n$	Brazil	[45]

1.4. The Impressively Wide Range of Medicinal Effects of SFs and SGs

Besides the natural actions of SFs and SGs in their own organisms of occurrence, these glycans have shown potential medicinal effects. In fact, they have been targets of research by many international groups, strictly due to their therapeutic promise. As the SFs and SGs consist of highly negatively-charged molecules, they have biophysical capacity to electrostatically interact with many health/disease-related basic proteins, or with positively-charged structural assemblies, such as the positive regions of virus particles. Although these electrostatic interactions depend essentially on a difference of net charge, such as sulfation density vs. basic amino acid content, it has been clearly proven in the past few years that the molecular interactions involving SFs and SGs are, in fact, mainly stereospecific (directly dependent on sulfation and glycosylation sites, anomericity, and monosaccharide-type composition), and not a mere consequence of the bulk charges related to sulfation degrees [15]. We briefly present below the general biochemical mechanisms of SFs and SGs in their main studied medical actions. It is noteworthy that the definitive mechanisms underlying the biomedical properties of these glycans are still under investigation.

1.4.1. Inflammation

In anti-inflammatory actions, possible mechanisms have been postulated by which SFs and SGs may affect the leukocyte recruitment to sites of injury where inflammation normally evolves [46]. The ability of SFs and SGs to prevent selectin-mediated cell-cell interactions was successfully evidenced by some in vitro experiments using brown algal SFs [46]. These molecules have proved to bind directly to purified and membrane-exposed P- and L-selectins [46,47], but curiously not to E-selectins [48]. The interaction of SFs and SGs with specific selectins points toward the concept of a selective beneficial action of these glycans in inflammation, and consequently supports the advantages of selectivity in drug candidates. The binding of these glycans at selectins therefore impairs the consequential selectin-mediated cell migration of the activated leukocytes to go further to the specific sites of injury. There is also a current hypothesis that exogenous SPs, like marine SFs and SGs, might compete with GAG molecules from cell surface proteoglycans for cytokine/chemokine bonds. This consequently arrests the correct formation of cytokine/chemokine gradients necessary for leukocyte activation and migration.

1.4.2. Hemostasis and Vascular Biology

Disorders related to the cardiovascular system and hematology, are the major cause of human death, accounting for around one-third of the total.

Despite many downsides, GAG-type heparin is the most exploited biomolecule for treating these disorders. Heparin has a narrow therapeutic window and a highly variable dose-response relationship, necessitating frequent coagulation monitoring during its administration. The main side-effects of this drug are hemorrhage, thrombocytopenia and osteoporosis. Moreover, the heparin sources utilized for large-scale production are very limited. It is obtained from pig intestine or bovine lung, and contamination of samples with pathogens is a serious concern in the extraction procedure from these sources [49,50]. The situation was further complicated recently due to contamination of heparin preparations with over-sulfated chondroitin sulfate [51]. This contaminant is severe since it induces hypotension associated with kallikrein release when intravenously injected [52].

New alternatives for treatments of diseases related to the cardiovascular system are therefore urgently required, and, in addition, SFs and SGs also display efficient levels in anticoagulation [15,16,49,50]. In fact, it is these marine glycans' biomedical effects as anticoagulants and antithrombotics that have been the most studied so far [15,16,46,49,50]. Their mechanisms of action most likely resemble the anticoagulant action of heparin, which potentiates the catalytic rates of natural blood inhibitors: the serpins antithrombin and heparin cofactor II [15,16,49,50,53,54]. However, in addition to this action, a serpin-independent anticoagulant mechanism was recently discovered for a red algal SG [55]. This novel mechanism is still unclear, but points toward a concept of the ability of this SG to blocking intermolecular complexes involving blood co-factors in initial steps of the coagulation cascade, such as the intrinsic pro-coagulant tenase and protrombinase complexes [55].

1.4.3. Angiogenesis

SFs and SGs have shown the ability to inhibit the formation of neovascularization, mainly by interfering in the necessary binding of molecules responsible for angiogenesis, such as vascular endothelial growth factors (VEGFs) [46,56], and/or basic fibroblast growth factors (bFGFs) [46,57], to their respective receptors. These molecular interventions have proved of huge clinical interest, especially for inhibiting the consequential feeding of tumor-affected areas in cancer-committed patients. In the report of Cumashi and co-workers [46], through a systematic and comparative tubulogenic assay using human umbilical vein endothelial cell (HUVEC) in culture, incubated with fucoidans of different structures, it has been shown that the angiogenic effects of brown algal SFs are highly stereospecific. The antiangiogenic effects of a green algal SG (again genusCodium) have also been demonstrated [58].

1.4.4. Tumor Progression and Spreading

The clearest mechanisms of SPs as potential candidates in anti-tumor therapies are most likely related to the inhibitory action of these glycans on tumor vascular networks, as evidenced by reduced rates in hemoglobin content in the affected areas using fucoidans [59]. Besides this neovascularization inhibiting function, fucoidans have proved to synergically reduce tumor spreading. This was observed through in vitro assays of cell-adhesion using highly metastatic breast cancer cell lines capable of binding normally onto human platelets fixed in a Matrigel. In some tests in which differential fractions of fucoidans were added, the binding property of the tumor cells can be abolished or reduced [59]. The ability of certain fucoidan-derivatives to interfere with breast cancer cell adhesion onto platelet-coated surfaces favors the concept that specific internal bioactive structural motifs of brown algal SFs really exist, and are responsible for combating metastasis [59]. This same experiment type was carried out using nine different fucoidans and heparin [46]. The authors observed that SFs from five specific brown algal species were able to reduce approximately 80% of the cell adhesion of tumor cells onto human platelet-coated plates [46]. The effectiveness reached by these fucoidans in tumor cell-adhesion was significant, and even higher than that achieved by highly sulfated glycans such as heparin, corroborating again the structural specificity in such clinical actions, rather than just a consequence of simple net charges.

1.4.5. Antioxidation

Recent reports have pointed out the ability of algal SGs to reduce oxidation levels through in vitro assays that might resemble the mechanisms of oxidative stress in mammals [60,61,62,63,64]. These reports have shown that these marine SPs are capable of inhibiting hydroxyl and superoxide radical formation or to chelate oxidant iron ions such as Fe(II), thus impairing their prompt oxidative effects in metabolisms, and consequently preventing their damaging effects on health. The primary antioxidant action of these glycans seems to be notably dependent on their charge nature, and future work must be done to prove the possible structural or conformational behavior of these SPs in such activities. The description of antioxidative effects of these marine SPs heavily corroborates the concept of Fucanomics and Galactanomics as promising projects of glycomics in terms of their medicinal impact, but, above all, supports the consumption of marine algae as food supplements—mainly because of the beneficial nutraceutical components contained in these organisms, such as antioxidants.

1.4.6. Infections

Infections and parasitic diseases are the second largest cause of death in humans, accounting for around one-fourth of the total. SFs and SGs have shown effects in preventing infections of viral origin [21,65,66], and bacterial origin, including those caused by Staphylococcus aureus [67]. The main mechanisms behind the antiviral activity of the reported algal SFs have been speculated to be through the binding properties of the sulfated glycans to positive regions of the viral particles during the adsorption period [21,65,66]. This SP-bonded virus particle loses its virulence in attaching to the host cells. Thus the consequent internalization of the virus particles, which would ultimately leave the infection progression, is blocked. As far as we know, no clear structure-function relationships have been achieved regarding this pharmacological activity. Nonetheless, it has been reported that certain structures may show more effectiveness in viral inhibition than others [21]. Although there is some speculation that marine SFs and SGs can act as antiprotozoal or antifungal agents, no solid evidence exists so far supporting these activities. Their antibacterial action is also unclear as well.

2. CONCEPTIONS IN FUCANOMICS AND GALACTANOMICS

Fucanomics and Galactanomics would be defined specifically as international subprojects of glycomics, comprising a systematic, compiled body of research about the structures, functions and metabolic paths involving SFs and SGs, respectively. This should be done using as many different species as possible. The characterization, notation and deposition of the structures would result in a large library of bioactive glycans from the sea. Based on the current definition of glycomics, this project about structural notation related to scientific names of the respective species adequately fits into the field of glycomics [1].

As detailed below, the intense research and publication of biomedical properties and mechanisms of SFs and SGs would represent a large, if not the most valuable, task of Fucanomics and Galactanomics. Additionally, an understanding about the regulative mechanisms involved in the biosynthesis of SFs and SGs, and the consequent impact of such mechanisms on the structural features, and in turn on the resultant biomedicinal properties of these glycans, would comprise another important objective of research. In fact, some reports concerning structural and functional changes influenced by either environmental or biological conditions have already started to emerge [68,69,70,71].

Annual seasons (winter vs. summer) have been demonstrated to be influential in sulfation patterns of sea urchin SFs, and thus a hypothetical controlling season-related mechanism in reproducing this invertebrate was raised [69]. Seasonal changes were also documented to be influential in structures of algal SFs and SGs, but still with no clear reasons [70]. The changes in sulfation density and consequently in both structural and functional properties of SGs in marine angiosperms were recently reported to be directly coupled to variations in salinity levels of the estuary regions which these organisms naturally inhabit [68]. Metal contamination in polluted areas has been shown to be influential in the expression of algal SPs, and hence in their respective abundance in cell walls [71]. These two last observations raise speculation about sulfation content of these glycans, somewhat related to the balance of counter-ions in the environment in which these polymers have been synthesized. In any event, unveiling these influential mechanisms associated with the task of deciphering the unknown biosynthetic routes of SFs and SGs would consequently give to Fucanomics and Galactanomics the proper knowledge to better handle and explore the beneficial properties of SFs and SGs. This is the rational research conduct in drug development based on naturally occurring carbohydrate-based molecules that might show possible structural variations dependent on external factors. The achievements from delineated researches might provide a boost to the current medicinalstatus-quo of SFs and SGs.

3. WHAT MAKE FUCANOMICS AND GALACTANOMICS SPECIAL GLYCOMICS SUBPROJECTS?

At first glance, the full structural assignment of an over 400 residue-complex glycan would appear to be an impossible mission for any glycoscientist. However, structural patterns based on oligomeric repetitions (Table 2) can greatly reduce the obstacles in structural studies involving complex carbohydrates. This regular pattern makes the structures of long glycans more manageable to common analytical practices, thus facilitating data interpretation [15,17]. Even though the glycans listed in Table 2 apparently show some relative structural simplicity, the diversity in structural forms and sulfation patterns is still high. Such variety still keeps both the species-specific identity and the capacity of these glycans to achieve biological actions through their structural “codes” [15]. This was initially and successfully accomplished in data interpretation of sea urchin fertilization through species-specific acrosome reactions [15,16,36,37] It seems to be similarly useful to understand the differential effective levels in medicinal properties of SFs and SGs [15,16].

Among naturally occurring bioactive high-MW SPs described so far in glycobiology, GAGs for example, SFs and SGs are even more unique in terms of structural patterns, together with their potent medicinal properties, from the viewpoint of achieving accurate structure-function correlations [15,16]. The consequent accuracy in structural-biofunction relationships can be exemplified by the amplified anticoagulant effect of SFs carrying lower proportions of Fucp residues, exclusively 2-O-sulfonated [15,16].

It is worth highlighting that the majority of the work undertaken so far to study the biomedical properties of marine SFs and SGs (section 1.4), has used molecules extracted from algae with a high degree of heterogeneity [46,47,48,49,50,51,52,53,54,55,56,57,58,59,60,61,62,63,64,65,66,67]. Although algal molecules may reach great levels of effectiveness in certain assays, their structure-function correlations are more difficult to determine. On the other hand, the use of well-defined SFs and SGs from invertebrates, and a few red algal species (Table 2), have enabled more confident propositions [15,16]. This places Fucanomics and Galactanomics as quite distinct subprojects as compared to other "glycan-omics", especially for those that may show therapeutic action [3,5,6,7,8,9,10,11]. It also comprises a huge step of glycomics towards the next pre-clinical testing of carbohydrate-based drug candidates, once potential bioactive glycans can be presented together with their respective biomedical mechanisms, and accurately studied through reliable structure-biofunction relationships.

4. THE IMPACT OF DRUG DEVELOPMENT IN GLYCOMICS

Glycomics would lose some of its major impact if the medicinal uses of exogenous glycans were left aside. In the case of GAG molecules, however, the importance of such therapeutic actions, of carbohydrate-based molecules has already been discussed previously [7]. It is worth remembering that GAG heparin is the second most used (to peptide insulin) natural macromolecule in medicine. Heparin naturally occurs in mast cells, and seems to have a primary key role in inflammation and allergic reactions [72]. However, due to its great affinity to the thrombin-inhibitor, antithrombin [73], unfractionated heparin and its low-MW derivatives are clinically explored for stopping the clotting processes in patients undergoing extracorporeal circulation during surgery or renal dialysis [74]. The low molecular variants of heparins also serve as the primary treatment of deep vein thrombosis, since these molecules contain more antithrombin-binding motifs per chain unit. As a consequence of the high incidence of thromboembolic disorders, especially in developed countries, there is a large-scale production of heparins. But due to some side-effects of this drug, new alternative anticoagulant agents are needed (Section 1.4.2) [15,16]. Anticoagulant effects have also been observed for some SFs and SGs, although at lower potencies than heparins, but clearly with lower side-effects also (Section 1.4.2). In addition, as opposed to heparin that can exhibit a high risk of contamination because of its mammalian origin, the SFs and SGs present an additional advantage over this GAG type: they are basically isolated from marine organisms which results in a much lower incidence of contaminants [15,16,50]. Such clinical benefits of SFs and SGs in anticoagulation likely in other biomedical functions (Section 1.4) are relevant contributing elements from Fucanomics and Galactanomics as from glycomics. The development of drugs based on natural occurring carbohydrates should be one of the greatest objectives of these projects, mainly because of their direct implications to human health.

5. FUTURE APPROACHES OF FUCANOMICS AND GALACTANOMICS

Because of the relatively recent description of SFs and SGs, many tasks have yet to be accomplished in Fucanomics and Galactanomics. One would be the prompt depositing of structures already characterized (Table 2) in an informative public carbohydrate databank. The deposition of structures related to their scientific names would end up in a valuable library of potential biomedical glycans to be shared with many laboratories in the world, thus speeding up data generation and collaborative networks regarding the science of SFs and SGs.

Another task would be research on the biosynthetic mechanisms of SFs and SGs. As opposed to GAGs, in which all steps are well-known at such a level that enables even proper manipulation of the biosynthetic systems for different purposes [75], the biosynthesis pathways of SFs and SGs are virtually unknown. Although some information about sea urchin (sulfo)transferases [76] and precursor nucleotide-sugars may exist, the biosynthetic routes of these glycans seem still to be unknown. Clarification of the abiotic and biotic determinants in molecular biology of SFs and SGs comprises another fruitful research line, as described in section 2.

Another task would be an attempt to establish some concise phylogenetic relationships between SFs and SGs. For instance, up to now, only linear SFs or SGs have occurred in echinoderms. In echinoidea (sea urchins), the structures are limited to a very short set of characteristics: only 2-O- and/or 4-O-sulfonation, together with a few types of glycosidic bonds (either α(1→3), or α(1→4), or β(1→3)) [15,16]. This observation raises the hypothesis of a possible short number of (sulfo)transferases in these invertebrates and the concept of very organized or simpler biosynthetic machinery as opposed to those of mammal SPs, like GAGs. In the latter case, a higher number of anabolic enzymes (including isoforms), is the largest contributing source for the consequent structural heterogeneities [77]. In a previous report, we have established a tentative phylogenetic view for SGs [15,16,21]. We noticed that chains of 3-linked β-Galp are heavily conserved throughout marine taxonomic groups (including red and green algae, sea angiosperm, clams, sea urchins, and tunicates), with a strong tendency toward 4-O-sulfonation in algae and marine angiosperm, 2-O-sulfonation in invertebrates, and 6-O-sulfonation, more randomly distributed among various organisms. This indicates that galactosyltransferases seem to be preserved through marine species, but sulfotransferases are likely to be more specifically distributed. Although 4- and 2-sulfotransferases are hypothetically considered to be preserved in plants and in invertebrates, respectively, 6-sulfotransferases are broadly dispersed among marine organisms. This phylogenetic assumption is quite preliminary and no correlation with SF structures has been undertaken so far. It is worth mentioning that these phylogenetic points are just speculative because they were based solely on the structural characteristics of glycans. The structural features of SFs or SGs taken alone cannot become the main tools to propose a definitive phylogenetic relationship if gene expression, transcription levels and amounts of nucleotide-sugar precursors and anabolic enzymes are not considered. Hence, this phylogenetic study is intrinsically dependent on the researches about the biosynthetic routes of SFs and SGs, discussed in the previous paragraph.

The studies concerning the spatial geometry (3D structural view) of SFs and SGs, their thermodynamic behaviors, and their conformational changes upon molecular interactions with the main proteins involved in the diseases documented in section 1.4 (coagulation (co)-factors, selectins, cytokines/chemokines, growth factors, endothelial adhesion-molecules, surface receptors of virus particles and bacteria) would represent outstanding breakthroughs of Fucanomics and Galactanomics. However, the high-MW of these glycans is the main issue for these advanced approaches, and would enhance complexity in both data acquisition and interpretation. Therefore, the production of SF- and SG-derived oligosaccharides is a fundamental primary step. In this direction, we have already established a protocol to efficiently produce low-MW SF-derivatives that still retain both well-defined molecular structures and biological properties [78,79]. Specific experiments of NMR spectroscopy (relaxation rates, NOE-based or residual dipolar coupling techniques) should be able to provide information regarding conformation (with or without the presence of binding proteins) as well as flexibility of these specific marine SPs. Degrees of freedom in studies of molecular motions related to sulfation patterns would help to understand the structural influences in molecular dynamics, as well as the binding capacities in protein interactions during their biomedical roles.

6. INCLUSION OF SFS AND SGS AS TOPICS IN FORUMS, CONSORTIUMS, AND MEETINGS OF GLYCOMICS

In the USA, the working group convened by the Division of Blood Diseases and resources of the National Heart, Lung, and Blood Institute (NHLBI), was created with the primary objective of identifying scientific opportunities and priorities emerging from the recent explosion of technological and biological advances from the glycosciences [80]. Fucanomics and Galactanomics would fit perfectly as topics of this objective. This reunion was a consequence of specific glycoscientific objectives raised by the Consortium for Functional Glycomics, funded by the National Institute of General Medical Science (NIGMS). This consortium was initiated and funded by the US National Institute of Health (NIH). In this report about the roles of glycans in hemostasis, inflammation and vascular biology [80], the following is said: "there is a need for funding mechanisms that will bring together the existing experts in the study of glycans with investigators in blood and vascular diseases, to foster fruitful collaboration that will bridge the existing gulf. Importantly, this frontier area is also ripe for translational research and drug development. It is also noteworthy that this situation is different in many European and Asian countries, e.g., Sweden, Australia, and Japan, where there are major nationwide commitments to, and interest in, the study of glycans. Thus, the USA lags behind the rest of the world in taking advantage of these opportunities".

Here, we want to reiterate that Brazil might even lag behind the USA in terms of creating these glyco-organizations and in taking the proper advantages of its own national glycoscientific opportunities. This can be seen by the lack of an official consortium in Brazil specifically concerning glycoscience. As documented here though, it is noteworthy that the majority of the results obtained with SFs and SGs of well-defined structures (Table 2), and with promising medicinal activities, were conducted by Brazilian scientists in Brazil! Furthermore, most of the marine species from which these novel SPs were extracted, were collected along the sea-shores of this country (Table 2).

Through this publication, we want to express the high importance in including these new glycans (Table 2) into either National or International Organization(s), in order to prompt the progress of Fucanomics and Galactanomics via possible collaborative and growing research networks. This should be undertaken with high priority, always keeping in mind the main objective of these glycomics subprojects, which is to provide to human society the right and proper medicinal benefits from SFs and SGs, especially those involved in hemostasis, inflammation and vascular biology, as pointed out by the committee of the above-mentioned glyco-forum report [80]. SFs and SGs exhibit great effectiveness in these three systems mentioned (section 1.4), and must be a topic in this or other glyco-forums as well.

7. CONCLUDING REMARKS

Here, we have made clear the importance of Fucanome and Galactanome to glycomics. This subdivision would make SFs and SGs more well known and participative within glycobiology. By definition, Fucanomics and Galactanomics comprise the full science related to the specific marine glycans composed mostly of sulfated Fucp and Galp units. These glycans show a broad range of medicinal functions and thus would serve as valuable carbohydrate-based therapeutic candidates in the science of drug discovery and development. Their pharmacological actions cover many systems, including inflammation, vascular biology, oncology, oxidative stress, virosis, pathogenesis, among other possible systems not discussed here. The significance of these therapeutic applications of Fucanomics and Galactanomics is highly relevant in order to enhance the glycomics impact, and its international recognition and importance. The rare structural features of some SFs and SGs (those composed of well-defined sulfation patterns and regular repeating oligomeric units, Table 2) make Fucanomics and Galactanomics unique glycomics subprojects because of the facilitated correlation of the biomedicinal levels achieved with the structural properties of SFs and SGs. The establishment of these structure-biofunction relationships has been more difficult to accomplish using other native glycans, including GAG polymers, but conversely have proved to be quite feasible using SFs and SGs.

It has also become clear that even though the division of glycomics into subprojects is really demanding and necessary, the complexity of studying such subprojects still remains high. There is a high chance in the future for assuming Agaranomics and Carrageenomics as hierarchically-related subprojects of Galactanomics here proposed. This would raise the complexity of Galactanomics, but would make easier the organization of large amounts of data. Possible other "glycan-omics" about to show up are Chondroitinomics, Heparinomics, Dermatanomics. All these are hierarchically included in Glycosaminoglycanomics, and in turn, in Proteoglycanomics as well.

Empirically speaking, Glycomics turned out to be undertaken more as a set of isolated research lines than just a single ongoing project. Even taking this subdivision trend, the subprojects themselves are quite challenging. By analogy to what was previously mentioned: "The Sialome—Far more than the sum of its parts" [11], here we reiterate: "The Glycome—Bigger than the division of its total". This division has been shown necessary for Glycomics' own evolution!

ACKNOWLEDGMENTS

The author acknowledges Paulo A.S. Mourão for his great contribution in Fucanomics and Galactanomics, Antonio Trincone for his kind invitation to the special issue "Marine Glycoconjugates", and the editorial office of Marine Drugs (including Lisa Han) for the outstanding editing and publishing work.

REFERENCES

1. Hart, G.W.; Copeland, R.J. Glycomics hits the big time. Cell **2010**, 143, 672–676.

2. Bertozzi, C.R.; Sasisekharan, R. Glycomics. In Essentials of Glycobiology, 2nd; Varki, A., Cummings, J.D., Esko, J.D., Freeze, H.H., Stanley, P., Bertozzi, C.R., Hart, G.W., Etzler, M.E, Eds.; Cold Spring Harbor Laboratory Press: Cold Spring Harbor, NY, USA, 2009; Chapter 48.

3. Pomin, V.H. Fucanome and galactanome: Marine glycomics contribution. J. Glycobiol. **2011**, 1.

4. Raman, R.; Raguram, S.; Venkataraman, G.; Paulson, J.C.; Sasisekharan, R. Glycomics: An integrated systems approach to structure-function relationships of glycans. Nat. Methods **2005**, 2, 817–824.

5. Pomin, V.H. Current glycomics' approaches: Subprojects and journals. J. Glycomics Lipidomics **2011**, 2, 1–3.

6. Gesslbauer, B.; Rek, A.; Falsore, F.; Rajkovic, E.; Kungl, A.J. Proteoglycanomics: Tools to unravel the biological function of glycosamynoglycnas. Proteomics **2007**, 7, 2870–2880.

7. Gesslbauer, B.; Kungl, A. Glycomics approaches toward drug development: Therapeutically exploring the glycosaminoglycanome. Curr. Opin. Mol. Ther. **2006**, 8, 521–528.

8. Lammana, W.C.; Kalus, I.; Padva, M.; Baldwin, R.J.; Merry, C.R.L.; Dierks, T. The heparanome—The enigma encoding and decoding heparan sulfate sulfation. J. Biotechnol. **2007**, 129, 290–307.

9. Turnbull, J.E. Heparan sulfate glycomics: Towards systems biology strategies. Biochem. Soc. Trans. **2010**, 38, 1356–1360.

10. Tran, V.M.; Nguyen, T.K.N.; Raman, K.; Kuberan, B. Applications of isotopes in advancing structural and functional heparanomics. Anal. Bioanal. Chem. **2011**, 399, 559–570.

11. Cohen, M.; Varki, A. The Sialome—Far more than the sum of its parts. OMICS **2010**, 14, 455–464.

12. Tissot, B.; North, S.J.; Ceroni, A.; Pang, P.C.; Panico, M.; Rosati, F.; Capone, A.; Haslam, S.M.; Dell, A.; Morris, H.R. Glycoproteomics: Past, present and future. FEBS Lett. **2010**, 583, 1728–1735.

13. Zarei, M.; Müthing, J.; Peter-Katalinić, J.; Bindila, L. Separation and identification of GM1b pathway Neu5Ac- and Neu5Gc gangliosides by on-line nanoHPLC-QToF MS and tandem MS: Toward glycolipidomics screening of animal cell lines. Glycobiology **2010**, 20, 118–126.

14. Pomin, V.H. Recent demands of Glycomics: Subprojects and the role of new journals. J. Glycobiol. **2011**, 1.

15. Pomin, V.H. Review: An overview about the structure-function relationship of marine sulfated homopolysaccharides with regular chemical structures. Biopolymers **2009**, 91, 601–609.

16. Pomin, V.H.; Mourão, P.A. Structure, biology, evolution and medical importance of sulfated fucans and galactans.Glycobiology **2008**, 18, 1016–1027.

17. Mulloy, B.; Ribeiro, A.C.; Vieira, R.P.; Mourão, P.A. Structural analysis of sulfated fucans by high-field NMR. Braz. J. Med. Biol. Med. **1994**, 27, 515–521.

18. Daniel, R.; Chevolot, L.; Carrascal, M.; Tissot, B.; Mourão, P.A.S.; Abian, J. Electrospray ionization mass spectrometry of oligosaccharides derived from fucoidan of Ascophyllum nodosum. Carbohydr. Res. **2007**, 342, 826–834.

19. Jiao, G.; Yu, G.; Zhang, J.; Ewart, S. Chemical structures and bioactivities of sulfated polysaccharides from marine algae. Mar. Drugs **2011**, 9, 196–223.

20. Berteau, O.; Mulloy, B. Sulfated fucans, fresh perspectives: Structures, functions, and biological properties of sulfated fucans and an overview of enzymes active toward this class of polysaccharide. Glycobiology **2003**, 13, 29–40.

21. Pomin, V.H. Structural and functional insights into sulfated galactans: A systematic review. Glycoconjugate J. **2010**, 27, 1–12.

22. Delattre, C.; Fenoradosoa, T.A.; Michaud, P. Galactans: An overview of their most important sourcing and applications as natural polysaccharides. Braz. Arch. Biol. Technol. **2011**, 54, 1075–1092.

23. Aquino, R.S.; Landeira-Fernandez, A.M.; Valente, A.P.; Andrade, L.R.; Mourão, P.A. Occurrence of sulfated galactans in marine angiosperms: Evolutionary implications. Glycobiology **2005**, 15, 11–20.

24. Chevolot, L.; Mulloy, B.; Ratiskol, J.; Foucault, A.; Colliec-Jouault, S. A disaccharide repeat unit is the major structure in fucoidans from two species of brown algae. Carbohydr. Res. **2001**, 330, 529–535.

25. Bilan, M.I.; Grachev, A.A.; Ustuzhanina, N.E.; Shashkov, A.S.; Nifantiev, N.E.; Usov, A.I. Structure of fucoidan from the brown seaweed Fucus evanescens. Carbohydr. Res. **2002**, 337, 719–730.

26. Nishino, T.; Naguno, T. Structural characterization of a new anticoagulant fucan sulfate from the brown seaweedEcklonia kurone. Carbohydr. Res. **1991**, 211, 77–90.

27. Chizhov, A.O.; Dell, A.; Morris, H.R.; Haslam, S.M.; McDowell, R.A.; Shashkov, A.S. A study of fucoidan from the brown seaweed Chorda filum. Carbohydr. Res. **1999**, 320, 108–119.

28. Matsubara, K.; Matsuura, Y.; Bacic, A.; Liao, M.; Hori, K.; Miyazawa, K. Anticoagulant properties of a sulfated galactan

preparation from a marine green alga, Codium cylindricum. Int. J. Biol. Macromol. **2001**, 28, 395–399.

29. Bilan, M.I.; Vinogradova, E.V.; Shashkov, A.S.; Usov, A.I. Structure of a highly pyruvylated galactan sulfate from the Pacific green alga Codium yezoense (Bryopsidales, Chlorophyta). Carbohydr. Res. **2007**, 342, 586–596.

30. Farias, E.H.; Pomin, V.H.; Valente, A.-P.; Nader, H.B.; Rocha, H.A.; Mourão, P.A. A preponderantly 4-sulfated, 3-linked galactan from the Green alga Codium isthmocladum. Carbohydr. Res. **2008**, 18, 250–259.

31. Kolender, A.A.; Matulewicz, M.C. Sulfated polysaccharides from the red seaweed Georgiella confluens. Carbohydr. Res.**2002**, 337, 57–68.

32. Zhang, Q.; Li, N.; Liu, X.; Zhao, Z.; Li, Z.; Xu, Z. The structure of a sulfated galactan from Porphyra haitanensis and itsin vivo antioxidant activity. Carbohydr. Res. **2004**, 339, 105–111.

33. Salehi, P.; Dashiti, Y.; Tajabadi, F.M.; Safidkon, F.; Rabei, R. Structural and compositional characteristics of a sulfated galactan from the red alga Gracilariopsis persica. Carbohydr. Res. **2011**, 83, 1570–1574.

34. Mulloy, B.; Ribeiro, A.-C.; Alves, A.P.; Vieira, R.P.; Mourão, P.A. Sulfated fucans from echinoderms have a regular tetrasaccharide repeating unit defined by specific patterns of sulfation at the 0-2 and 0-4 positions. J. Biol. Chem. **2004**,269, 22113–22123.

35. Alves, A.P.; Mulloy, B.; Moy, G.W.; Vacquier, V.D.; Mourão, P.A. Females of the sea urchin Strongylocentrotus purpuratus differ in the structures of their egg jelly sulfated fucans. Glycobiology **1998**, 8, 939–946.

36. Vilela-Silva, A.C.E.S.; Alves, A.P.; Valente, A.-P.; Vacquier, V.D.; Mourão, P.A. Structure of the sulfated alpha-L-fucan from the egg jelly coat of the sea urchin Strongylocentrotus franciscanus: patterns of preferential 2-O- and 4-O-sulfation determine sperm cell recognition. Glycobiololy **1999**, 9, 927–233.

37. Vilela-Silva, A.C.E.S.; Castro, M.O.; Valente, A.-P.; Biermann, C.H.; Mourão, P.A. Sulfated fucans from the egg jellies of the closely related sea urchins Strongylocentrotus droebachiensis and Strongylocentrotus pallidus ensure species-specific fertilization. J. Biol. Chem. **2002**, 277, 379–387.

38. Alves, A.P.; Mulloy, B.; Diniz, J.A.; Mourão, P.A. Sulfated polysaccharides from the egg jelly layer are species-specific inducers of acrosomal reaction in sperms of sea urchins. J. Biol. Chem. **1997**, 272, 6965–6971.

39. Castro, M.O.; Pomin, V.H.; Santos, L.L.; Vilela-Silva, A.C.; Hirohashi, N.; Pol-Fachin, L.; Verli, H.; Mourão, P.A. A unique 2-sulfated beta-galactan from the egg jelly of the sea urchin Glyptocidaris crenularis: Conformation flexibility versus induction of the sperm acrosome reaction. J. Biol. Chem. **2009**, 284, 18790–18800.

40. Farias, W.R.; Valente, A.-P.; Pereira, M.S.; Mourão, P.A. Structure and anticoagulant activity of sulfated galactans. Isolation of a unique sulfated galactan from the red algae Botryocladia occidentalis and comparison of its anticoagulant action with that of sulfated galactans from invertebrates. **2000**, 275, 29299–29307.

41. Pereira, M.G.; Benevides, N.M.; Melo, M.R.; Valente, A.-P.; Melo, F.R.; Mourão, P.A. Structure and anticoagulant activity of a sulfated galactan from the red alga, Gelidium crinale. Is there a specific structural requirement for the anticoagulant action? Carbohydr. Res. **2005**, 340, 2015–2023.

42. Carlucci, M.J.; Pujol, C.A.; Ciancia, M.; Noseda, M.D.; Matulevicz, M.C.; Damonte, E.B.; Cerezo, A.S. Antiherpetic and anticoagulant properties of carrageenans from the red seaweed Gigartina skottsbergii and their cyclized derivatives: Correlation between structure and biological activity. Int. J. Biol. Macromol. **1997**, 20, 97–105.

43. Penman, A.; Rees, D.A. Carrageenans. IX. Methylation analysis of galactan sulphates from Furcellaria fastigiata,Gigartina canaliculata, Gigartina chamissoi, Gigartina atropurpurea, Ahnfeltia durvillaei, Gymnogongrus furcellatus,Eucheuma isiforme, Eucheuma uncinatum, Aghardhiella tenera, Pachymenia hymantophora, and Gloiopeltis cervicornis. Structure of xi-carrageenan. J. Chem. Soc. **1973**, 19, 2182–2187.

44. Mourão, P.A.; Perlin, A.S. Structural features of sulfated glycans from the tunic of Styela plicata (Chordata-Tunicata). A unique occurrence of L-galactose in sulfated polysaccharides. Eur. J. Biochem. **1987**, 166, 431–436.

45. Santos, J.A.; Mulloy, B.; Mourão, P.A. Structural diversity among sulfated alpha-L-galactans from ascidians (tunicates). Studies on

the species Ciona intestinalis and Herdmania monus. Eur. J. Biochem. **1992**, 204, 669–677.

46. Cumashi, A.; Ushakova, N.A.; Preobrazhenskaya, M.E.; Piccoli, A.; Totani, L.; Ustyuzhanina, N.E.; Bilan, M.I.; Usov, A.I.; Grachev, A.A.; Morozevich, G.E.; et al. A comparative study of the anti-inflammatory, anticoagulant, antiangiogenic, and antiadhesive activities of nine different fucoidans from brown seaweeds. Glycobiology **2007**, 17, 541–552.

47. Foxal, C.; Watson, S.R.; Dowbenko, D.; Lasky, L.A.; Kiso, M.; Hasegawa, A.; Asa, D.; Brandley, B.K. The three members of the selectin receptor family recognize a common carbohydrate epitope, the sialyl Lewis(x) oligosaccharide. J. Cell. Biol. **1992**, 117, 895–902.

48. Game, S.M.; Rajapurohit, P.K.; Clifford, M.; Bird, M.I.; Priest, R.; Bovin, N.V.; Nifantiev, N.E.; O'Beirne, G.; Cook, N.D. Scintillation proximity assay for E-, P-, and L-selectin utilizing polyacrylamide-based neoglycoconjugates as ligands.Anal. Biochem. **1998**, 258, 127–135.

49. Mourao, P.A.; Pereira, M.S. Searching for alternatives to heparin: Sulfated fucans from marine invertebrates. Trends Cardiovasc. Med. **1999**, 9, 225–232.

50. Mourão, P.A. Use of sulfated fucans as anticoagulant and antithrombotic agents: Future perspectives. Curr. Pharm. Des. **2004**, 10, 967–981.

51. Guerrini, M.D.; Beccati, Z.; Shriver, Z.; Naggi, A.; Viswanathan, K.; Bisio, A.; Capila, I.; Lansing, J.C.; Guglieri, S.; Fraser, B.; et al. Oversulfated chondroitin sulfate is a contaminant in heparin associated with adverse clinical events.Nat. Biotechnol. **2008**, 26, 669–675.

52. Kishimoto, T.K.; Viswanathan, K.; Ganguly, T.; Elankumaran, S.; Smith, S.; Pelzer, K.; Lansing, J.C.; Sriranganathan, N.; Zhao, G.; Galcheva-Gargova, Z.; et al. Contaminated heparin associated with adverse clinical events and activation of the contact system. N. Engl. J. Med. **2008**, 358, 2457–2467.

53. Melo, F.R.; Pereira, M.S.; Foguel, D.; Mourão, P.A. Antithrombin-mediated anticoagulant activity of sulfated polysaccharides. J. Biol. Chem. **2004**, 279, 20824–20835.

54. Pomin, V.H.; Pereira, M.S.; Valente, A.P.; Tollefsen, D.M.; Pavão, M.S.G.; Mourão, P.A. Selective cleavage and anticoagulant activity of a sulfated fucan: stereospecific removal of a 2-sulfate ester

from the polysaccharide by mild acid hydrolysis, preparation of oligosaccharides, and heparin cofactor II-dependent anticoagulant activity.Glycobiology **2005**, 15, 369–381.

55. Glauser, B.F.; Rezende, R.M.; Melo, F.R.; Pereira, M.S.; Francischetti, I.M.; Monteiro, R.Q.; Rezaie, A.R.; Mourão, P.A. Anticoagulant activity of a sulfated galactan: Serpin-independent effect and specific interaction with factor Xa.Thromb. Haemostasis. **2009**, 102, 1183–1193.

56. Koyanagi, S.; Tanigawa, N.; Nakagawa, H.; Soeda, S.; Shimeno, H. Oversulfation of fucoidan enhances its anti-angiogenic and antitumor activities. Biochem. Pharmacol. **2003**, 65, 173–179.

57. Soeda, S.; Kozako, T.; Iwata, K.; Shimeno, H. Oversulfated fucoidan inhibits the basic fibroblast growth factor-induced tube formation by human umbilical vein endothelial cells: Its possible mechanism of action. Biochim. Biophys. Acta**2000**, 1497, 127–134.

58. Matsubara, K.; Mori, M.; Matsumoto, H.; Hori, K.; Miyazawa, K. Antiangiogenic properties of a sulfated galactan isolated from a marine green alga, Codium cylindricum. J. Appl. Phycol. **2003**, 15, 87–90.

59. Croci, D.O.; Cumashi, A.; Ushakova, N.A.; Preobrazhenskaya, M.E.; Piccoli, A.; Totani, L.; Ustyuzhanina, N.E.; Bilan, M.I.; Usov, A.I.; Grachev, A.A.; et al. Fucans, but not fucomannoglucuronans, determine the biological activities of sulfated polysaccharides from Laminaria saccharina brown seaweed. PLoS One **2001**, 6.

60. Rupérez, P.; Ahrazem, O.; Leal, J.A. Potential antioxidant capacity of sulfated polysaccharides from the edible marine brown seaweed Fucus vesiculosus. J. Agric. Food Chem. **2002**, 50, 840–845.

61. de Souza, M.R.; Marques, C.T.; Dore, C.M.G.; da Silva, F.R.F.; Rocha, H.A.O.; Leite, E.L. Antioxidant activities of sulfated polysaccharides from brown and red seaweeds. J. Appl. Phycol. **2007**, 19, 153–160.

62. Costa, L.S.; Fidelis, G.P.; Telles, C.B.; Dantas-Santos, N.; Camara, R.B.; Cordeiro, S.L.; Costa, M.S.; Almeida-Lima, J.; Melo-Silveira, R.F.; Oliveira, R.M.; et al. Antioxidant and antiproliferative activities of heterofucans from the seaweed Sargassum filipendula. Mar. Drugs **2011**, 9, 952–966.

63. Barahona, T.; Encinas, M.V.; Mansilla, A.; Matsuhiro, B.; Zúñiga, E.A. A sulfated galactan with antioxidant capacity from the green

variant of tetrasporic Gigartina skottsbergii (Gigartinales, Rhodophyta). Carbohydr. Res. **2012**, 347, 114–120.

64. Barahona, T.; Chandía, N.P.; Encinas, M.V.; Matsohiro, B.; Zúñiga, E.A. Antioxidant capacity of sulfated polysaccharides from seaweeds. A kinetic approach. Food Hydrocol. **2011**, 25, 529–535.

65. Adhikari, U.; Mateu, C.G.; Chattopadhyay, K.; Pujol, C.A.; Damonte, E.B.; Ray, B. Structure and antiviral activity of sulfated fucans from Stoechospermum marginatum. Phytochemicals **2006**, 67, 2474–2482.

66. Romanos, M.T.V.; Andrada-Serpa, M.J.; Mourão, P.A.S.; Yoneshigue-Valentin, Y.; Pereira, M.S.; Santos, N.; Wigg, M.D. Sulfated fucan from marine alga inhibits HeLa cells infection by HTLV-1 free particles: Semi-quantitative analysis.Rev. Bras. Pharmacogn. **2011**, 21, 229–233.

67. Pierre, G.; Sopena, V.; Juin, C.; Mastouri, A.; Graber, M.; Maugard, T. Antibacterial activity of a sulfated galactan extracted from the marine alga Chaetomorpha aerea against Staphylococcus aureus. Biotechnol. Bioprocess Eng. **2011**, 16, 937–945.

68. Aquino, R.S.; Grativol, C.; Mourão, P.A. Rising from the sea: correlations between sulfated polysaccharides and salinity in plants. PLoS One **2011**, 28.

69. Cinelli, L.P.; Castro, M.O.; Santos, L.L.; Garcia, C.R.; Vilela-Silva, A.C.; Mourão, P.A. Expression of two different sulfated fucans by females of Lytechinus variegatus may regulate the seasonal variation in the fertilization of the sea urchin. Glycobiology **2007**, 17, 877–885.

70. Honya, M.; Morim, M.; Anzai, M.; Araki, Y.; Nisizawa, K. Monthly changes in the content of fucans their constituent sugars and sulphate in cultured Laminaria japonica. Hydrobiologia **1999**, 398, 411–416.

71. Andrade, L.R.; Leal, R.N.; Noseda, M.; Duarte, M.E.; Pereira, M.S.; Mourão, P.A.; Farina, M.; Filho, G.M.A. Brown algae overproduce cell wall polysaccharides as a protection mechanism against the heavy metal toxicity. Mar. Pollut. Bull. **2010**, 60, 1482–1488.

72. Oschatz, C.; Maas, C.; Lecher, B.; Jansen, T.; Björkqvist, J.; Tradler, T.; Sedlmeier, R.; Burfeind, P.; Cichon, S.; Hammerschmidt, S.; et al. Mast cells increase vascular permeability by heparin-initiated bradykinin formation in vivo.Immunity **2011**, 34, 258–268.

73. Maaroufi, R.M.; Jozefowicz, M.; Tapon-Bretaudière, J.; Fischer, A.M. Mechanism of thrombin inhibition by antithrombin and heparin cofactor II in the presence of heparin. Biomaterials **1997**, 18, 203–211.

74. Blossom, D.B.; Kallen, A.J.; Patel, P.R.; Elward, A.; Robinson, L.; Gao, G.; Langer, R.; Perkins, K.M.; Jaeger, J.L.; et al. Outbreak of adverse reactions associated with contaminated heparin. N. Engl. J. Med. **2008**, 359, 2674–2684.

75. Esko, J.D.; Stewart, T.E.; Taylor, W.H. Animal cell mutants defective in glycosaminoglycan biosynthesis. Proc. Natl. Acad. Sci. USA **1985**, 82, 3197–3201.

76. Ageenko, N.V.; Kiselev, K.V.; Odintsova, N.A. Expression of pigment cell-specific genes in the ontogenesis of the sea urchin Strongylocentrotus intermedius. Evid. Based Complement. Altern. Med. **2011**, 2011.

77. Pomin, V.H.; Sharp, J.S.; Li, X.; Wang, L.; Prestegard, J.H. Characterization of glycosaminoglycans by ^{15}NNMR spectroscopy and in vivo isotopic labeling. Anal. Chem. **2010**, 82, 4078–4088.

78. Pomin, V.H.; Pereira, M.S.; Valente, A.-P.; Tollefsen, D.M.; Pavão, M.S.; Mourão, P.A. Selective cleavage and anticoagulant activity of a sulfated fucan: Stereospecific removal of a 2-sulfate ester from the polysaccharide by mild acid hydrolysis, preparation of oligosaccharides, and heparin cofactor II-dependent anticoagulant activity.Glycobiology **2005**, 15, 369–381.

79. Pomin, V.H.; Valente, A.-P.; Pereira, M.S.; Mourão, P.A. Mild acid hydrolysis of sulfated fucans: A selective 2-desulfation reaction and an alternative approach for preparing tailored sulfated oligosaccharides. Glycobiology **2005**,15, 1376–1385.

80. Varki, A.P.; Baum, L.G.; Bellis, S.L.; Cummings, R.D.; Esko, J.D.; Hart, G.W.; Linhardt, R.J.; Lowe, J.B.; McEver, R.P.; Srivastava, A.; et al. Working group report: The roles of glycans in hemostasis, inflammation and vascular biology.Glycobiology **2008**, 18, 747–749.

11

CHAPTER

Production of Hyaluronic Acid by Streptococcus Zooepidemicus on Protein Substrates Obtained from Scyliorhinus Canicula Discards

José A. Vázquez [1,*], Lorenzo Pastrana [2], Carmen Piñeiro [3], José A. Teixeira [4], Ricardo I. Pérez-Martín [5] and Isabel R. Amado [2,4]

[1]*Grupo de Reciclado y Valorización de Materiales Residuales (REVAL), Instituto de Investigacións Mariñas (IIM-CSIC), r/Eduardo Cabello, 6. Vigo-36208 Galicia, Spain*

[2]*Departamento de Química Analítica y Alimentaria, Facultad de Ciencias de Ourense (Universidad de Vigo), Campus As Lagoas s/n, Ourense-32004 Galicia, Spain*

[3]*Servicio de Instrumentación Científica (SICIM), Instituto de Investigacións Mariñas (IIM-CSIC), r/Eduardo Cabello, 6. Vigo-36208 Galicia, Spain*

[4]*Centre of Biological Engineering, University of Minho, Campus Gualtar, 4710-057 Braga, Portugal*

[5]*Grupo de Bioquímica de Alimentos, Instituto de Investigacións Mariñas (IIM-CSIC), r/Eduardo Cabello, 6. Vigo-36208 Galicia, Spain*

ABSTRACT

This work investigates the production of hyaluronic acid (H) by Streptococcus equi subsp. zooepidemicus in complex media formulated with peptones obtained from Scyliorhinus canicula viscera by-products. Initially, in batch cultures, the greatest productions were achieved using commercial media (3.03 g/L) followed by peptones from alcalase hydrolyzed viscera (2.32 g/L) and peptones from non-hydrolyzed viscera (2.26 g/L). An increase of between 12% and 15% was found in subsequent fed-batch cultures performed on waste peptones. Such organic nitrogen sources were shown to be an excellent low-cost substrate for microbial H, saving more than 50% of the nutrient costs.

Keywords: viscera waste valorization; hyaluronic acid production; Streptococcus zooepidemicus; marine peptones; Scyliorhinus canicula by-products; logistic equation

1. INTRODUCTION

One of the most important macromolecules for cosmetic and pharmacological formulations is hyaluronic acid (H). It is a high-molecular-mass glycosaminoglycan polysaccharide that can be found in different animal tissues such as skin, cartilage, vitreous humor, synovial liquid, bacterial cell capsule, etc. [1,2,3]. Its rheological and biological features including pseudoplasticity, biocompatibility material, and water holding capacity are especially adequate and essential for multiple types of commercial applications in different fields [4,5,6]. Although the production of H has been, during past decades, based on its extraction from bovine vitreous humor, umbilical cord, or rooster comb, microbial H from Streptococcus strains is currently the main alternative. Marine sources of H, such as fish vitreous humor, have also been explored, but lower yields than from the microbial sources were found [7,8].

Nevertheless, the cost of H production is basically dependent on the price of the commercial culture media. Streptococci are facultative anaerobe bacteria that show fastidious nutrient requirements in relation to the organic nitrogen [9,10] which cannot be replaced by inorganic nitrogen salts. The formulation of low-cost media is therefore essential for the industrial production of H. In this context, peptones generated from fishing by-products and discards are valuable and effective substrates [11]. In addition, the new policy of zero discards and the exhaustive reduction of marine wastes by the European Union (EU) requires the use of valorizing solutions to solve them. This microbial bioconversion strategy is also framed into the biorefinery concept which is one of the most important pillars to develop in the next decade from the blue biotechnology issue.

The small-spotted catshark (Scyliorhinus canicula) is an abundant fish in the northeastern Atlantic Ocean and Mediterranean Sea which is considered as a discard in some fisheries [12,13]. Recently, the processes for the recovery and purification of another glycosaminoglycan (chondroitin sulfate) from cartilage by-products of this species have been optimized [14]. However, to our knowledge, no valorized alternatives for viscera wastes have yet been developed. The present work studies the feasibility of peptones obtained from three treatments of S. canicula viscera as nitrogen sources for H production by Streptococcus equi subsp. zooepidemicus ATCC 35246.

2. RESULTS AND DISCUSSION

In order to evaluate the viability of the three nitrogen sources obtained from small-spotted catshark wastes, commercial tryptone in the control medium was substituted by the peptones prepared using thermal autohydrolysis and enzyme catalysis of viscera. The remaining nutrients were similar in all media. Time-courses of batch cultures are displayed inFigure 1. The experimental data of the bioproductions quantified (biomass: X, lactic acid: L, and hyaluronic acid: H) were modeled on the logistic Equation (1). The agreement between predicted and experimental data was excellent (Table 1), with values of R^2 higher than 0.986. The p-values from Fisher's F-test also indicated the consistency of model (1) to describe the cultivation data.

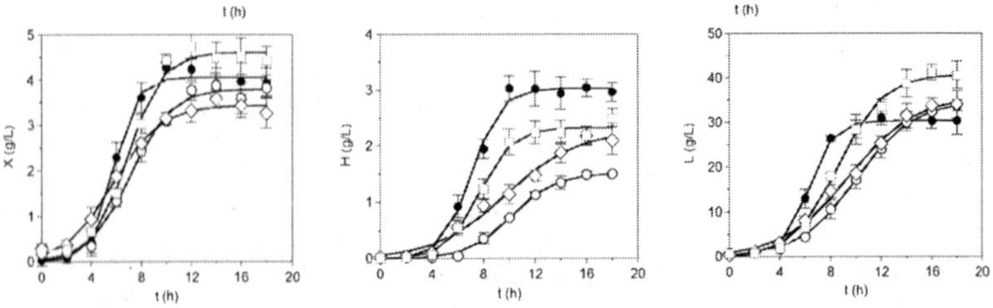

Figure 1. Batch cultivations of S. zooepidemicus on different nutritive media. G: glucose, Pr: total proteins, X: biomass, H: hyaluronic acid, L: lactic acid. ◊: Medium A, ○: Medium B, □: Medium C, •: CM. In the case of bioproductions (X, L, and H), experimental data were fitted to the logistic Equation (1) (continuous line). Error bars are the confidence intervals (n = 2, α = 0.05).

All the kinetic parameters defined by Equations (1)–(4) for the three bioproductions were statistically significant (Student'st-test, Table 1). According to these parameters, the highest maximum biomass (X_m = 4.61 ± 0.26 g/L) and lactic acid production (L_m = 40.74 ± 1.82 g/L) were found in Medium C, but CM medium conducted the highest maximum hyaluronic acid production (H_m = 3.03 ± 0.14 g/L) and the maximum and specific maximum rate for X, L, and H productions (v_x, v_h, v_l,μ_x, μ_h, and μ_l). In concordance, the times to reach the asymptotic or plateau phase of sigmoid trends in CM (t_P) were shorter than in alternative media. The production of the glycosaminoglycan, taking into account the values of H_m, followed this sequence between broths: CM > Medium C = Medium A > Medium B. The H_m results in Media C and A were somewhat inferior (2.32 ± 0.13 g/L) to the tryptone ones, but were nevertheless remarkable and similar to those derived from tuna viscera (2.41 ± 0.02 g/L) [15]. Although the slight and not significant difference between the H productions in Media A and C indicates that the use of alcalase in the process of viscera hydrolysis could be excluded, the greater concentration of protein extracted by means of enzyme application could justify this.

Table 1. Parametric estimations corresponding to the logistic Equation (1) applied to S. zooepidemicus batch cultures. Numerical values of the parameters are shown with their confidence intervals. R^2 is the determination coefficients for the mathematical fittings to Equation (1) and p-values from Fisher's F-test verify the robustness of the equation. The different production yields are also calculated.

Parameters	Medium A	Medium B	Medium C	CM
X_m (g/L)	3.43 ± 0.15	3.80 ± 0.20	4.61 ± 0.26	4.05 ± 0.16
v_x (g·L^{-1}·h^{-1})	0.461 ± 0.009	0.587 ± 0.131	0.835 ± 0.230	1.11 ± 0.33
λ_x (h)	2.03 ± 0.77	3.92 ± 0.81	4.15 ± 0.86	3.99 ± 0.62
τ_x (h)	5.75 ± 0.44	7.15 ± 0.47	6.91 ± 0.48	5.80 ± 0.28
μ_x (h^{-1})	0.538 ± 0.109	0.618 ± 0.151	0.725 ± 0.488	1.10 ± 0.32
t_x (h)	9.46 ± 0.80	10.39 ± 0.67	9.66 ± 0.71	7.62 ± 0.34
$Y_{X/G}$	0.058	0.071	0.075	0.080
$Y_{X/Pr}$	1.632	1.254	0.991	1.275
R^2	0.994	0.994	0.993	0.995
p-Value	<0.001	<0.001	<0.001	<0.001

H_m (g/L)	2.26 ± 0.33	1.51 ± 0.06	2.32 ± 0.13	3.03 ± 0.14
v_h (g·L^{-1}·h^{-1})	0.207 ± 0.053	0.225 ± 0.028	0.456 ± 0.126	0.667 ± 0.169
λ_h (h)	4.23 ± 1.46	6.79 ± 0.45	5.14 ± 0.79	4.81 ± 0.66
τ_h (h)	9.69 ± 1.33	10.15 ± 0.29	7.68 ± 0.44	7.09 ± 0.36
μ_h (h^{-1})	0.366 ± 0.128	0.596 ± 0.085	0.785 ± 0.232	0.879 ± 0.235
t_h (h)	15.15 ± 2.01	13.50 ± 0.56	10.23 ± 0.89	9.36 ± 0.40
$Y_{H/G}$	0.040	0.029	0.044	0.061
$Y_{H/Pr}$	1.143	0.512	0.584	0.967
$Y_{H/X}$	0.700	0.408	0.589	0.759
R^2	0.986	0.998	0.993	0.995
p-Value	<0.001	<0.001	<0.001	<0.001
L_m (g/L)	35.78 ± 2.28	34.60 ± 1.14	40.74 ± 1.82	30.25 ± 0.77
v_l (g·L^{-1}·h^{-1})	3.46 ± 0.54	3.90 ± 0.29	5.73 ± 0.89	7.81 ± 1.39
λ_l (h)	4.31 ± 0.84	5.61 ± 0.35	5.09 ± 0.61	4.34 ± 0.38
τ_l (h)	9.48 ± 0.71	10.05 ± 0.27	8.65 ± 0.38	6.28 ± 0.20
μ_l (h^{-1})	0.387 ± 0.079	0.451 ± 0.042	0.562 ± 0.099	1.03 ± 0.19
t_l (h)	14.65 ± 1.34	14.48 ± 0.59	12.21 ± 0.70	8.21 ± 0.25
$Y_{L/G}$	0.651	0.643	0.713	0.622
$Y_{L/Pr}$	18.44	11.29	9.42	9.87
R^2	0.995	0.999	0.997	0.998
p-Value	<0.001	<0.001	<0.001	<0.001

The consumption of glucose was faster in CM and the protein time-course was similar in Media B, C, and CM. The final protein uptake was inferior in Medium A. The best productivities in relation to protein uptake were also generated by Medium A. In the remaining cases, the highest yields were mainly defined by CM. The molecular weights of H sampled at the end of the fermentations (18 h) were 1800, 1730, 1900, and 1850 kDa for Media A, B, C, and CM, respectively.

The concentrations of H and corresponding molecular weights obtained with the bacteria S. zooepidemicus ATCC 35246 in the present study were similar or, in some cases, higher than those reported in the literature [10,16]. Anaerobic and aerobic cultures in a base medium formulated with glucose, mineral salts ($MgSO_4$ and Na_2HPO_4), and yeast extract as unique sources of nitrogen showed productions around 1.5 and 2.5 g/L of H, respectively [17]. The molecular weights of such H were of 2 and 3 MDa, respectively. These authors also found values of μ_m (0.45–0.75 h^{-1}), working under different experimental conditions, in agreement with the outcomes in the alternative media (Table 1). However, our yields of hyaluronic production/glucose consumption were lower in comparison with those previously reported in the mentioned work [17]. In this context, Johns et al. [16] previously established the best conditions of agitation (>400 rpm) and pH (6.6–6.9) for S. zooepidemicus ATCC 35246 in a complex medium (similar to the base medium supplemented with guanine, adenine, and uracil). The values of H and L productions were similar to those described herein.

In recent times, some culture media based on nutrients recovered from agrifood wastes have been studied as new options for the bacterial production of H. Fruit bagasse (cashew apple) was studied as a substrate for S. zooepidemicus fermentations in a solid-state modality generating 6 mg/g of H [18]. Low levels of H (~0.2 g/L) were achieved when whey or soy protein were utilized as nitrogen sources in Erlenmeyer flask experiments [19,20]. In no case were such bioproductions comparable to those obtained with marine peptones, mainly because batch cultures were performed at flask scale and without pH control. It is well known that the lack of constant pH conditions (among 6.0–7.0) and the subsequent spontaneous evolution of the acidic environment in S. zooepidemicus culture provoke the inhibition of H production [16,17,21].

Selecting the best residual media (A and C), fed-batch cultures were performed in order to increase the biosynthesis production of the polysaccharide by S. zooepidemicus. Figure 2 shows the case of Medium A which included a glucose fed-batch every 2 h from 10 h of culture. The numerical analyses of all cultures are summarized in Table 2. Logistic Equation (1) was again a perfect tool for kinetics modeling ($R^2 > 0.978$ and $p < 0.001$). CM was the most effective medium in terms of maximum concentration bioproductions and rates of productions. However, the increment of H_m comparing batch and fed-batch resources was (in %) 12, 15, and 7 for Media A, C, and CM, respectively. Moreover, the highest yields of biomass, hyaluronic, and lactic acids as functions of protein consumed were defined by Medium A. The M_w of the biopolymer in that medium was larger (2110 kDa) than previously detected in the batch.

These outcomes of hyaluronic acid formation on residual media were in agreement with those previously found using peptones extracted from Isirus oxyrhinchus and Raja clavata by-products [11]. However, the clear biphasic sigmoid patterns described in these fed-batch cultivations were not observed in the current study. Other authors have also investigated the capacity of the fed-batch strategy to increase the microbial hyaluronic acid production [22] or to perform successive batch fermentations [23]. In general, lactic acid bacteria have also shown excellent responses to the glucose-intermittent fed in terms of enhancing cultivation time and biomass, and lactic acid and bacteriocin productions [24,25,26].

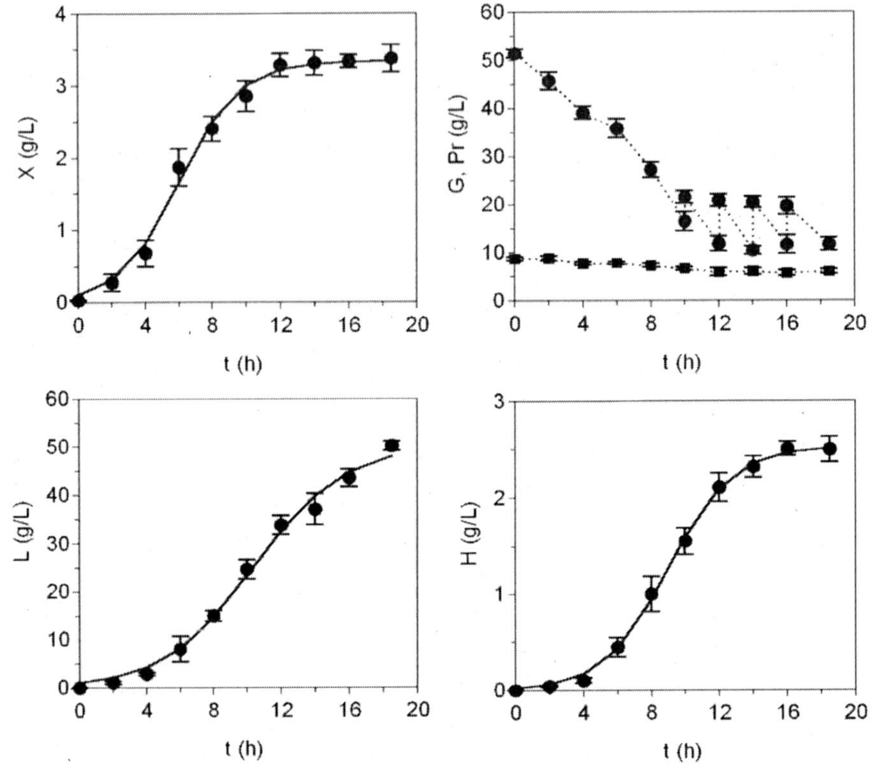

Figure 2. Fed-batch cultivation of S. zooepidemicus on medium formulated with peptones obtained from S. canicula viscera without hydrolysis (Medium A). Concentrated glucose solutions were employed for fed-batch operatory. Experimental data of bioproductions were fitted to the logistic Equation (1) (continuous line). Error bars are the confidence intervals ($n = 2$, $\alpha = 0.05$).

Based on the commercial prices of the tryptone, yeast extract glucose, and mineral salts ingredients, the costs of H production in the Media A, B, C, and CM were calculated (Figure 3). The H_m data used for calculation were selected fromTable 1. The three alternative peptones proved to be an excellent source of nitrogen in terms of reducing cost. Media B, A, and C reduced the price of a gram of hyaluronic acid produced in batch fermentations by 31%, 54%, and 55%, respectively, in comparison with CM. Similar cost reductions were observed in fed-batch cultures (data not shown). In addition, current results were concordant with those reported when medium for S. zooepidemicus was formulated with tuna peptone and mussel-processing effluents [15].

Table 2. Parametric estimations corresponding to the logistic Equation (1) applied to S. zooepidemicus fed-batch cultures. Numerical values of the parameters are shown with their confidence intervals. R^2 is the determination coefficient for the mathematical fittings to Equation (1) and p-values from Fisher's F-test verify the robustness of the equation. The different production yields are also calculated.

Parameters	Medium A	Medium C	CM
X_m (g/L)	3.34 ± 0.16	3.27 ± 0.17	4.56 ± 0.29
v_x (g·L^{-1}·h^{-1})	0.467 ± 0.098	0.464 ± 0.103	0.930 ± 0.350
λ_x (h)	2.47 ± 0.83	2.69 ± 0.87	3.46 ± 1.03
τ_x (h)	6.05 ± 0.48	6.22 ± 0.50	8.34 ± 0.77
μ_x (h^{-1})	0.559 ± 0.128	0.567 ± 0.138	0.820 ± 0.321
t_x (h)	9.63 ± 0.60	9.74 ± 0.87	7.62 ± 0.34
$Y_{X/G}$	0.047	0.049	0.052
$Y_{X/Pr}$	1.201	0.976	1.027
R^2	0.994	0.994	0.988
p-Value	<0.001	<0.001	<0.001
H_m (g/L)	2.53 ± 0.09	2.66 ± 0.17	3.23 ± 0.25
v_h (g L^{-1}·h^{-1})	0.332 ± 0.035	0.321 ± 0.054	0.589 ± 0.234
λ_h (h)	5.19 ± 0.44	5.42 ± 0.75	3.76 ± 1.22
τ_h (h)	9.00 ± 0.27	9.56 ± 0.52	6.50 ± 0.65
μ_h (h^{-1})	0.526 ± 0.061	0.483 ± 0.095	0.730 ± 0.299
t_h (h)	12.80 ± 0.99	13.70 ± 0.61	9.24 ± 0.81
$Y_{H/G}$	0.035	0.039	0.039
$Y_{H/Pr}$	0.896	0.772	0.767
$Y_{H/X}$	0.746	0.791	0.747
R^2	0.999	0.996	0.986
p-Value	<0.001	<0.001	<0.001
L_m (g/L)	50.50 ± 5.44	51.13 ± 4.16	81.88 ± 19.65
v_l (g·L^{-1}·h^{-1})	4.68 ± 0.83	5.08 ± 0.76	6.53 ± 1.92
λ_l (h)	5.01 ± 0.99	5.11 ± 0.78	4.49 ± 1.89
τ_l (h)	10.42 ± 0.94	10.14 ± 0.69	10.76 ± 2.22
μ_l (h^{-1})	0.370 ± 0.090	0.398 ± 0.079	0.319 ± 0.143
t_l (h)	15.82 ± 2.79	15.16 ± 2.19	17.03 ± 4.01
$Y_{L/G}$	0.700	0.762	0.895
$Y_{L/Pr}$	17.99	15.04	17.77
R^2	0.994	0.996	0.978
p-Value	<0.001	<0.001	<0.001

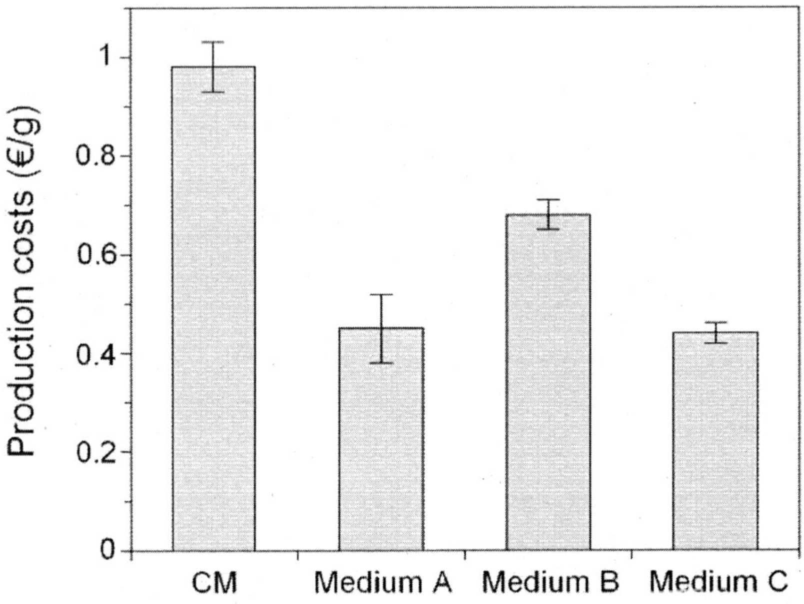

Figure 3. Hyaluronic production costs in each medium tested for batch culture conditions. Error bars are the confidence intervals ($n = 2$, $\alpha = 0.05$).

3. EXPERIMENTAL SECTION

3.1. Preparation of Protein Substrate (Peptones)

S. canicula by-products (stomachs and intestines) were collected from fish-processing industries located in Vigo (Spain). Fresh material was mixed with distilled water (40% w/v), grinded, and divided to perform the following procedures: (a) autoclaving at 101 °C/1 h, centrifugation at 12,000× g/30 min, and supernatant recovery (Peptone A); (b) autohydrolysis of homogenate by incubation at 30 °C/200 rpm/5 h, autoclaving of the autohydrolysate at 101 °C/1 h, centrifugation at 12,000×g/30 min, and supernatant recovery (Peptone B); (c) enzymatic hydrolysis by alcalase addition (2% v/p, 48 AU/kg) and incubation at 50 °C/200 rpm/3 h, autoclaving of the enzymatic hydrolysate at 101 °C/1 h, centrifugation at 12,000× g/30 min, and supernatant recovery (Peptone C). The composition of peptones was (in g/L): 15.3 ± 0.2 of protein and 1.8 ± 0.1 of reducing sugars (Peptone A), 15.5 ± 0.3 of protein and 2.4 ±

0.3 of reducing sugars (Peptone B), 23.5 ± 0.4 g/L of protein and 2.2 ± 0.2 of reducing sugars (Peptone C).

3.2. Bacterial Strain

Growth experiments and hyaluronic acid production were carried out using the Lancefield group C, β-hemolytic and mucoid bacterium: Streptococcus equi subsp. zooepidemicus ATCC 35246 obtained from the American Type Culture Collection (Manassas, VA, USA). Stocks of this bacterium were stored at −80 °C in complex medium (defined below) with 25% glycerol. Inocula were conducted according the steps reported in the previous work [11].

3.3. Microbiological Culture Media and Growth Conditions

The complex (CM) and alternative growth media for S. zooepidemicus contained the following nutrients (in g/L of distilled water): glucose, 50; yeast extract, 5; K_2HPO_4, 2.0; KH_2PO_4, 2.0; $MgSO_4$, 0.5; $(NH_4)SO_4$, 0.5; and either tryptone (Cultimed, Panreac Química, Spain), 15, or the peptones (A, B, C) obtained from each S. canicula visceral treatment at the protein-Lowry concentration of 8. These media were: Medium A, Medium B, and Medium C formulated with Peptone A, Peptone B, and Peptone C, respectively. The pH in all cultures was adjusted to 6.7 after autoclaving and sterilized at 121 °C/15 min.

S. zooepidemicus was cultivated in a glass 2 L bioreactor (with a working volume of 1.8 L) at 37 °C, 500 rpm of agitation, without aeration, and pH controlled with sterile solutions of NaOH and HCl (both 5 M). In the fed-batch cultivations, glucose was added every 2 h (from 8 to 10 h of culture) up to a concentration of 20 g/L using a sterile glucose solution of 500 g/L.

3.4. Analytical Methods

The procedures of sampling from the bioreactor, preparation of samples, and purification of H are extensively described in a previous report [15]. Briefly, samples from the bioreactor were initially blended with a 10% volume of 5% (w/v) SDS for 10 min and centrifuged at 5000× g/30 min. H was precipitated by mixing the supernatant with three volumes of ethanol and then centrifuged at 5000× g/10 min. The sediment was redissolved with one volume of NaCl (1.5 M) and three volumes of ethanol and subsequently centrifuged again at 5000× g/10 min. Finally, this last sediment was resuspended in distilled water for HA determination. The rest of the determinations performed (in duplicate) were: (1) optical density (OD) at 700 nm and subsequent calculation of biomass (dry weight) using a corresponding calibration curve; (2) total soluble proteins according the method of Lowry et al. [27]; (3) H analyzed using the method of Blumenkrantz and Asboe-Hansen [28] with the modifications of Murado et al. [29]; (4) reducing sugars measured by the 3,5-dinitrosalicylic reaction [30]; (5) glucose, lactic acid, and molecular weight (Mw) of H using HPLC methodology [11]. Thus, ION-300 column (Transgenomic, San José, CA, USA) with 6 mM H_2SO_4 as mobile phase (flow = 0.4 mL/min) at 65 °C and a refractive-index detector were used for lactic acid and glucose quantification. Mw of H was determined by size-exclusion chromatography on HPLC equipped with an Ultrahydrogel linear column (Waters, Milford, MA, USA) using 0.1 M $NaNO_3$ as mobile phase (flow = 0.6 mL/min) and a refractive-index detector. Standards of dextran (Sigma, St. Louis, MO, USA) with different molecular weights (5, 12, 25, 50, 80, 150, 270, 410, 670, 1400, 2000, and 3500 kDa) were used for calibration.

3.5. Modeling of S. zooepidemicus Cultivation

The logistic equation was chosen to model the kinetic profiles of the different bioproductions (P) as biomass (X), hyaluronic acid (H), and lactic acid (L) productions [31]:

$$P = \frac{P_m}{1 + \exp\left[2 + \frac{4v_P\left(\lambda_P - t\right)}{P_m}\right]} \tag{1}$$

Complementarily, we have calculated other parameters with clear biological means to help the characterization of the cultures in the media studied:

$$\mu_P = \frac{4v_P}{P_m} \tag{2}$$

$$\tau_P = \lambda_P + \frac{2}{\mu_P} \tag{3}$$

$$t_{mP} = \tau_P + \frac{P_m}{2v_P} \quad (4)$$

where P is the product determined (X, L, or H); P_m is the maximum product formation (g/L); v_P is the maximum production rate ($g \cdot L^{-1} \cdot h^{-1}$); λ_P is the product's lag phase (h); μ_P is the specific maximum production rate (h^{-1}); τ_P is the time required to achieve half the maximum production (h); and t_{mP} is the time required to reach the maximum production (h). In addition, the yields of productions on glucose (Y_P/Y_G), protein uptakes (Y_P/Y_{Pr}), and hyaluronic acid relative to biomass (Y_H/Y_X) were also determined.

3.6. Numerical and Statistical Analyses

Fitting procedures and parametric estimations calculated from the results were carried out by minimizing the sum of quadratic differences between the observed and model-predicted values, using the non-linear least-squares (quasi-Newton) method provided by the macro-"Solver" of the Microsoft Excel spreadsheet. Confidence intervals from the parametric estimates (Student's t-test) and consistence of mathematical models (Fisher's F-test) were assessed by "SolverAid" macro (Levie's Excellaneous
website: http://www.bowdoin.edu/~rdelevie/excellaneous).

4. CONCLUSIONS

Protein material derived from fishing discards (S. canicula viscera) proved to be an appropriate alternative to replace commercial tryptone in the microbial production of H. Peptones obtained by alcalase and thermal treatment of by-products yielded more than 2 g/L of the polysaccharide with higher than 1800 kDa in batch fermentations. Furthermore, fed-batch operatory significantly improved the productions but not the molecular weights of H. In all cases, the reduction of the prime cost of H using marine-based media was remarkable.

ACKNOWLEDGMENTS

We appreciate the excellent technical support of Margarita Nogueira and Ana I. Durán (IIM-CSIC). We also thank to Ramiro Martínez (Novozymes

A/S, Spain) for supplying us with alcalase. English used in this manuscript has been reviewed by Concorde Language School S.L. (Vigo, Spain). Isabel Rodríguez Amado was awarded with a postdoctoral contract from the Xunta de Galicia (Plan I2C, 2012). This research was funded by the projects MARMED 2011-1/164 (Atlantic Area Programme, EU), 0687_NOVOMAR_1_P (POCTEP Programme, EU) and iSEAS LIFE13 ENV/ES/000131 (LIFE+ Programme, EU).

AUTHOR CONTRIBUTIONS

J.A.V. performed the experiments, developed the mathematical models and wrote the manuscript. L.P., R.I.P.M. and J.A.T. have been involved in critical reading of the manuscript and in the design of the present study. C.P. helped in the chromatographic determinations. I.R.A. helped in the design of the present study, in the experiments carried out and in the manuscript preparation. All authors read and approved the manuscript.

CONFLICTS OF INTEREST

The authors declare no conflict of interest.

REFERENCES

1. Shiedlin, A.; Bigelow, R.; Christopher, W.; Arbabi, S.; Yang, L.; Maier, R.V.; Wainwright, N.; Childs, A.; Miller, R.J. Evaluation of hyaluronan from different sources: Streptococcus zooepidemicus, rooster comb, bovine vitreous, and human umbilical cord. Biomacromolecules **2004**, 5, 2122–2127.
2. Yamada, T.; Kawasaki, T. Microbial synthesis of hyaluronan and chitin: New approaches. J. Biosci. Bioeng. **2005**, 99, 521–528.
3. Collins, M.N.; Birkinshaw, C. Hyaluronic acid based scaffolds for tissue engineering—A review. Carbohyd. Polym. **2013**,92, 1262–1279.
4. Kogan, G.; Soltés, L.; Stern, R.; Gemeiner, P. Hyaluronic acid: A natural biopolymer with a broad range of biomedical and industrial applications. Biotechnol. Lett. **2007**, 29, 17–25.

5. Vázquez, J.A.; Rodríguez-Amado, I.; Montemayor, M.I.; Fraguas, J.; González, M.P.; Murado, M.A. Chondroitin sulphate, hyaluronic acid and chitin/chitosan production using marine waste sources: Characteristics, applications and eco-friendly processes: A review. Mar. Drugs **2013**, 11, 747–774.

6. Collins, M.N.; Birkinshaw, C. Hyaluronic acid solutions—A processing method for efficient chemical modification. J. Appl. Polym. Sci. **2013**, 130, 145–152.

7. Amagai, I.; Tashiro, Y.; Ogawa, H. Improvement of the extraction procedure for hyaluronan from fish eyeball and the molecular characterization. Fish. Sci. **2009**, 75, 805–810.

8. Murado, M.A.; Montemayor, M.I.; Cabo, M.L.; Vázquez, J.A.; González, M.P. Optimization of extraction and purification process of hyaluronic acid from fish eyeball. Food Bioprod. Proc. **2012**, 90, 491–498.

9. Kim, S.J.; Park, S.Y.; Kin, C.W. A novel approach to the production of hyaluronic acid by Streptococcus zooepidemicus. J. Microbiol. Biotechnol. **2006**, 16, 1849–1855.

10. Cooney, M.J.; Goh, L.T.; Lee, P.L.; Johns, M.R. Structured model based analysis and control of the hyaluronic acid fermentation by Streptococcus zooepidemicus: Physiological implications of glucose and complex nitrogen-limited growth. Biotechnol. Prog. **1999**, 15, 898–910.

11. Vázquez, J.A.; Montemayor, M.I.; Fraguas, J.; Murado, M.A. High production of hyaluronic and lactic acids byStreptococcus zooepidemicus in fed-batch cultures using commercial and marine peptones from fishing by-products.Biochem. Eng. J. **2009**, 44, 125–130.

12. Olaso, I.; Velasco, F.; Pérez, N. Importance of discarded blue whiting (Micromesistius poutassou) in the diet of the lesser spotted dogfish (Scyliorhinus canicula) in the Cantabrian Sea. ICES J. Mar. Sci. **1998**, 55, 331–341.

13. Rodríguez-Cabello, C.; Fernández, A.; Olaso, I.; Sánchez, F.; Gancedo, R.; Punzón, A.; Cendrero, O. Overview of continental shelf elasmobranch fisheries in the Cantabrian Sea. J. Northwest Atl. Fish. Sci. **2005**, 35, 375–385.

14. Blanco, M.; Fraguas, J.; Sotelo, C.G.; Pérez-Martín, R.I.; Vázquez, J.A. Production of chondroitin sulphate from head, skeleton and fins of Scyliorhinus canicula by-products by combination of

enzymatic, chemical precipitation and ultrafiltration methodologies. Mar. Drugs **2015**, 13, 3287–3308.

15. Vázquez, J.A.; Montemayor, M.I.; Fraguas, J.; Murado, M.A. Hyaluronic acid production by Streptococcus zooepidemicusin marine by-products media from mussel processing wastewaters and tuna peptone viscera. Microb. Cell Factories**2010**, 9, 46.

16. Johns, M.R.; Goh, L.T.; Oeggerli, A. Effect of pH, agitation and aeration on hyaluronic acid production by Streptococcus zooepidemicus. Biotechnol. Lett. **1994**, 16, 507–512.

17. Armstrong, D.C.; Johns, M.R. Culture conditions affect the molecular weight properties of hyaluronic acid produced by Streptococcus zooepidemicus. Appl. Environ. Microbiol. **1997**, 63, 2759–2764.

18. De Macedo, A.C.; Santana, M. Hyaluronic acid depolymerization by ascorbate-redox effects on solid state cultivation of Streptococcus zooepidemicus in cashew apple fruit bagasse. World J. Microbiol. Biotechnol. **2012**, 28, 2213–2219.

19. Pires, A.; Macedo, A.; Eguchi, S.; Santana, M. Microbial production of hyaluronic acid from agricultural resource derivatives. Bioresour. Technol. **2010**, 101, 6506–6509.

20. Pan, N.C.; Vignoli, J.A.; Baldo, C.; Pereira, H.C.B.; da Silva, R.S.S.F.; Celligoi, M.A.P.C. Agroindustrial byproducts for the production of hyaluronic acid by Streptococcus zooepidemicus ATCC 39920. Int. J. Sci. Technol. Res. **2015**, 4, 114–118.

21. Fong Chong, B.; Blank, L.M.; Mclaughlin, R.; Nielsen, L.K. Microbial hyaluronic acid production. Appl. Microbiol. Biotechnol. **2005**, 66, 341–351.

22. Liu, L.; Du, G.; Chen, J.; Wang, M.; Sun, J. Enhanced hyaluronic acid production by a two-stage culture strategy based on the modeling of batch and fed-batch cultivation of Streptococcus zooepidemicus. Biores. Technol. **2008**, 99, 8532–8536.

23. Huang, W.C.; Chen, S.J.; Chen, T.L. Production of hyaluronic acid by repeated batch fermentation. Biochem. Eng. J.**2008**, 40, 460–464.

24. Guerra, N.P.; Agrasar, A.T.; Macías, C.L.; Pastrana, L. Modelling the fed-batch production of pediocin using mussel processing wastes. Proc. Biochem. **2005**, 40, 1071–1083.

25. Vázquez, J.A.; Mirón, J.; González, M.P.; Murado, M.A. Bacteriocin production and pH gradient. Some mathematical models and their problems. Enzyme Microb. Technol. **2005**, 37, 54–67.

26. Guerra, N.P.; Bernárdez, P.F.; Pastrana, L. Fed-batch pediocin production on whey using different feeding media.Enzyme Microb. Technol. **2007**, 41, 397–406.

27. Lowry, O.H.; Rosebrough, N.J.; Farr, A.L.; Randall, R.J. Protein measurement with the folin phenol reagent. J. Biol. Chem. **1951**, 270, 27299–27304.

28. Blumenkrantz, N.; Asboe-Hansen, G. New method for quantitative determination of uronic acids. Anal. Biochem. **1973**,54, 484–489.

29. Murado, M.A.; Vázquez, J.A.; Montemayor, M.I.; Cabo, M.L.; González, M.P. Two mathematical models for the correction of carbohydrate and protein interference in the determination of uronic acids by the m-hydroxydiphenyl method. Biotechnol. Appl. Biochem. **2005**, 41, 209–216.

30. Bernfeld, P. Enzymes of starch degradation and synthesis. Adv. Enzymol. **1951**, 12, 379–427.

31. Vázquez, J.A.; Lorenzo, J.M.; Fuciños, P.; Franco, D. Evaluation of non-linear equations to model different animal growths with mono and bi-sigmoid profiles. J. Theor. Biol. **2012**, 314, 95–105.

12

CHAPTER

Purification and Characterization of a Fucoidanase (FNase S) from a Marine Bacterium Sphingomonas paucimobilis PF-1

Woo Jung Kim [1,2], Joo Woong Park [1], Jae Kweon Park [3], Doo Jin Choi [1] and Yong Il Park [1,*]

[1]*Department of Biotechnology and The Biomaterial Engineering Research Center, The Catholic University of Korea, Bucheon, Gyeonggi-do 420-743, Korea*

[2]*Biocenter, Gyeonggi Institute of Science and Technology Promotion (GSTEP), Suwon, Gyeonggi-do 443-270, Korea*

[3]*Department of Pharmaceutical Science, University of Gachon, Yeonsu-dong, Yeonsu-gu, InCheon 406-799, Korea*

ABSTRACT

The Search for enzyme activities that efficiently degrade marine polysaccharides is becoming an increasingly important area for both structural analysis and production of lower-molecular weight oligosaccharides. In this study, anendo-acting fucoidanase that degrades Miyeokgui fucoidan (MF), a sulfated galactofucan isolated from the sporophyll (called Miyeokgui in Korean) of Undaria pinnatifida, into smaller-sized galactofuco-oligosaccharides (1000–4000 Da) was purified from a marine bacterium, Sphingomonas paucimobilis PF-1, by ammonium sulfate precipitation, diethylaminoethyl (DEAE)-Sepharose column chromatography, and chromatofocusing. The specific activity of this enzyme was approximately 112-fold higher than that of the crude enzyme, and its molecular weight was approximately 130 kDa (FNase S), as determined by native gel electrophoresis and 130 (S1), 70 (S2) and 60

(S3) kDa by sodium dodecyl sulfate-polyacrylamide gel electrophoresis (SDS-PAGE). The optimum pH and temperature of FNase S were pH 6.0–7.0 and 40–45 °C, respectively. FNase S activity was enhanced by Mn^{2+} and Na^{+} (115.7% and 131.2%), but it was inhibited by Ca^{2+}, K^{+}, Ba^{2+}, Cu^{2+} (96%, 83.7%, 84.3%, and 89.3%, respectively), each at 1 mM. The K_m, V_{max} and K_{cat} values of FNase S on MF were 1.7 mM, 0.62 mg·min^{-1}, and 0.38·S^{-1}, respectively. This enzyme could be a valuable tool for the structural analysis of fucoidans and production of bioactive fuco-oligosaccharides.

Keywords: fucoidan; Sphingomonas sp.; fucoidanase; galactofuco-oligosaccharides

1. INTRODUCTION

Fucoidans are a group of sulfated polysaccharides that exhibit various biological activities including anti-viral, anti-bacterial, anticoagulant and anti-tumoral activities [1,2,3,4]. The biological activities of fucoidans vary depending on the sources. The fucoidans of brown algae represent a family of homo- and hetero-polysaccharides that are mainly composed of fucose residues that are sulfated at positions 2 and/or 4 [5] and linked through α-1,2-, α-1,3-, or α-1,4-O-glycosidic bonds [6,7]. Fucoidans have also been demonstrated to contain minor amounts of mannose, xylose, galactose, glucose, uronic acids, and rhamnose [8]. It is generally believed that monosaccharide composition and sulfate contents are related to the different biological activities of fucoidans.

Despite their diverse pharmacologic activities, the structural variation, high molecular masses, and viscous nature of polysaccharides including marine-derived ones may limit their successful applications, especially as therapeutic agents, in terms of problems in polysaccharide standardization, solubility, and bioavailability [9,10]. Bioactive oligosaccharides with lower-molecular weights would help overcome these problems. Indeed, recently, significant efforts have been made to prepare oligosaccharides from marine polysaccharides, aiming at the pharmaceutical usage of bioactive oligosaccharides [8,9,10,11,12]. For example, the fuco-oligosaccharides produced by the hydrolysis of fucoidan with fucoidanases have been applied as pharmaceutical materials in medical sciences [13]. On the other hand, fucoidanases can be either exo- or endo-acting enzymes. However, no commercial endo-fucosidases [5] are currently available.

Therefore, searching for new enzyme activities that degrade marine polysaccharides becomes increasingly important area for both academic researchers and industry people. Especially, finding new fucoidanases and their subsequent kinetic characterizations would provide not only insight into the relationships between the structures and the biological activities of fucoidans but also improved technologies for the production of industrially important bioactive fuco-oligosaccharides [11]. Thus, enzymes having fucoidan-degrading activities have been sought for enzymatic modifications of fucoidans for the preparation of lower-molecular weight fuco-oligosaccharides. The major advantage for the enzymatic digestion over the chemical cleavage is the preservation of sulfate groups on fucoidans, which are recognized as the major factors implicated in the various biological activities of sulfated polysaccharides (SPS).

Previously, we reported on the isolation and identification of a marine bacterial strain, Sphingomonas paucimobilis PF-1, that degrades the Miyeokgui fucoidan (MF), which was purified in house from the sporophyll (commonly called Miyeokgui in Korean and Mekabu in Japanese) of Korean Undaria pinnatifida [14]. When intact cells of this strain was used as an enzyme source and incubated with MF as the substrate, seven distinct low-molecular weight fuco-oligosaccharides (LMFOs) with masses lower than approximately 4 kDa, were produced [14]. On the other hand, we also reported that the MF fucoidan is actually an O-acetylated sulfated galactofucan polysaccharide with a molecular mass of 1246 kDa, as characterized by various methods, such as organic elemental analysis, high performance liquid chromatography (HPLC) analysis of neutral sugars, Fourier transform infrared spectroscopy (FTIR), FT-Raman (Fourier transform-Raman), ^{1}H and ^{13}C nuclear magnetic resonance (NMR) [15].

In the present study, we first report on the purification and characterization of a fucoidanase (FNase S) from Sphingomonas paucimobilis PF-1, which degrades the high-molecular weight fucoidan (MF) into low-molecular weight galactofuco-oligosaccharides.

2. RESULTS AND DISCUSSION

2.1. Enzyme Purification

Fucodianase was purified from the supernatant that was prepared from the cell harvest via ammonium sulfate fractionation, diethylaminoethyl (DEAE)-Sepharose Fast-Flow anion-exchange chromatography and chromatofocusing. Consequently, the elution profile of the DEAE Sepharose chromatography indicated that among the five peaks with fucoidanase activities, fractions 2 to 8 (peak I) exhibited relatively higher peak than the others (Figure 1A). The fractions in peak I were collected and concentrated with 80% $(NH_4)_2SO_4$. The elution profile from the chromatofocusing revealed a single peak with fucoidanase activity that appeared from fractions 71 to 75 (Figure 1B). These results, summarized in Table 1, indicate that the enzyme was purified to homogeneity, and a 112.8-fold increase in the specific fucoidanase activity (relative to the activity of the cell disruption supernatant) with a yield of approximately 3.2%. Notably, all of the processes employed in the purification step may provide useful tools for the purification of those displayed on the cell surface since we previously reported that this fucoidan-degrading enzyme activity of S. paucimobilis PF-1 is observed with the harvested intact cells, implying its localization on the surface of the cell [14].

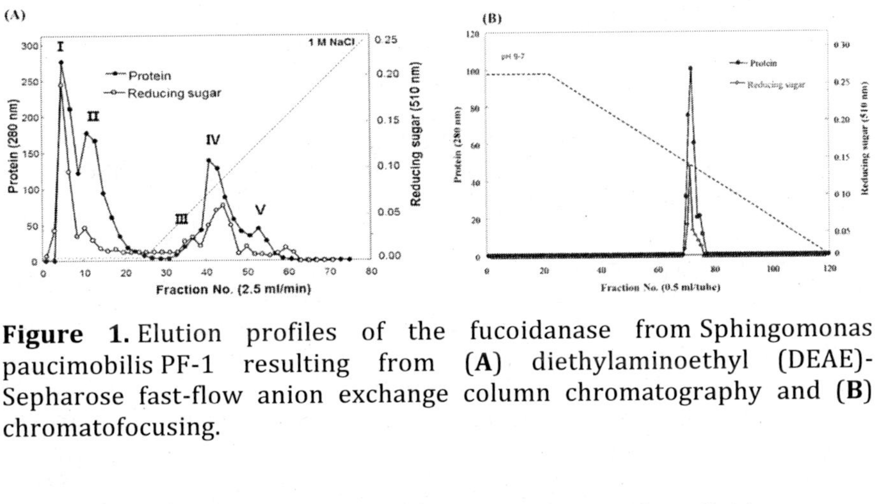

Figure 1. Elution profiles of the fucoidanase from Sphingomonas paucimobilis PF-1 resulting from **(A)** diethylaminoethyl (DEAE)-Sepharose fast-flow anion exchange column chromatography and **(B)** chromatofocusing.

Table 1. Isolation and purification of fucoidanase from Sphingomonas paucimobilis PF-1.

Step	Total volume (mL)	Protein (mg/mL)	Total protein (mg)	Activity (mg/mL)	Specific activity ($U \cdot mg^{-1}$ of protein)	Total activity (unit)	Yield (%)
Cell disruption supernatant	100	44.287	4428.7	0.841	0.019	84.08	100
80% Ammonium sulfate	70	19.582	1370.74	0.538	0.027	37.66	44.79
DEAE-Sepharose	15	0.147	2.205	0.210	1.429	3.15	3.75
Chromatofocusing	3	0.042	0.126	0.090	2.143	0.27	3.2

2.2. Estimation of the Molecular Mass and N-Terminal Sequence of the Fucoidanase

To determine the protein purity and molecular mass of the purified enzyme, native-polyacrylamide gel electrophoresis (PAGE) and sodium dodecyl sulfate (SDS)-PAGE was used according to the methods described by Laemmli [16]. The results shown in Figure 2A indicated that there was one single protein band from the final concentrated elutant, and the relative molecular mass of the purified fucoidanase was estimated to be 130 kDa (FNase S) with native-PAGE. It has been reported that the majority of Flavobacteriaceae sp. fucoidanases have molecular masses of 105 kDa, based on SDS-PAGE [17]. In general, the high-molecular weight enzyme (e.g., FNase S) can be dissociated by treatment with SDS and boiling for 10 min into several components that can be separated with SDS-PAGE. To determine whether the protein was monomeric, the purified enzyme was analyzed with SDS-PGAE. As shown in Figure 2B, the isolated multi-complex enzyme exhibited three subunits that were derived from the purified enzyme, appeared as a single band in the native-PAGE, and were determined to have molecular weights of 130 (S1), 70 (S2), and 60 kDa (S3) by SDS-PAGE. A similar result was reported by another research group. Furukawa et al. [18] reported that the molecular masses of the fucoidanase E1, E2, and E3 of the Vibrio sp. N-5 were 39 kDa, 68 kDa, and 68 kDa, respectively, based on SDS-PAGE analysis. Although the evidence is insufficient to reach a clear conclusion, these results indicated that the purified fucoidanases from both of the bacterial strains appeared to be similar to the large extracellular enzyme complex, called the cellulosome, of the plant-cell-wall-degrading anaerobic microorganisms [19]. The cellulosome consists of a scaffolding protein and many bound enzymes. Similarly, the subunits of FNase S of Sphingomonas paucimobilis PF-1 are probably involved in the enzyme's activity, stability and/or binding to substrate in physiological conditions.

The N-terminal 10-amino acid sequences of the S1, S2, and S3 subunits were analyzed and determined to be SXPEAASLPG, SPQFDVVXIG, and SLQFDVVVIG, respectively. A BLASTp search of GENBANKnr using the three domains revealed homologies primarily with hypothetical bacterial conserved proteins, with various annotations, such as protein-containing ATPase, core domain (EGF 27867), carbohydrate kinase, PfkB family protein (YP003854323) and dihydrolipoyl dehydrogenase (ZP08017938) (Table 2), which are not related to any glycohydrolases or glycosidases. Based on these results, we suggest that the purified protein might be a unique enzyme with fucoidanase activity, although full amino acid sequences of these subunits need to be further determined in the future.

Table 2. Determinations of the N-terminal amino acid sequences of (**A**) S1, (**B**) S2, and (**C**) S3. The isolated multi-complex enzyme was composed of three subunits with molecular weights of 130 (S1), 70 (S2), and 60 kDa (S3), as determined with SDS-PAGE.

Matched protein	Mol. Mass (kDa)	Positives (%)	Matched species	Functional category	Matched Peptide	Accession No.
(A) The homology comparison of *N*-terminal 10-amino acid sequence of the S1 subunit to the hypothetical bacterial conserved proteins by a BLASTp search.						
Protein containing ATPase, AAA+ type, core domain	244	8/10 (80)	*Rhodopirellula baltica* WH47	Protein containing ATPase, core domain	maedas**peaaslpg**aagdgg	EGF 27867
Hypothetical protein RB6669	202	8/10 (80)	*Rhodopirellula baltica* SH 1	ATP binding site	maedas**peaaslpg**aagdgg	NP 867389
Chitinase	127	8/10 (80)	*Arthroderma otae* CBS 113480		wp**d**v**ldaaslp**tltletagg	XP002842640
ABC sugar (glycerol) transporter, inner membrane subunit	32	8/10 (80)	*Sagittula stellata* E-37	Inner membrane subunit	madtt**sapgaaslpg**dvtak	ZP01747232
Glycoside hydrolase family protein	103	8/10 (80)	*Roseiflexus castenholzii* DSM 13941	Sugar binding domain	sls**dgpesaslpg**rfplret	YP001433316

(**B**) The homology comparison of *N*-terminal 10-amino acid sequence of the S2 subunit to the hypothetical bacterial conserved proteins by a BLASTp search.						
Carbohydrate kinase, PfkB family protein	34	8/10 (80)	*Parvularcula bermudensis* HTCC2503	Adenosine_kinase	ma**pqfdviaig**naivdllah	YP003854323
Hypothetical protein DFA_11777	78	9/10 (90)	*Dictyostelium fasciculatum*	Hypothetical protein	mvs**phydvvvig**agiaglsq	EGG14015
Dihydrolipoamide dehydrogenase	50	8/10 (80)	*Xanthomonas axonopodis* pv. *citri str. 306*	Pyridine nucleotide-disulphide oxidoreductase	m**seqfdvvvig**agpagyhaa	NP641866
Sugar kinase	34	8/10 (80)	*Erythrobacter* sp SD-21	Adenosine kinase	mt**dprydvvaig**naivdvma	ZP01863524

Matched protein	Mol. Mass (kDa)	Positives (%)	Matched species	Functional category	Matched Peptide	Accession No.
(C) The homology comparison of *N*-terminal 10-amino acid sequence of the S3 subunit to the hypothetical bacterial conserved proteins by a BLASTp search.						
Dihydrolipoyl dehydrogenase	51	9/10 (90)	*Lautropia mirabilis* ATCC 51599	Dihydrolipoyl dehydrogenase	**mslefdvvvig**agpggyiaa	ZP08017938
Anaerobic glycerol-3-phosphate dehydrogenase subunit B	45	9/10 (90)	*Serratia odorifera* DSM 4582	ANAEROBIC glycerol-3-phosphate dehydrogenase	**mqfdvvvigg**	ZP06638221
Transmembrane protein	42	9/10 (90)	*Chromobacterium violaceum* ATCC 12472	Transmembrane protein	**mqfdviviga**	NP902173

AAA+, ATPases associated with various cellular activities; ABC, ATP-binding cassette.

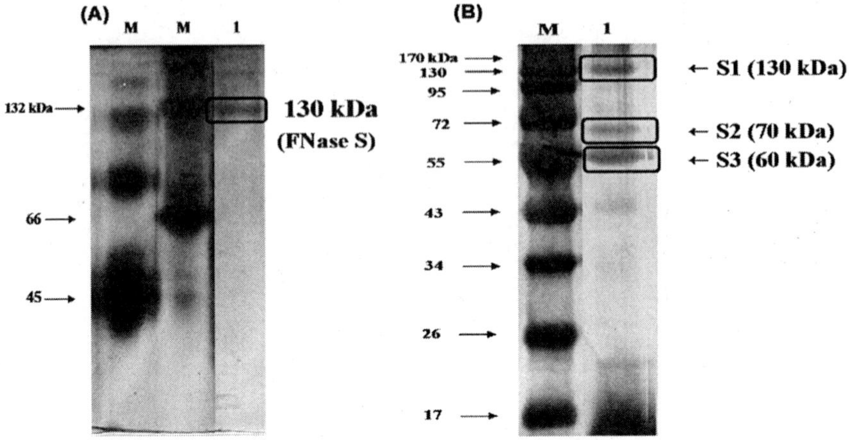

Figure 2. Gel-electrophoresis analyses of the purified enzyme. The purified fucoidanase was analyzed with (**A**) 12% native-polyacrylamide gel electrophoresis (PAGE) and (**B**) sodium dodecyl sulfate (SDS)-PAGE, and stained with silver. M, molecular weight marker (kDa); lane 1, purified fucoidanase on the native-PAGE (**A**); lane 1, purified fucoidanase on the SDS-PAGE performed without mercaptoethanol in the sample buffer (**B**).

2.3. Optimum pH and Temperature for Enzyme Activity and Stability

The FNase S activity was measured at various pH values (pH 4.0~9.0) in buffers with the same ionic concentrations to determine the optimal conditions. The pH stability was tested via 24-h preincubations of the FNase S in appropriate buffers at 45 °C that had the same ionic concentrations but different pH values that ranged from 4.0 to 9.0. The maximum activity was observed between pH 6.0 and 7.0 (Figure 3A). The FNase S activity was measured as a function of temperature from 25 °C to 55 °C, and the activity was found to be greatest at 45 °C (Figure 3B). The thermostability was tested after preincubation at various temperatures (25–55 °C) for one day at pH 6.0 and then by measuring the residual fucoidanase activity at 45 °C. As shown in Figure 3B, this enzyme was shown to be stable at 40–45 °C. The optimum pH and temperature for this enzyme were very close to the enzyme of the marine Vibrio sp., which has optimum pH and temperature ranges of pH 6.0 to 8.0 and 35 to 45 °C, respectively [18].

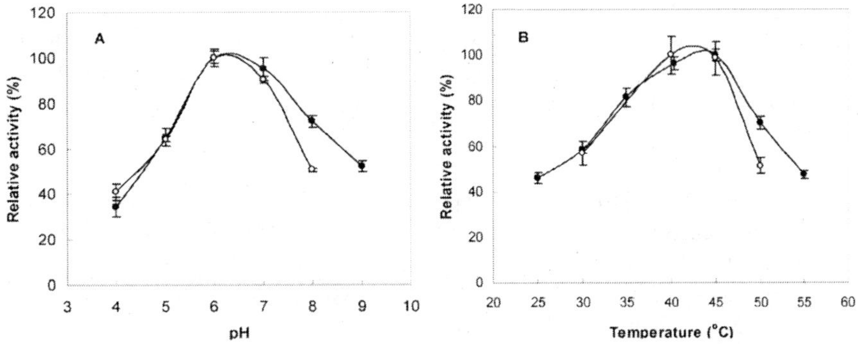

Figure 3. Effects of pH and temperature on enzyme activity and stability. (**A**) The enzyme activities were measured at various pH levels at 45 °C for three days (●). The enzyme solutions were preincubated at various pH levels for one day, and the remaining activities were measured at pH 6.0 and 45 °C for three days (○). (**B**) The enzyme activities were measured at various temperatures in 10 mM sodium acetate buffer at a pH of 6.0 for three days (●). The enzyme solutions were preincubated at various temperatures for one day in 10 mM sodium acetate buffer (pH 6.0), and the remaining activities were measured at 45 °C for three days (○).

2.4. Effects of Meal Ions and Substrate Specificities on the Enzyme Activities

The effects of metal ions on activity of FNase S were investigated (Table 3A). Mn^{2+} and Na^{+} (concentrations of 1 mM) caused significant increases of 115% and 131%, respectively, while Ca^{2+} (86%), K^{+} (83%), Ba^{2+} (84%) and Cu^{2+} (89%) caused significant decreases in enzyme activity. FNase S may require mono- or divalent metal ions (Mn^{2+} and Na^{+}) for effective enzyme activity or digestion of Miyeokgui fucoidan (MF) via the metabolic processes that occur in marine environments. The substrate specificities of FNase S on various carbohydrates were investigated (Table 3B). A commercial fucoidan (FF, Sigma, St. Louis, MO, USA) and alginate stimulated the FNase S enzyme activity by 134% and 129%, respectively, when compared with that (100%) on MF as the control. However, heparin (79%), starch (25%), laminarin (54%) and dextran (15%) acted to decrease the activity of FNase S. Therefore, we concluded that FNase S exhibits substrate specificities for fucoidan and alginate. However, to more clearly define the relationships between the molecular structures of these polysaccharides and the enzyme activities, additional comparison studies with other fucoidanases are needed. Unfortunately, to date, less is known about the enzymatic properties of fucoidanases. It is very important to achieve a detailed understanding of the key factors, such as the sulfur content, the sugar composition and the molecular weight, of substrates involved in the enzymatic characteristics.

Table 3. Effects of metal ions (**A**) and substrate specificities (**B**) on fucoidanase activities. The data are given as the means ± SD (standard deviation), n = 3. (**A**) Control, the fucoidanase activity in the absence of metal ions was regarded as 100%. (**B**) Control, the activity of the purified enzyme (FNase S) on MF was regarded as 100%.

Metal ions (1 mM)	Relative activity (%)
(**A**) Effects of metal ions on fucoidanase activities.	
Control	100
Na^{+}	131.2 ± 0.83
Mn^{2+}	115.7 ± 0.93
K^{+}	83.7 ± 0.26
Cu^{2+}	89.3 ± 2.06
Ca^{2+}	86 ± 1.68
Ba^{2+}	84.3 ± 2.8
(**B**) Effects of substrate specificities on fucoidanase activities.	
Substrate	**Relative activity (%)**
Control	100
FF *	134.5 ± 1.9
Heparin	79.2 ± 1.4
Alginate	129.4 ± 2.3
Starch	25.2 ± 0.8
Laminarin	54.9 ± 2.2
Dextran	15.5 ± 3.1

* FF, a commercial fucoidan (Sigma).

2.5. Kinetic Parameters

Initially, the kinetic parameters of the FNase S against the natural substrates, FF and MF, were determined to be K_m = 1.7 and 1.8 mM, and V_{max} = 0.62 and 0.64 mg/min, respectively (Figure 4). These results also indicated that the K_{cat} values of the FNase S were 0.376 S^{-1} and 0.343

S^{-1} (V_{max}/K_m), respectively (Table 4). Interestingly, MF and FF were hydrolyzed at similarly fast rates (K_{cat}) and high K_{cat}/K_m values. As demonstrated in Figure 3 and Figure 4, the determination of the kinetic parameters and substrate specificities of FNase S are important for understanding its enzymatic properties. However, our ability to find valuable information from other comparable studies was limited due to the paucity of such studies. Therefore, our current date could be the most valuable in terms of its contributions to kinetic studies of the activities of fucoidanases and other related enzymes.

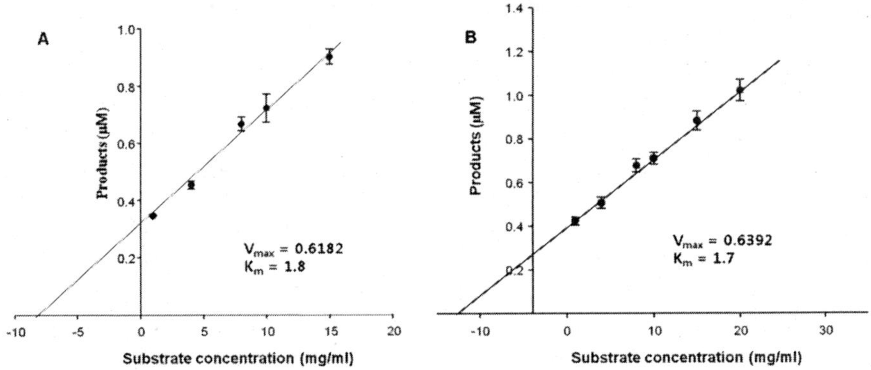

Figure 4. Assay of the enzyme kinetics at different substrate concentrations. The concentration of the enzyme was fixed at 1 mg/mL. (**A**) Miyeokgui fucoidan (MF) and (**B**) FF (commercial fucoidan, Sigma) were used as the substrates. The values of the products were calculated by measuring the reducing sugars released from the substrate fucoidans, (**A**) MF and (**B**) FF, and were expressed in molar concentrations.

Table 4. Kinetic parameters of the purified enzyme (FNase S) on Miyeokgui fucoidan (MF) and commercial fucoidan (FF) as the substrates.

Substrate	K_{cat} (S^{-1})	K_m (mM)	V_{max} (mg/min)	K_{cat}/K_m ($S^{-1}\cdot mM^{-1}$)
FF	0.376 ± 0.04	1.7 ± 0.36	0.62	0.221
MF	0.343 ± 0.1	1.8 ± 0.25	0.64	0.19

FF, commercial fucoidan (Sigma); MF, Miyeokgui fucoidan.

2.6. Production and Characterization of Galactofuco-Oligosaccharides by FNase S

The fucoidanase activity was determined by measuring the level of reducing sugars and confirmed by HPLC analyses of the degradation products (Figure 5). The fucoidan-degrading activity was also clearly confirmed by the fractionation of the degradation products through Bio-Gel P-4 size-exclusion chromatography and the analysis of the monosaccharide compositions of the resulting oligosaccharides (Figure 6). The oligosaccharides generated were resolved into seven distinct low-molecular mass fractions with peaks that ranged from 1 to 7. Based on comparisons with the masses of the standard malto-oligosaccharides (1–7 glucose units, Sigma), the relative molecular masses of these fucoidan oligosaccharides were determined to be 3312 Da (peak 1), 2494 Da (peak 2), 1699 Da (peak 3), 1543 Da (peak 4), 1312 Da (peak 5), 817 Da (peak 6), and 318 Da (peak 7) (Table 5). Based on the total sugar levels, one major fraction (peak 2) represented approximately 72% of the initial oligosaccharide mixture. Based on the high performance anion-exchange chromatography with pulsed amperometric detection (HPAEC/PAD) analysis that followed trifluoroacetic acid (TFA) hydrolysis, the constituent monosaccharides of these seven major oligosaccharides were determined to be the following: peak 1, fucose, galactose, mannose and xylose at 8:7:2:2 (given as the mole ratios); peak 2, 3:8:2:2; peak 3, 5:3:1:1; peak 4, 3:3:2:1; peak 5, fucose and galactose at 7:1; peak 6, fucose and galactose, 4:1; peak 7, fucose only (dimer) (Table 5). The uronic acid and sulfate contents of these fractions were not determined at this time. These results clearly indicated that fucose and galactose were the major monosaccharides in these oligosaccharides that were derived from MF fucoidan.

This fucoidanase activity of FNase S was not related to α-L-fucosidase. While the commercial α-L-fucosidase released p-nitrophenol from the artificial substrate, p-nitrophenyl-α-L-fucoside, our fucoidanase enzyme preparation (FNase S) did not produce p-nitrophenol, which demonstrated that the fucoidanase activity of this enzyme in the degradation of MF was not related to the α-L-fucosidase activity (data not shown). Taken collectively, these results strongly suggested that FNase S produces low-molecular weight galactofuco-oligosaccharides from MF, a sulfated galactofucan of the sporophyll (calledMiyeokgui in Korean) of Korean Undaria pinnatifida, and that this FNase S is not an α-L-fucosidase but is rather an endo-acting fucoidanase that randomly attacks the fucoidan chains as other reports have demonstrated [8,20].

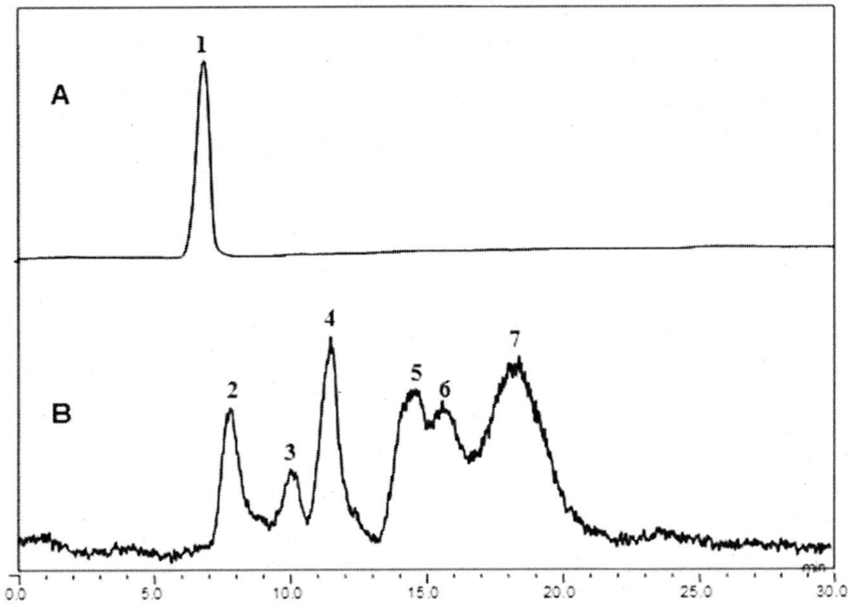

Figure 5. High performance liquid chromatography (HPLC) analysis of Miyeokgui fucoidan (MF) and its enzymatic hydrolysates produced by FNase S. (**A**) MF and (**B**) its hydrolysates were analyzed on a Shodex OHpak SB-806HQ column using HPLC system equipped with evaporative light scattering detector (ELSD). The elution was performed with distilled water at a rate of 0.8 mL/min.

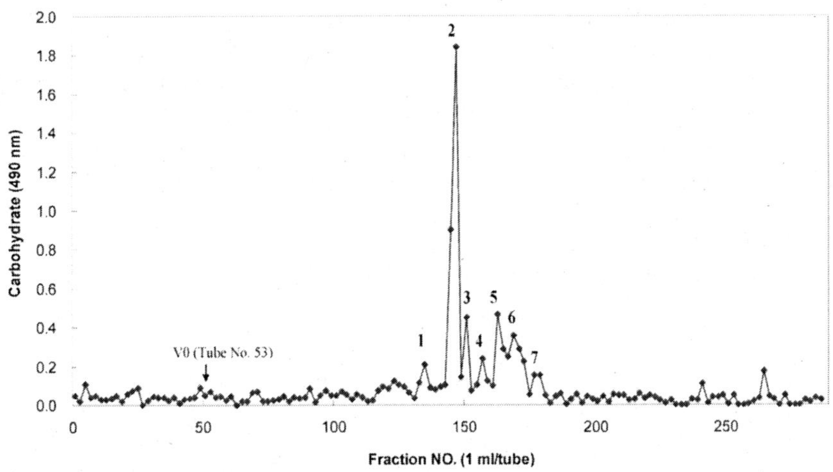

Figure 6. Bio-Gel P-4 column chromatography of the fragmented Miyeokgui fucoidan (MF) after digestion by FNase S.

Table 5. Monosaccharide compositions and relative molecular masses of the galactofuco-oligosaccharides generated from MF by FNase S.

Peak No [a]	M_r [b] (Da)	Monosaccharide composition [c]	Relative amount (%) [d]
1	3312	Fucose, Galactose, Mannose, Xylose (mole ratio, 8:7:2:2)	4.5
2	2494	Fucose, Galactose, Mannose, Xylose (mole ratio, 3:8:2:2)	72
3	1699	Fucose, Galactose, Mannose, Xylose (mole ratio, 5:3:1:1)	6.5
4	1543	Fucose, Galactose, Mannose, Xylose (mole ratio, 3:3:2:1)	2.7
5	1312	Fucose, Galactose (mole ratio, 7:1)	5.3
6	817	Fucose, Galactose (mole ratio, 4:1)	6.1
7	318	Fucose (2 moles)	2.9

[a] The oligosaccharides were fractionated through a Bio-Gel P-4 column after digestion of MF by FNase S; [b] The relative molecular mass of each oligosaccharide was estimated using malto-oligosaccharides (1–7 glucose units, Sigma) as standard molecular markers; [c] The monosaccharide compositions were determined with high performance anion-exchange chromatography with pulsed amperometric detection (HPAEC-PAD) analyses; [d] The carbohydrate contents of each peak were the total neutral sugars as determined with the phenol-sulfuric acid method and the relative amounts of each oligosaccharide were estimated by setting the sum of the major oligosaccharides (peak 1–7) to 100%.

This fucoidanase activity of FNase S was not related to α-L-fucosidase. While the commercial α-L-fucosidase released p-nitrophenol from the artificial substrate, p-nitrophenyl-α-L-fucoside, our fucoidanase enzyme preparation (FNase S) did not produce p-nitrophenol, which demonstrated that the fucoidanase activity of this enzyme in the degradation of MF was not related to the α-L-fucosidase activity (data not shown). Taken collectively, these results strongly suggested that FNase S produces low-molecular weight galactofuco-oligosaccharides from MF, a sulfated galactofucan of the sporophyll (calledMiyeokgui in Korean) of Korean Undaria pinnatifida, and that this FNase S is not an α-L-fucosidase but is rather an endo-acting fucoidanase that randomly attacks the fucoidan chains as other reports have demonstrated [8,20].

Bakunina [21] reported on the 25 strains of epiphytic marine bacteria isolated from the brown algae Fucus evanescens andChorda filum, and 53 bacterial strains isolated from the sea cucumber Apostichopus japonicus were screened for having fucoidanase activity using fucoidans prepared from the brown algae F. evanescens and Laminaria cichorioides (Table 6). The bacterial epiphytes Cytophaga sp. and some bacterial isolates of the genera Alteromonas and Pseudoalteromonas from the sea cucumber were also found to possess fucoidanase activity [13]. However, the enzyme activity was comparable but lower than that of the already known microbial fucoidanases [22,23]. Although very little is known about the genes that encode fucoidanases, the molecular cloning and biochemical characterization of a special fucanase FcnA termed fucan α-1,4-endohydrolase isolated from the family Flavobacteriaceae have decoded the structural basis of sulfated polysaccharides [17,24]. The results of this study demonstrated that our FNase S produces low-molecular weight galactofuco-oligosaccharides from fucoidan (actually a sulfated galactofucan). However, these results raise the questions on how bacteria that produce fucoidanase can take up dimer and larger oligosaccharides into the cytoplasmic space through the cell wall for use as nutrients. Additional studies of the catabolic cascade of MF and its oligomers are underway. In addition, mining and molecular cloning of corresponding genes for FNase S in the future would stimulate the applicability of this enzyme to the various industries for the development of new nutraceuticals, cosmeceuticals, as well as pharmaceuticals.

Table 6. Fucoidan-degrading enzymes and their properties.

Producing microorganism	MW (Da)	Habitat	Substrate	Cleaving pattern	Refs.
Cytophaga, *Alteromonas*, *Pseudoalteromonas*	n.d.	Intra-cellular	Fucoidan from *Fucus evanescens*, *Laminaria cichorioides* and *Laminaria japonica*	Exo	[21]
Dendryphiella arenaria	n.d.	Intra-cellular	Fucoidan from *Fucus vesiculosus* and *Laminaria digitata*	n.d.	[25]
Flavobacteriaceae	n.d.	Extra-cellular	Fucoidan from *Pevetia canaliculata*	Endo	[24]
Flavobacteriaceae	67,000	Extra-cellular	Fucoidan from *Kjellmaniella crassifolia*	Endo	[26]
Littorina kurila	n.d.	Intra-cellular	Fucoidan from *Fucus distichus*	Disaccharides	[27]
Pseudoalteromonas atlantica	n.d.	Extra-cellular	Fucoidan from brown seaweed	Exo	[28]
Patinopecten yessoensis	85,000	Intra-cellular	Fucoidan from *Nemacystus decipieus*	Endo	[20]
Pecten maximus	200,000	Extra-cellular	Fucoidan from *Ascophyllum nodosum.*	endo	[29]
Pecten maximus	n.d.	Intra-cellular	Fucoidan from *Ascophyllum nodosum*	Releases L-fucose	[30]
Pecten maximus	n.d.	Intra-cellular	Sulfated L-fucopyranoside	Sulfoesterase	[31]
Vibrio sp. N5	40,000–68,000	Extra-cellular	Fucoidan from *Kjellmaniella crassifolia*	Endo	[18]
Fucophilus fucidanolyticus	n.d.	Intra-cellular	Fucoidan from *Cladosiphon okamuranus*	Endo	[26]
Gram-negative bacterium	67,000	Extra-cellular	Fucoidan from *Kjellmaniella crassifolia*	Endo	[32]
Sphingomonas paucimobilis PF-1	≥132,000	Extra-cellular	Fucoidan *from Undaria pinnatifida* sporophyll	Endo	This study

n.d., not determined.

FNase S was incubated with 1% fucoidan for three days at 45 °C in 50 mM sodium acetate buffer (pH 5.6) containing 0.01% sodium azide. The reaction products were eluted with 50 mM sodium nitrate at a flow rate of 0.5 mL/min. Each fraction was monitored for carbohydrates with the phenol-H_2SO_4 method at 490 nm using L-fucose as a reference. The arrow indicates the void volume at which the unhydrolyzed fucoidan and hydrolyzed products with larger molecular masses (over approximately 5000 Da) reached the upper exclusion limit of the gels. Only major peaks (1 to 7) were collected and examined for their monosaccharide compositions with high performance anion-exchange chromatography with pulsed amperometric detection (HPAEC-PAD) analyses.

3. EXPERIMENTAL SECTION

3.1. Preparation of Fucoidan

Miyeokgui fucoidan (MF) was extracted and purified as described previously [33]. Briefly, fucoidan was extracted from the dried sporophylls (called Miyeokgui in Korean) of Undaria pinnatifida (called Miyeok in Korean and Wakame in Japanese) that were collected from the southern coastal area of Wando, Korea, mainly through 0.1 N HCl extraction and 75% ethanol precipitation. The crude fucoidan obtained was further purified through DEAE-Cellulose column chromatography. Commercial fucoidan (tentatively named FF in this study) from alga Fucus vesiculosus was purchased from Sigma Chemical (St. Louis, MO, USA).

3.2. Microorganism and Culture Conditions

The growth of Sphingomonas paucimobilis PF-1 (KCTC 11130BP) was performed as previously reported by our group [14]. The cells were cultured in a minimal medium consisting of 0.5% MF with 2% Bacto peptone in dH_2O (pH 7.0) at 30 °C for 4 days on a shaking incubator (180 rpm, JEIO TECH Co., Seoul, Korea) and then centrifuged at 6000 rpm, 4 °C for 30 min. The cells harvested were stored at −20 °C and used for the purification of the fucoidanolytic enzymes.

3.3. Production and Purification of Fucoidanase

All experiments were performed at 4 °C unless otherwise stated. The cells (50 g, wet weight) were homogenized on ice using a Sonifier 450 (Branson, Danbury, CT, USA) for 1 h in 50 mL of 50 mM sodium acetate buffer (pH 5.6). The homogenate was centrifuged at 16,000 rpm for 40 min to remove the insoluble materials. The supernatant solution was gradually brought up to 40% (w/v) saturation with $(NH_4)_2SO_4$ and then centrifuged at 16,000 rpm for 30 min to remove the precipitates. Continuously, the supernatant solution was further brought to 80% (w/v) saturation with $(NH_4)_2SO_4$ and precipitated overnight at 4 °C. The precipitate was collected by centrifugation at 16,000 rpm for 30 min, dissolved in 10 mL of 50 mM sodium acetate buffer (pH 5.6), and dialyzed (10 kDa molecular weight cut off) extensively against 2 L of the same buffer for 3 days at 4 °C and stored at −20 °C. The dialyzed solution (20 mL) was loaded onto a column (2.8 × 30 cm) of DEAE-Sepharose (GE Healthcare, Marlborough, MA, USA) and equilibrated with 10 mM Tris-

HCl buffer containing 10 mM sodium chloride (pH 7.2). The column was washed with the same buffer and then was eluted with a linear gradient of 0–1 M sodium chloride. The non-binding fractions on the DEAE-Sepharose were collected and dialyzed against 2 L 0.025 M ethanolamine buffer (pH 9.4). The concentration was applied onto a Mono P (GE Healthcare) column equilibrated with 0.025 M ethanolamine buffer (pH 9.6) and then eluted with a linear gradient of Polybuffer 96 buffer (pH 7.0) (GE Healthcare). The apparent molecular weight of the purified enzyme was estimated with native PAGE and SDS-PAGE analyses using a 12% polyacrylamide gel and Coomassie Brilliant Blue R-250 staining [34]. The molecular masses of the protein standard ranged from 10 to 170 kDa and were prestained on a pro-stain Ladder (Fermentas, Hanover, MD, USA).

3.4. Determination of the N-terminal Amino Acid Sequence

The N-terminal amino acid residues of fucoidanase were determined using an Applied Biosystems 476A automatic protein sequencer (Amino Acid Sequencing Facility, Institute of Korea Basic Science, Daejeon, Korea). The amino acid sequence was compared with those available in SWISS-PROT and/or the Protein Data Bank.

3.5. Effects of pH and Temperature on Fucoidanase Activity and Stability

The optimal temperature for fucoidanase activity was determined over the range of 25 to 55 °C in 50 mM sodium phosphate buffer (pH 6.0) containing 0.5% of MF as the substrate. The thermal stability of the purified enzyme was tested by pre-incubating the enzyme at different temperatures (25 to 55 °C) for 1 day. The remaining activities of fucoidanase were measured immediately after this treatment with the standard method mentioned previously. The activity of fucoidanase was determined by measuring the amount of reducing sugars released from the fucoidan with the Somogyi-Nelson method, using fucose as the standard [35]. One unit of fucoidanase activity was defined as the amount of enzyme (mg) that released 1 nmol of reducing sugar per min. To determine the optimal pH, the fucoidanase activity of the enzyme was studied at 45 °C in the following buffers: 50 mM sodium acetate buffer (pH 4.0 and 5.0), 50 mM sodium phosphate buffer (pH 6.0), and 50 mM Tris-HCl (pH 7.0, 8.0, 9.0). The pH stability of the enzyme was measured based on the residual activity after pre-incubating the enzyme in the above buffer (pH 4.0–9.0) at 45 °C for 1 day.

3.6. Effects of Metal Ions on Fucoidanase Activity

The effects of several metal ions on fucoidanase activity were examined in reaction mixtures supplemented with NaCl (Na^{+}), $MnSO_4$ (Mn^{2+}), KCl (K^{+}), $CuSO_4$ (Cu^{2+}), CaCl (Ca^{2+}), and $BaCl_2$ (Ba^{2+}) at final concentrations of 1 mM. The reactions were performed at the optimal pH and temperature over 72 h with MF (0.5%) as the substrate. The fucoidanase activity was determined with the Somogyi-Nelson method as described above.

3.7. Determination of the Substrate Specificity

The substrate specificity of the purified enzyme was determined using the following natural substrates: commercial fucoidan (FF) from the alga Fucus vesiculosus (Sigma), heparin, alginate, starch, laminarin, and dextran. The purified enzyme (10 µg) was added to 1 mL of 50 mM sodium phosphate buffer (pH 6.0) containing 0.5% substrates. The reactions were performed at the optimal pH and temperature over 72 h. The fucoidanase activity was quantitatively determined by measuring the amount of reducing sugar released from the polysaccharide substrates with the Somogyi-Nelson method, using fucose or glucose as the standard [35].

3.8. Determination of the Kinetic Parameters

The hydrolysis kinetics of the purified enzyme on the intact fucoidan (MF) and commercial fucoidan (FF, Sigma) were determined at pH 6.0 and 45 °C with substrate concentrations that varied from 0 to 20 mM and/or 0 to 30 mM, respectively; this step was followed by the standard enzyme assay described above. The K_m, V_{max} and K_{cat} values were obtained based on a Lineweaver-Burk plot and were expressed as the means of the three different experiments. This plot is widely used to determine important terms in enzyme kinetics, such as K_m and V_{max} for determining the type of enzyme inhibition by using competitive, non-competitive and uncompetitive inhibitors. In this plot, competitive inhibitors have the same y-intercept as uninhibited enzyme, since V_{max} is unaffected by competitive inhibitors. However, different slopes and x-intercepts between the data sets can be seen in the same plot. On the other hand, non-competitive inhibition has plots with the same x-intercept with different slopes and y-intercepts. Subsequently, uncompetitive inhibition can cause different intercepts on both the y- and x-axes. Herewith, our data simply showed in the case of uncompetitive inhibition, since no inhibitor was used to determine the kinetics of fucoidanase. These assays were set up in triplicates. If the plot of 1/V and 1/(substrate) is linear, then the slope is equal to $1/V_{max}$, $1/K_m$, and K_{cat}/K_m.

3.9. Size-Exclusion HPLC Analysis for Fucoidanase Activity

Fucoidanase activity was determined by measuring the level of reducing sugars and confirmed with size-exclusion HPLC (Dionex Co., Sunnyvale, CA, USA) using a Shodex OHpak column (SB-806HQ, 8.0 × 300 mm, Showa Denko Co., Tokyo, Japan) under the established conditions, following 75% ethanol precipitation. Next, the supernatant was vacuum-dried using a Speed-Vac (Module spin 40, Biotron, Seoul, Korea), 100 μL dH_2O was added to dissolve the pellet. Ten microliters of the sample was injected, eluted with water at a flow rate of 0.8 mL/min at 60 °C and detected with an evaporative light scattering detector (ELSD, Deerfield, IL, USA).

3.10. Identification of the Galactofuco-Oligosaccharides

The reaction mixture of MF and purified enzyme was fractionated through a Bio-Gel P-4 column (1.5 × 95 cm), and each oligomer eluted from the column was analyzed for its monosaccharide composition and molecular mass. The reaction mixture of 10 mL of 50 mM sodium phosphate buffer (pH 6.0) containing 1% fucoidan and 10 mL of purified enzyme (10 μg/min) was incubated for 3 days at 45 °C in a shaking water bath (120 rpm). After incubation, the enzyme was removed by centrifugation and the supernatant was concentrated by freeze drying. The dried material was dissolved in 5 mL of dH_2O and precipitated with 75% ethanol. Next, the supernatant was vacuum-dried using a Speed-Vac. The dried material was dissolved in 1 mL of 50 mM sodium nitrate and 0.5 mL of the material was applied to the Bio-Gel P-4 column and eluted at 0.5 mL/min with the same solvent. The fractions containing carbohydrates were pooled together and freeze-dried to determine the monosaccharide compositions of the galactofuco-oligosaccharides. Pooled fractions were dissolved in 0.1 mL of dH_2O, an equal volume of 4 M trifluoroacetic acid (TFA) was added, and the mixtures were allowed to stand for 4 h at 100 °C with gentle stirring. After the reaction, the mixture was filtered through a 0.45-μm syringe filter and vacuum-dried using a Speed-Vac. The dried material was re-dissolved in 0.1 mL of dH_2O. The monosaccharide compositions of the TFA-hydrolyzed galactofuco-oligosaccharides were assessed with HPAEC-PAD (Dionex Co.) as described by Lee et al. [36].

3.11. Statistical Analysis

The statistical significances of the differences between the mean values were determined with Student's t-tests. The experimental values represent the means ± the standard deviations.

4. CONCLUSIONS

In this study, we purified an enzyme (FNase S) that degrades Miyeokgui fucoidan (MF), which is a sulfated galactofucan, into smaller-sized galactofuco-oligosaccharides (1000–4000 Da), from a strain of Sphingomonas paucimobilis. Experiments involving the over-production and the purification of fucoidanases are thus of great interest, as is the processing of these enzymes to make them usable in industrial processes. The naturally occurring fucoidans require further structural elucidation to create a source of pharmaceutical drugs or any value-added raw materials for industrial processes. Additional knowledge about fucoidan-degrading enzymes will facilitate studies that aim to elucidate the structures of fucoidans and other fucoidan type polysaccharides and enable the production of lower-molecular weight functional galactofuco-oligosaccharides.

ACKNOWLEDGMENTS

This work was supported by a grant from the Gyeonggi-do GRRC program (Project No. 2014-B03) of the Catholic University of Korea, for which the authors are thankful.

AUTHOR CONTRIBUTIONS

Woo Jung Kim performed most of the purification and kinetic parameter analysis experiments as the first author; Joo Woong Park performed cell culture; Jae Kweon Park extracted and prepared fucoidan; Doo Jin Choi performed monosaccharide composition analysis; and Yong Il Park designed the experiments and wrote the paper as the corresponding author.

CONFLICTS OF INTEREST

The authors declare no conflict of interest.

REFERENCES

1. Hoshino, T.; Hayashi, T.; Hayashi, J.; Lee, J.B.; Sankawa, U. An antivirally active sulfated polysaccharide fromSargassum horneri (TURNER) C. AGARDH. Biol. Pharm. Bull. **1998**, 21, 730–734.

2. Cumashi, A.; Ushakova, N.A.; Preobrazhenskaya, M.E.; D'Incecco, A.; Piccoli, A.; Totani, L.; Tinari, N.; Morozevich, G.E.; Berman, A.E.; Bilan, M.I.; et al. A comparative study of the anti-inflammatory, anticoagulant, anti-angiogenic and anti-adhesive activities of nine different fucoidans from brown seaweeds. Glycobiology **2007**, 17, 541–552.

3. Alekseyenko, T.V.; Zhanayeva, S.Y.; Venediktova, A.A.; Zvyagintseva, T.N.; Kuznetsova, T.A.; Besednova, N.N.; Korolenko, T.A. Antitumor and antimetastatic activity of fucoidan, a sulfated polysaccharide isolated from the Okhotsk sea Fucus evanescens brown alga. Bull. Exp. Biol. Med. **2007**, 147, 730–732.

4. Hahnenberger, R.; Jakobson, A.M. Antiangiogenic effect of sulfated glycosaminoglycans and polysaccharides in the chick embryo chorioallantoic membrane. Glycoconj. J. **1991**, 8, 350–353.

5. Chevolot, L.; Foucault, A.; Chaubet, F.; Kervarec, N.; Sinquin, C.; Fisher, A.M.; Boisson-Vidal, C. Further data on the structure of brown seaweed fucans: Relationships with anticoagulant activity. Carbohydr. Res. **1999**, 319, 154–165.

6. Usov, A.I.; Smirnova, G.P.; Bilan, M.I.; Shashkov, L.S. Polysaccharides of brown algae. Bioorg. Khim. **1998**, 24, 437–445.

7. Mulloy, B.; Ribeiro, A.C.; Alves, A.P.; Vieira, R.P.; Mourão, P.A.S. Sulfated fucans from echinoderms have a regular tetrasaccharide repeating unit defined by specific patterns of sulfation at the O-2 and O-4 positions. J. Biol. Chem. **1994**,269, 22113–22123.

8. Berteau, O.; Mulloy, B. Sulfated fucans, fresh perspectives: Structures, functions, and biological properties of sulfated fucans

and an overview of enzymes active toward this class of polysaccharides. Glycobiology **2003**, 13, 29R–40R.

9. Silchenko, A.S.; Kusaykin, M.I.; Kurilenko, V.V.; Zakharenko, A.M.; Isakov, V.V.; Zaporozhets, T.S.; Gazha, A.K.; Zvyagintseva, T.N. Hydrolysis of Fucoidan by Fucoidanase Isolated from the Marine Bacterium, Formosa algae. Mar. Drugs **2013**, 11, 2416–2430.

10. Ji, J.; Wang, L.C.; Wu, H.; Luan, H.M. Bio-function Summary of Marine Oligosaccharides. Int. J. Biol. **2011**, 3, 74–86.

11. Holtkamp, A.D.; Kelly, S.; Ulber, R.; Lang, S. Fucoidans and fucoidanases-focus on thchniques for molecular structure elucidation and modification of marine polysaccharides. Appl. Microbiol. Biotechnol. **2009**, 82, 1–11.

12. Courtois, J. Oligosaccharides from land plants and algae: Production and applications in therapeutics and biotechnology. Curr. Opin. Microbiol. **2009**, 12, 261–273.

13. Bakunina, I.; Nedashkovskaya, O.I.; Alekseeva, S.A.; Ivanova, E.P.; Romanenko, L.A.; Gorshkova, N.M.; Isakov, V.V.; Zvyagintseva, T.N.; Mikhailov, V.V. Degradation of fucoidan by the marine proteobacterium Pseudoalteromonas citrea.Mikrobiologiia **2002**, 71, 49–55.

14. Kim, W.J.; Kim, S.M.; Lee, Y.H.; Kim, H.G.; Kim, H.K.; Moon, S.H.; Suh, H.H.; Jang, K.H.; Park, Y.I. Isolation and characterization of marine bacterial strain degrading fucoidan from Korean Undaria pinnatifida sporophylls. J. Microbiol. Biotechnol. **2008**, 18, 616–623.

15. Synytsya, A.; Kim, W.J.; Kim, S.M.; Pohl, R.; Synytsya, A.; Kvasnicˇka, F.; Čopíková, J.; Park, Y.I. Structure and antitumour activity of fucoidan isolated from sporophyll of Korean brown seaweed Undaria pinnatifida. Carbohydr. Polym. **2010**, 81, 41–48.

16. Laemmli, U.K. Cleavage of structural proteins during assembly of head of bacteriophage T4. Nature **1970**, 227, 680–685.

17. Colin, S.; Deniaud, E.; Jam, M.; Descamps, V.; Chevolot, Y.; Kervarec, N.; Yvin, J.C..; Barbeyron, T.; Michel, G.; Kloareg, B. Cloning and biochemical characterization of the fucanase FcnA: Definition of a novel glycoside hydrolase family specific for sulfated fucans. Glycobiology **2006**, 16, 1021–1032.

18. Furukawa, S.; Fujikawa, T.; Koga, D.; Ide, A. Purification and some properties of exo-type fucoidanases from Vibrio sp. N-5. Biosci. Biotechnol. Biochem. **1992**, 56, 1829–1834.

19. Doi, R.H.; Kosugi, A. Plant-cell-wall-degrading enzyme complexes. Nat. Rev. **2004**, 2, 541–551.

20. Kitamura, K.; Matsuo, M.; Yasui, T. Enzymatic degradation of fucoidan by fucoidanase from the hepatopancreas ofPatinopecten yessoensis. Biosci. Biotechnol. Biochem. **1992**, 56, 490–494.

21. Bakunina, I.Yu.; Shevchenko, L.S.; Nedashkovskaya, O.I.; Shevchenko, N.M.; Alekseeva, S.A.; Mikhailov, V.V.; Zvyagintseva, T.N. Screening of marine bacteria for fucoidanases. Microbiology **2000**, 69, 303–308.

22. Sakai, T.; Kimura, H.; Kojima, K.; Ikai, K.; Akiyoshi, S.; Nakanishi, Y.; Kato, I. Oligosaccharides manufacture by hydrolysis of fucoidan. Chem. Abstr. **1997**, 126, 113.

23. Gonzalez, J.M.; Weiner, R.M. Phylogenetic characterization of marine bacterium strain 2-40, a degrader of complex polysaccharides. Int. J. Syst. Evol. Microbiol. **2000**, 50, 831–834.

24. Descamps, V.; Colin, S.; Lahaye, M.; Jam, M.; Richard, C.; Potin, P.; Barbeyron, T.; Yvin, J.C.; Kloareg, B. Isolation and culture of a marine bacterium degrading the sulfated fucans from marine brown algea. Mar. Biotechnol. **2005**, 8, 1–13.

25. Kelly, S.; Holtkamp, A.; Poth, S.; Lang, S.; Ulber, R. Untersuchungen zur potenziellen fucoidanase-Aktivität von Dendryphiella Arenaria. Chem. Ing. Tech. **2008**, 80, 399–403.

26. Sakai, T.; Ishizuka, K.; Shimanaka, K.; Ikai, K.; Kato, I. Structures of oligosaccharides derived from Cladosiphon okamuranus fucoidan by digestion with marine bacterial enzymes. Mar. Biotechnol. **2003**, 5, 536–544.

27. Bilan, M.I.; Kusaykin, M.I.; Grachev, A.A.; Tsvetkova, E.A.; Zvyagintseva, T.N.; Nifantiev, N.E.; Usov, A.I. Effect of enzyme preparation from the marine mollusk Littorina kurila on fucoidan from the brown alga Fucus distichus.Biochemistry (Mosc.) **2005**, 70, 1321–1326.

28. Yaphe, W.; Morgan, K. Enzymic hydrolysis of fucoidin by Pseudomonas atlantica and Pseudomonas carrageenovora.Nature **1959**, 183, 761–762.

29. Berteau, O.; McCort, I.; Goasdoue, N.; Tissot, B.; Daniel, R. Characterization of a new alpha-L-fucosidase isolated from the marine mollusk Pecten maximus that catalyzes the hydrolysis of alpha-L-fucose from algal fucoidan (Ascophyllum nodosum). Glycobiology **2002**, 12, 273–282.

30. Daniel, R.; Berteau, O.; Jozefonvicz, J.; Goasdoue, N. Degradation of algal (Ascophyllum nodosum) fucoidan by an enzymatic activity contained in digestive glands of the marine mollusk Pecten maximus. Carbohyd. Res. **1999**, 322, 291–297.

31. Daniel, R.; Berteau, O.; Chevolot, L.; Varenne, A.; Gareil, P.; Goasdoue, N. Regioselective desulfation of sulfated L-fucopyranoside by a new sulfoesterase from the marine mollusk Pecten maximus: Application to the structural study of algal fucoidan (Ascophyllum nodosum). Euro. J. Biochem. **2001**, 268, 5617–5626.

32. Sakai, T.; Kawai, T.; Kato, I. Isolation and characterization of a fucoidan-degrading marine bacterial strain and its fucoidanase. Mar. Biotechnol. **2004**, 6, 335–346.

33. Kim, W.J.; Kim, S.M.; Kim, H.G.; Oh, H.R.; Lee, K.B.; Lee, Y.K.; Park, Y.I. Purification and anticoagulant activity of a fucoidan from Korean Undaria pinnatifida sporophyll. Algae **2007**, 22, 247–252.

34. George, V.; Diwan, A.M. Simultaneous staining of proteins during polyacrylamide gel electrophoresis in acidic gels by countermigration of Coomassie brilliant blue R-250. Anal. Biochem. **1983**, 132, 481–483.

35. Somogyi, M. Notes on sugar determination. J. Biol. Chem. **1952**, 195, 19–23.

36. Lee, Y.K.; Lim, D.J.; Lee, Y.H.; Park, Y.I. Variation in fucoidan contents and monosaccharide compositions of Korean Undaria pinnatifida (Harvey) suringar (Phaeophyta). Algae **2006**, 21, 157–160.

Index